Linear Measures *(Continued)*

1 centimeter = 0.3937 inch
1 meter = 100 centimeters = 39.37 inches = 3.28083 feet = 1.0936 yards
1 kilometer = 1000 meters = 0.62137 mile

Velocity:

1 foot per second (ft./sec.) = 0.3048 meter per second (m./sec.) = 18.29 meters per minute (m./min.) = 1.0973 kilometers per hour = 0.6818 mile per hour
1 mile per hour = 88 feet per minute = 1.4667 feet per second = 0.447 meter per second = 26.82 miters per minute = 1.6093 kilometers per hour
1 meter per second = 3.28083 feet per second = 2.2369 miles per hour
1 centimeter per second (cm./sec.) = 0.02237 mile per hour
1 kilometer per hour = 16,6667 meters per minute = 0.2778 meter per second = 0.9113 foot per second = [illegible] per hour

Weights:

1 ounce = 28.3495 grams
1 pound = 16 ounces = 0.454 kilogram
1 kilogram = 35.27 ounces = 2.2 pounds

Area:

1 square inch = 6.452 square centimeters
1 square foot = 0.0929 square meter
1 square yard = 0.8361 square meter
1 square centimeter = 0.155 square inch
1 square meter = 10.764 square feet = 1.1960 square yards

Volume:

1 cubic inch = 16.3872 cubic centimeters
1 cubic centimeter = 0.061 cubic inch   1 quart = 0.946 cubic centimeter   1 liter = 1.057 quarts
1 cubic foot = 28.317 liters = 0.02832 cubic meter

Pressure:

1 atmosphere = 34.0 feet of water = 760 millimeters or 29.92 inches of mercury = 14.7 pounds per square inch

Physiology of Muscular Activity

Seventh Edition, Illustrated

PETER V. KARPOVICH, M.D., M.P.E.
Research Professor of Physiology, Emeritus
Springfield College, Springfield, Massachusetts

WAYNE E. SINNING, Ph.D.
Professor of Physical Education
Springfield College, Springfield, Massachusetts

W. B. SAUNDERS COMPANY
Philadelphia · London · Toronto

W. B. Saunders Company West Washington Square
Philadelphia, Pa. 19105

12 Dyott Street
London, WC1A 1DB

1835 Yonge Street
Toronto 7, Ontario

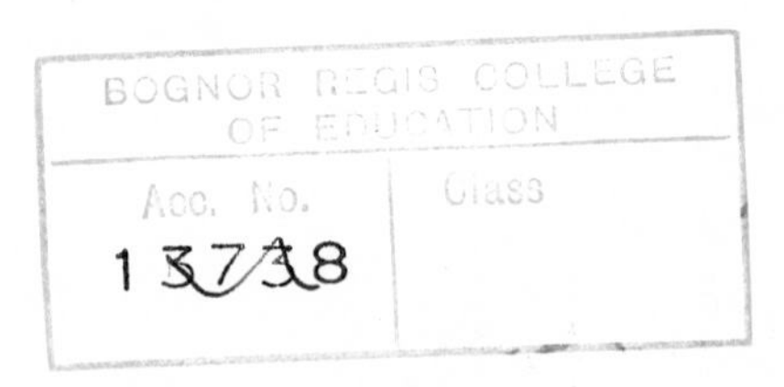

Listed here is the latest translated edition of this book together with the language of the translation and the publisher.

Italian (5th Edition) Edizione (Leonardo) Scientifiche
Rome, Italy

Japanese (5th Edition) Baseball Magazine Sha
Tokyo, Japan

Physiology of Muscular Activity SBN 0-7216-5297-2

 Made in the United States of America. Press of W. B. Saunders Company. Library of Congress catalog card number 73-139428.

Print No.: 9 8 7 6 5 4 3 2

DEDICATED TO THE MEMORY OF

DR. ARTHUR H. STEINHAUS

PREFACE

This edition of *Physiology of Muscular Activity,* like previous editions, is written for physical education students taking their first course in exercise physiology. It is assumed that students have had a prior course in elementary physiology.

It is not practicable to list all of the changes in a revision such as this. However, those familiar with previous editions will note that the chapter on muscle has been extensively revised to include more information on the interaction between muscle and nerve as well as muscle contraction *in vivo* as revealed through electromyography. Chapter 3, which is entirely new, deals with the organization of the nervous system and its function in the control of motor activities. It was added in response to requests by numerous instructors who stated a need for such information. Chapter 19 has also been completely revised to include newer material on somatotyping and a discussion of body composition and exercise. Another major change is an expansion of the material on exercise at higher altitudes.

New material on physiological activity at the cellular level has also been emphasized. Perhaps the most significant development in exercise physiology during the last half-decade is the increase in the use of electron microscopy and microbiochemical techniques to study the effects of exercise on the cell. Although the reviews of completed studies are not all-inclusive, it is hoped that there is enough exposure to provide the student with an appreciation of the significance of such research and a realization of its potential in helping us to understand more clearly the bodily response to exercise. No modern exercise physiology course can be complete without at least an introduction to this newer work.

Many people have contributed to this revision. Special thanks, however, must be extended to Dr. Clifford Keeney and Dr. Margaret Thorsen of Springfield College for review of parts of the manuscript and to Dr. Charles Tipton of the University of Iowa for his help in preparing the section on determining minimal weights for high-school wrestlers by the use of anthropometric measurements. The assistance and cooperation of the staff of Marsh Memorial Library, Springfield College, in obtaining resource material is also acknowledged, especially the efforts of Mr. Gerald Davis, Assistant Librarian in Charge of Reader Service. Finally, appreciation is expressed to Mrs. Harriet Chaiken for endless hours of typing as well as checking the accuracy of the manuscript; and to Mrs. Karpovich and Mrs. Sinning for assistance with proofreading.

W. E. S.
P. V. K.

CONTENTS

Chapter One

SKELETAL MUSCLE 1

The Muscle Cell 1
Muscle Structure 4
The Chemical Composition of Muscle 6
The Blood Supply of Muscle 7
Muscle Innervation 8
Red and White Muscle 9
The Contractile Process 9
Muscular Contraction 13
Physical Properties of Muscle 17
Questions 19

Chapter Two

MUSCLE TRAINING AND RELATED PROBLEMS 20

Changes in Structure of Muscle 20
Gain in Strength 21
Body Weight and Strength 22
Weight Training and Athletics 25
What Happens When Weight Training Stops? 26
Is Tapering Off Necessary? 26
Gain in Endurance 26
Chemical Changes 27
End Plate Transmission 27
More Complete Use of All Fibers 28
Capillaries 28
Muscle Tone 29
Effect of Warming-Up 29
Muscle Soreness After Exercise 32
Muscle Cramps 33
Menstruation and Muscle Strength 35
Irregular Chromosome Arrangements 35
Questions 36

Chapter Three

NERVOUS CONTROL OF MUSCULAR CONTRACTION ... 37

Organization of the Nervous System ... 37
The Sensation of Movement ... 46
Neural Integration of Muscle Contraction ... 49
Questions ... 59

Chapter Four

PHENOMENA ASSOCIATED WITH NERVE CONTROL ... 60

The Guiding Role of the Head ... 60
Reaction and Reflex Time ... 61
Questions ... 71

Chapter Five

THE PROVISION OF ENERGY FOR MUSCULAR WORK ... 72

The Release of Energy in the Cell ... 72
Cell Function and Training ... 77
Protein As a Source of Energy During Exercise ... 80
Respiratory Quotient ... 80
Effect of Exercise on the Respiratory Quotient ... 82
Fat As a Source of Energy ... 83
Sugar and Endurance ... 84
The Glycogen of the Muscle and the Liver ... 85
Athlete's Diet ... 86
Meat-Eaters versus Vegetarians ... 86
The Pre-Game Meal ... 87
Effect of Lack of Food and Fasting Upon Work ... 87
Questions ... 88

Chapter Six

THE ROLE OF OXYGEN IN PHYSICAL EXERTION ... 90

The Demand for Oxygen ... 90
Oxygen Debt ... 92
The Recovery Process After Exercise ... 94
Oxygen Intake and Oxygen Debt As Limiting Factors in Exertion ... 97
Intermittent Exercise ... 99
Factors Determining the Rate of Oxygen Intake ... 100
Questions ... 101

Chapter Seven

WORK, ENERGY AND MECHANICAL EFFICIENCY 102

Work 102
Methods of Measuring the Amount of Energy Used in Physical Activities 103
Calculation of Energy 106
General Procedure for Measuring the Amount of Energy Used... 107
Mechanical Efficiency of the Body 113
Efficiency During Aerobic and Anaerobic Work 114
Apparatus Employed in Measuring Work Output 114
Maximal Muscle Force During Isotonic Contraction 117
Questions 120

Chapter Eight

ENERGY COST OF VARIOUS ACTIVITIES 121

Basal Energy Requirements 121
Posture 122
Walking on a Horizontal Plane 122
Pack Carrying 125
Climbing and Going Down Stairs 126
Running 126
Skiing and Snowshoeing 129
Swimming 130
Calisthenics 134
Rowing 134
Football 134
Bicycling 136
Weight Lifting 137
Wrestling 138
Snow Shoveling 138
Energy Spent on Housework 140
Effect of Training on Basal Metabolic Rate (B.M.R.) 141
Effect of Training Upon Work Output and Efficiency 141
Questions 143

Chapter Nine

RESPIRATION 145

General Considerations 145
Pulmonary Ventilation in the Sedentary Individual 146
Minute-Volume of Lung Ventilation During Physical Work 146
Frequency and Depth of Respiration During Work 148
Vital Capacity of the Lungs 151
Pain in the Side 152
Alveolar Air 153
Alveolar Air Changes During Work 154
Nasal versus Mouth Breathing 154

Regulation of Respiration 155
The Control of Breathing Through the Blood Supply of the Respiratory Center 155
Control of Respiration Through Reflexes from Working Muscles. 157
"Second Wind" 159
Effect of Training on Respiration 161
Respiratory Gymnastics 162
Questions 163

Chapter Ten

BLOOD COMPOSITION AND TRANSPORTATION OF GASES 164

Buffer Substances in the Blood 164
Organs Responsible for the Regulation of the Acid-Base Balance. 166
The Transport of Oxygen 166
Oxygen Pulse 169
The Temperature of the Blood 170
The Transport of Carbon Dioxide 170
Lactic Acid 172
The Red Blood Corpuscles 174
Changes in the Count of White Blood Corpuscles After Muscular Activity 177
Effect of Training Upon White Cell Count 178
Blood Platelets or Thrombocytes 178
Specific Gravity of the Blood in Exercise 179
Effect of Exercise Upon Blood Sugar 179
Phosphates in Exercise 180
Athletes as Blood Donors 180
Questions 182

Chapter Eleven

BLOOD CIRCULATION AND THE HEART 183

Measurement of the Cardiac Output 183
Influence of Posture on Heart Output 184
Effect of Exercise Upon the Minute-Volume and Stroke Volume. 186
Blood Circulation Time 189
Effect of Training on the Stroke Volume 190
Effect of Training Upon the Heart 191
Cardiovascular Disease and Exercise 195
The Heart in the Prepubescent Child 195
Questions 197

Chapter Twelve

THE PULSE RATE 198

Pulse Rate and Age 199
Postural Pulse Rate Change 199

Food Intake and Time of Day, and Pulse Rate 201
Emotions and Pulse Rate 201
Pulse Rate Before Exercise 202
Pulse Rate During Exercise 202
Pulse Rate and Participation in Special Physical Activities 204
Heart Rate and Step-Up Exercise 205
Return of Pulse Rate to Normal 207
Relation Between Resting and Post-Exercise Pulse Rates 209
Regulation of the Frequency of the Heart Beat 210
Reflex from Working Muscles 211
Effect of Training Upon Pulse Rate 212
Questions 214

Chapter Thirteen

ARTERIAL AND VENOUS BLOOD PRESSURE 215

Function of the Arteries 215
Arterial Blood Pressure 216
Postural Blood Pressure Changes 217
Anticipatory Rise in Blood Pressure 218
Arterial Blood Pressure During Muscular Exertion 218
Factors Influencing Arterial Blood Pressure During Exercise 221
Postexercise Blood Pressure 222
Valsalva Phenomenon 222
Weight Lifting and Isometric Contractions 223
Effect of Training on Arterial Blood Pressure 225
Effect of Muscular Activity Upon Venous Pressure 225
Effect of Respiration Upon Venous Pressure 226
Questions 227

Chapter Fourteen

COORDINATION OF FUNCTIONS OF VARIOUS ORGANS FOR MUSCULAR WORK 228

Local Control 228
Remote Control 228
Epinephrine 229
Effect of Heat 230
"Milking" Action of Muscles 230
Effectiveness of Reflex versus Chemical Control 231
Factors Limiting Athletic Performance 231
Respiration 232
Transportation of Oxygen 232
Effect of Compensatory Adjustments on Digestion 233
Effect on the Kidneys 233
Summary 234
Questions 235

Chapter Fifteen

FATIGUE AND STALENESS 236

Types of Fatigue 236
Symptoms of Fatigue 237
Causes of Fatigue 238
Probable Seats of Fatigue 239
Factors Contributing to Inefficiency and Fatigue in Industry 241
Boredom 242
Staleness 242
Prevention of Nervous Breakdown 243
Questions 244

Chapter Sixteen

EXERCISE UNDER UNUSUAL ENVIRONMENTAL CONDITIONS 245

Temperature 245
Effects of High Altitude 255
Underwater Swimming 262
Questions 265

Chapter Seventeen

HEALTH, PHYSICAL FITNESS AND AGE 266

Definition of Health 266
Physical Fitness 267
How Much Physical Fitness is Necessary? 268
Relation Between Health and Physical Fitness 269
Mens Sana in Corpore Sano 271
Physical Fitness and Immunity to Disease 272
Allergic Reaction to Exercise 273
Indisposition and Collapse After Strenuous Exertion 274
Longevity of Athletes 275
Athletic Contests for Children and Adolescents 276
The Age of Maximal Proficiency in Sports and Athletics 278
Sources of Fitness 279
Fitness of the American Child 279
Questions 280

Chapter Eighteen

TESTS OF PHYSICAL FITNESS 281

Classification of Tests 281
Muscular Strength Tests 282
Heart Tests 282
Pulse Rate 283
Blood Pressure 283

Respiratory Tests 284
Breath Holding 284
Lung Ventilation 285
Oxygen Use 285
The Tuttle Pulse-Ratio Test 288
The Harvard Step-Up Test 289
The McCurdy-Larson Test 292
The Kraus-Weber Test 293
What Type of Tests Should Be Used? 293
Questions 294

Chapter Nineteen

BODY CONSTITUTION AND COMPOSITION 295

Body Constitution 295
Body Composition 304
Making Weight 312
Questions 313

Chapter Twenty

PHYSICAL ACTIVITY FOR CONVALESCENTS 315

Questions 320

Chapter Twenty-One

ERGOGENIC AIDS IN WORK AND SPORTS 321

Alcohol 323
Alkalies 324
Amphetamine (Benzedrine) 325
Caffeine 326
Cocaine 326
Fruit Juices 327
Gelatin and Glycine 327
Hormones 329
Lecithin 329
Oxygen 330
Phosphates 331
Sodium Chloride 332
Sugar 332
Tobacco Smoking and Athletic Performance 333
Ultraviolet Rays 335
Vitamins 335
Conclusions 337
Questions 338

BIBLIOGRAPHY 340

INDEX 361

Chapter One

SKELETAL MUSCLE

A discussion of skeletal muscle is a most appropriate starting point for a book on exercise physiology. The bodily adjustments which will be of concern throughout the text are made to meet the demands of contracting muscles for nutrients, oxygen, waste removal, and general homeostatic balance. The properties of muscle tissue also impose physical performance limitations. The present discussion is confined to skeletal muscle; smooth and cardiac muscle types will be discussed where appropriate.

THE MUSCLE CELL

The structural unit of the muscle is the *muscle fiber* or *cell.* The fibers are bound together to form the entire muscle (Fig. 1). Each fiber is composed of bundles of *myofibrils* which, with numerous nuclei and other subcellular structures, are suspended in the fluid *sarcoplasm* and enclosed by a membranous covering called the *sarcolemma* (Fig. 1). The myofibril is the smallest contractile unit within the cell. The myofibrils, in turn, are made up of even smaller, elongated structures called *myofilaments,* of which there are two types, *actin* and *myosin.* Actin filaments and myosin filaments are named for the proteins from which they are made. Interspersed throughout the cell and between the bundles of myofibrils is the *sarcoplasmic reticulum,* a tubular, fluid-conducting organelle.

In addition to the sarcoplasmic reticulum, there is a second tubular system located within the confines of the sarcolemma. This system is referred to as the *T-system.* It apparently opens to extracellular spaces via pores through the sarcolemma and contains extracellular fluid.[407] It is in contact with, but not a part of, the sarcoplasmic reticulum.

The intracellular structural organization of a muscle fiber is shown microscopically and diagrammatically in Figure 2. The fiber is cylindri-

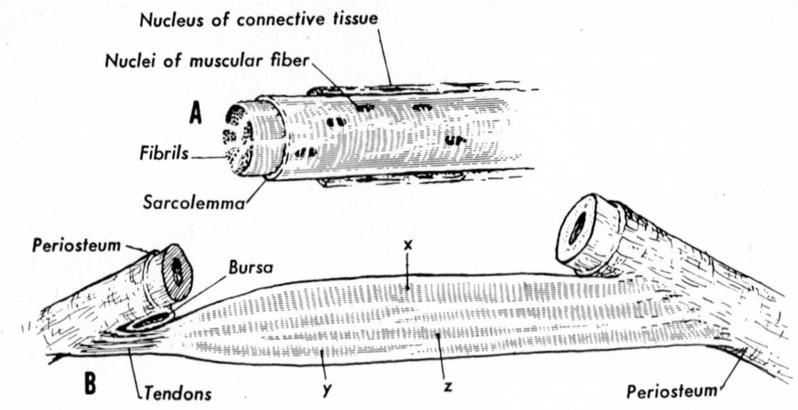

Figure 1. Skeletal muscle. *A*, Structure of a striated muscle fiber, showing part of the endomysium surrounding the sarcolemma, some nuclei, and a group of myofibrils. *B*, Gross structure of a muscle, showing some muscle fibers attached directly to the periosteum, some to a tendon, and others in an intermediary position which increases the length of the muscle. (Braus, P.: Anatomie des Menschen. J. Springer, 1924.)

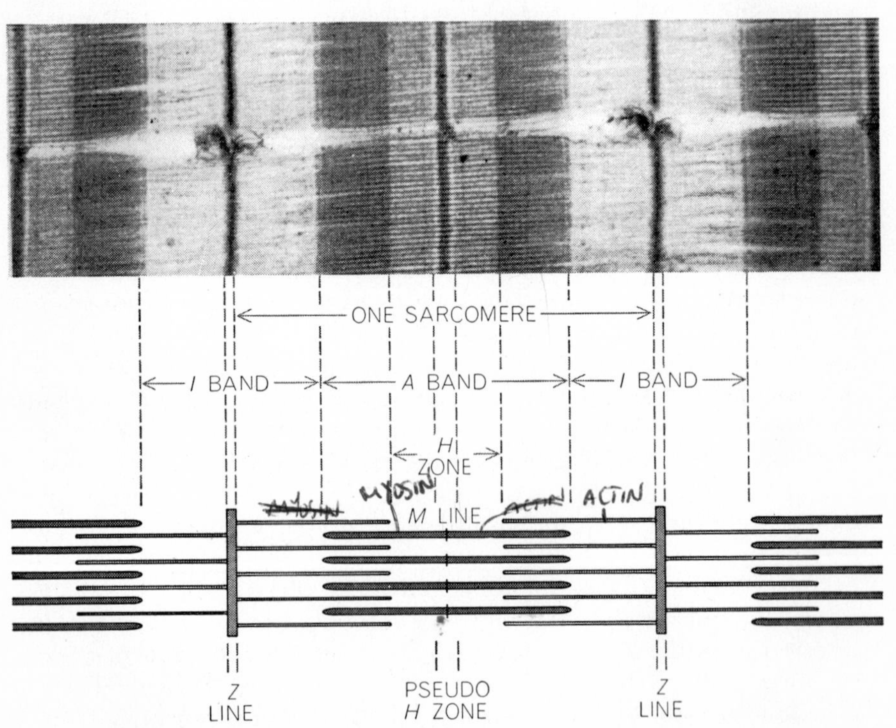

Figure 2. An electron micrograph of striated muscle from a frog. The structures and their functional significance are further described in the text. (From The Mechanism of Muscular Contraction, by H. E. Huxley. Copyright © 1965 by Scientific American, Inc. All rights reserved.)

cal in form and has tapered ends. It may be from 1 millimeter to 30 centimeters in length and from 10 to 100 microns in diameter. The relative cross-sectional dimensions of the component parts are shown in Figure 3. Because of its cross striations, skeletal muscle is frequently referred to as "striated muscle." These striations are due to alternate bands; the light bands are called *isotropic* or I-bands, the dark ones *anisotropic* or A-bands. The I-bands consist primarily of actin filaments, while the A-bands consist of myosin filaments with actin filaments interdigitated among them. The actin filaments are attached to membranous *Z-lines*. The portion of the fibril from one Z-line to the next is the *sarcomere*. The orderly arrangement of the myofibrils and their respective sarcomeres is responsible for the overall striated appearance of the fiber.

Figure 4 shows the schematic arrangement of the myofilaments during contraction, relaxation, and stretch. The myosin filaments have finger-like projections extending to the actin filaments. These projections are called *interfilamentous bridges*.

Other structures and some important inclusions are found within

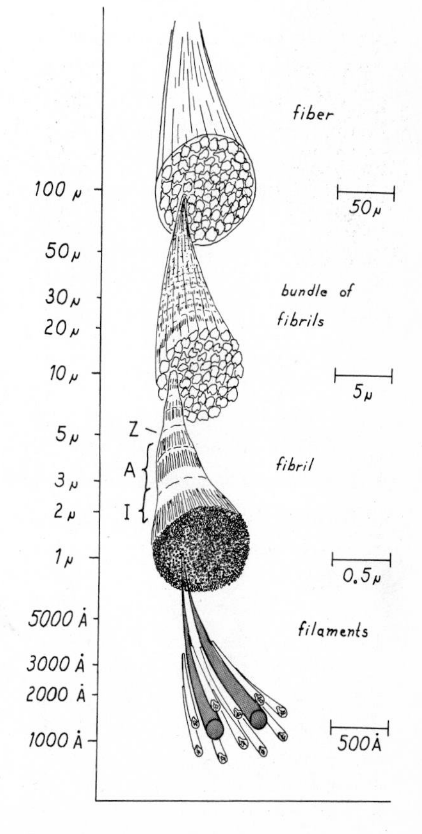

Figure 3. Cross-sectional dimensions of the structural components of the muscle fiber. Note that this is a logarithmic scale. (Buchtal and Kaiser: Dan. Biol. Medd. *21*:1, 1951.)

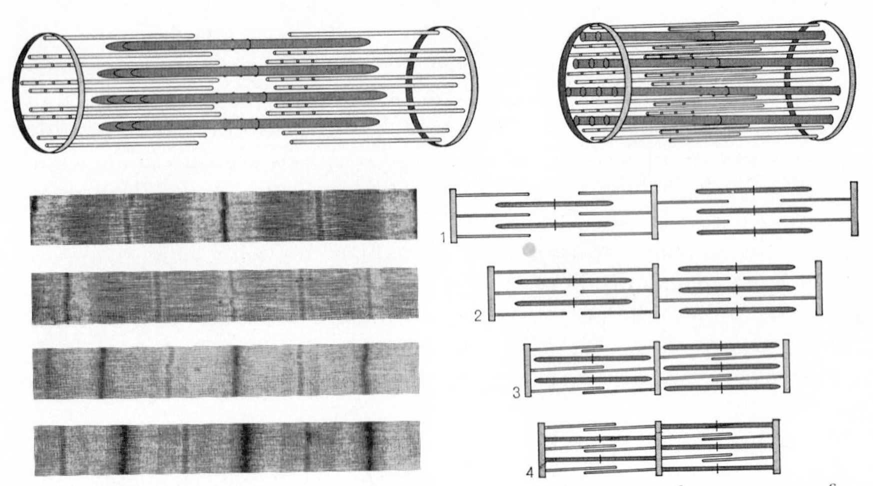

Figure 4. Changes in the relationships between the structural components of a myofibril during contraction, relaxation, and stretch. The actin filaments slide between the myosin filaments; note that the A-band remains at a constant length under all three conditions. (From The Mechanism of Muscular Contraction, by H. E. Huxley. Copyright © 1965 by Scientific American, Inc. All rights reserved.)

the sarcoplasm. The *mitochondria* are located between the myofibrils primarily in the I-band area, and are small organelles which contain enzymes essential for the utilization of oxygen in providing energy for muscle contraction. Scattered throughout the sarcoplasm are granules of *muscle glycogen,* which is a storage form of carbohydrate. Fat droplets are frequently seen in contact with the mitochondria.

MUSCLE STRUCTURE

A muscle is formed by a systematic binding of the muscle fibers by connective tissue, as shown in the cross-sectional diagram in Figure 5. Each fiber is wrapped in whole or in part by a delicate sheath of connective tissue called the *endomysium.* The endomysium winds around and between a dozen or more fibers, grouping them into a primary bundle or fasciculus. Collections of such bundles are also wrapped in a sheet of connective tissue, the *perimysium,* and form, with other primary bundles, a secondary bundle. Secondary bundles are bound into tertiary bundles by the *epimysium.* The epimysium is the connective tissue sheath surrounding the entire muscle. Large sheets of this connective tissue pass deep into the muscle from the epimysium, forming *septa.* Septa vary in size, and they may be further subdivided, but no special terminology is given to such subdivisions.[529]

Even though each type of binding connective tissue has distinctive

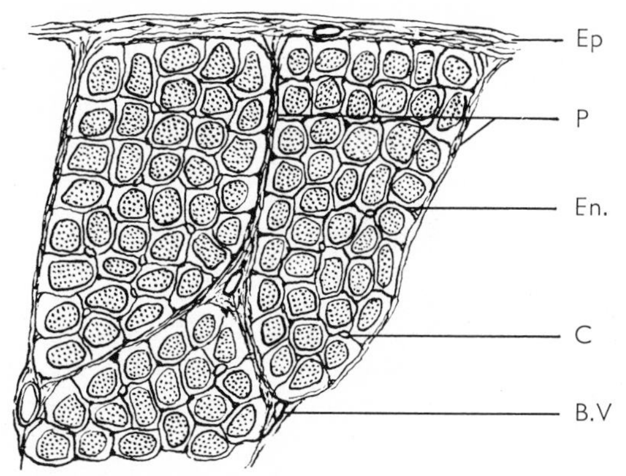

Figure 5. Organization of connective tissue binding of a skeletal muscle. The epimysium (*Ep*) surrounds the muscle and forms septa within the muscle. The perimysium (*P*) binds the bundles, while the endomysium (*En*) binds the individual fibers into bundles. Larger blood vessels (*BV*) are found as deep as the perimysium, while capillaries (*C*) extend as far as the muscle fibers. (Wells, E. W.: The Microanatomy of Muscle. *In* Structure and Function of Muscle, Volume I. Edited by Bourne, G. H. Academic Press, 1960.)

characteristics, there is a continuity of structure from one tissue type to another. At the muscle-tendon junction, some tendon fibers attach directly to the sarcolemma of the fibers, while others become continuous with the endomysium, perimysium, or epimysium. The tendon extends from the muscle to the bone and becomes continuous with the periosteum (Fig. 1). In some instances, muscles connect directly to the bone. As a result of this arrangement, contraction of the fiber is eventually passed to the bone, causing movement of the body segment.[529]

When a muscle is short (under two inches), the fibers run the entire length of the muscle. In longer muscles, bundles form chains which extend from one end to the other. As may be seen from the diagrams shown in Figure 6, some muscles are made of fibers lying parallel to the long axis of the muscle (*A*), while in other muscles, fibers form an angle with the long axis, producing pennate (*B*) and bipennate (*C*) arrangements. The latter arrangement greatly increases the strength of the muscle, because the force a muscle can develop is proportionate to the size of its cross section. The true cross section is perpendicular to the direction of the fibers, and is much greater in pennate and bipennate muscles than in parallel ones.

Although the relationship between potential force and cross-sectional area has been accepted for a long time, it has been difficult to measure the exact cross-sectional area of intact, human muscle. Ikai and Fukunaga[263] have recently reported a method of obtaining a cross-sectional view by use of ultra-sound waves and special recording equip-

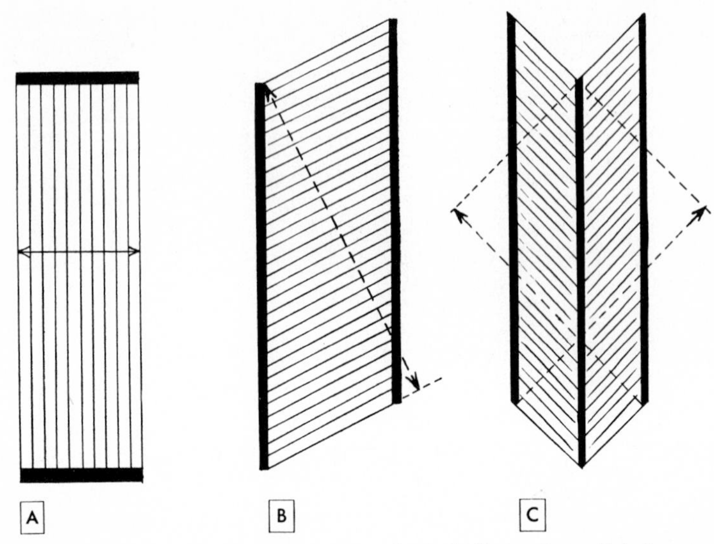

Figure 6. Different types of arrangement of muscle fibers: *A*, parallel; *B*, pennate; and *C*, bipennate. The true cross sections are indicated by interrupted lines. If the cross section of *A* is taken as 100 per cent, then that of *B* is 250 per cent and *C*, 450 per cent.

ment. They found an average value of 6.3 kilograms force per square centimeter of cross section in 245 healthy male and female subjects. There were no differences noted due to sex or amount of training. In another study, however, Ikai[262] reported increases in the cross section and force per unit of cross section as a result of training.

THE CHEMICAL COMPOSITION OF MUSCLE

The most abundant single constituent of muscle is water, which forms about 75 per cent of its mass. Proteins comprise about 20 per cent, and extractives and inorganic salts, 5 per cent. The constituents may be grouped as follows: (1) proteins, (2) carbohydrates and fats, (3) nitrogenous extractives (creatine, urea, and the like), (4) non-nitrogenous extractives (lactic acid, for example), (5) pigments, (6) enzymes, and (7) inorganic salts. Some of these substances are a part of the machinery of the muscle, others make up the fuel from which energy is derived, while others are no doubt only waste products produced from previous contracti e activity.

Among the proteins found in muscle are myosin (50 to 55 per cent), actin (20 to 25 per cent), and tropomyosin (10 to 15 per cent).[531] Actomyosin, the combined form of actin and myosin, is also found in small quantities. The sarcoplasm also contains a protein called myogen which is separate from the contractile proteins and the red pigment, myoglobin. This protein holds a small amount of oxygen for immediate needs during muscular contraction. Myoglobin has a greater affinity for oxygen than hemoglobin, but it dissociates oxygen five times faster: in 1/100 versus 1/20 of a second.

The most important carbohydrate is glycogen. It may be present in amounts varying from 0.5 to 1.5 per cent of the weight of the muscle, and is derived from the sugar of the blood. During muscle activity, this store of glycogen diminishes and, if the activity is sufficiently prolonged, may disappear entirely. The relation of glycogen to activity will be considered later.

The fat is located chiefly in the connective tissue of the muscle, but a certain amount is contained within the muscle fibers. Connective tissue, on analysis, yields the neutral fats, whereas the muscle fibers yield primarily cholesterol and phospholipids and small quantities of neutral fats.

THE BLOOD SUPPLY OF MUSCLE

Each muscle receives blood through one or more arteries. An artery branches freely to the perimysium, and from there supplies a profuse network of capillaries throughout the endomysium supplying the individual muscle fibers. The scheme of the general vascular distribution is shown in Figure 7. By this arrangement, the contractile material of the muscle cell is separated from the blood by only the surrounding connective tissue and the thin endothelial cells which constitute the capillary walls. The capillaries are abundant; there may be more than 4000 per square millimeter* of muscle cross section.

*About the area of cross section of the lead in a pencil.

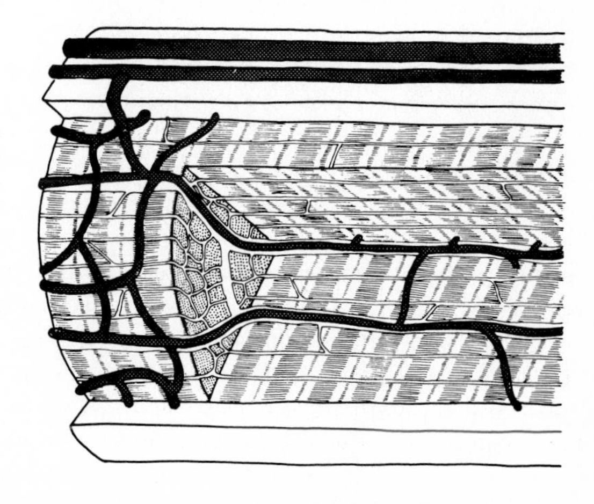

Figure 7. The distribution of vascular structures within the muscle. Each muscle fiber is close to a capillary, which facilitates exchange between the cell and the blood. (Ruch, T. C., and Patton, H. D. (editors): Physiology and Biophysics, 19th ed. W. B. Saunders Co., 1965.)

Veins follow the general distribution of the arteries. It is of interest to note that even their smallest branches are equipped with valves. When the veins are squeezed during muscle contraction, blood is pushed along in the direction of the heart, since the valves prevent a backflow. During relaxation the veins are filled again. Thus, one can speak of a "milking" action of muscles in reference to venous circulation.

There are no lymphatic ducts between the individual muscle fibers, but they are found in the fascia, and they penetrate as far as the perimysium. The intercellular fluid, therefore, drains slowly through connective tissue between the muscle fibers until it reaches the nearest lymph vessel.

MUSCLE INNERVATION

A muscle is also supplied with one or more branches from larger nerve trunks. The branches contain both afferent and efferent nerve fibers, each kind making up approximately one-half of the total number.

The afferent fibers arise from sensory endings found within the muscle and tendons. Some of these fibers arise from bare nerve endings and pressure-sensitive pacinian corpuscles. Others have their origins in one of three types of stretch-sensitive endings: the *annulospiral* and *flower spray* terminations of the muscle spindle or the *Golgi tendon organ.* The spindle and tendon organs are very important for the regulation of muscle contraction and will be discussed in detail relative to total nervous system function in Chapter 3.

The efferent fibers are of two types: either large-diameter or small-diameter. All these fibers have their origin in motor neurons located in the ventral horn of the gray matter of the spinal cord (see Chapter 3). The collection of motor neurons innervating a given muscle is called the *motor pool.* The larger nerves supply stimulation to the large muscle fibers (*extrafusal fibers*), leading to the production of contractile force (see Fig. 25). The smaller nerves supply small muscle fibers (*intrafusal fibers*) lying within the structure of the muscle spindle. Artificial stimulation of the small nerves provides no readily apparent contractile force.

The individual muscle fibers are, so to speak, insulated from each other. Each muscle fiber is supplied with at least one motor nerve ending which penetrates its sarcolemma. Inside the sarcolemma, this nerve branches into irregularly shaped terminations embedded in a granular substance which represents the remnants of undifferentiated muscular protoplasm. This combination of nerve fiber and granular substance constitutes the *motor end plate,* through which a motor nerve stimulus causes the fibers to contract (see Fig. 14, Chapter 3).

A motor neuron and the muscle fibers it innervates are referred to as a *motor unit.* The unit operates on an all-or-none principle in that all of the fibers will contract on stimulation by the motor nerve. The number of muscle fibers in a motor unit is related to the degree of precision of contraction required of the muscle. The intrinsic muscles of the eye, for example, may have only three fibers in a unit, while motor units of the thigh have several hundred. The *innervation ratio* is the numerical expression of the average number of muscle fibers per motor unit in a muscle. Motor units made of long fibers may have more than one end plate for each fiber. Muscle fibers belonging to one motor unit are usually interspersed among fibers belonging to other motor units.

Unmyelinated sympathetic fibers, for the control of smooth muscles in the blood supply system, are also found going to skeletal muscles. Some investigators have claimed that skeletal muscles are directly innervated by them, but positive histological proof is absent. The consensus is that sympathetic fibers which innervate the blood vessels of muscles may occasionally terminate in the muscles, or at least appear to do so.

RED AND WHITE MUSCLE

Skeletal muscles vary in color; some are pale, whereas others contain more pigment, which causes a dark color. The red muscles are found in places where sustained contractions are required. Histologically, the fibers of the red muscle differ from those of the white in that they contain more myoglobin. The red fibers are the less irritable, give a slower but greater contraction, and can maintain contraction for a longer period than can the white fibers.

In man, red and white fibers intermingle. Some muscles, however, have a predominance of either red or white fibers. For example, the soleus has more red fibers than the gastrocnemius. Therefore, it is thought that the gastrocnemius acts first and that the soleus then joins in sustaining contraction.

THE CONTRACTILE PROCESS

Muscle contraction is a complex process, dependent not only on the motor neurons and muscle fibers, but also on the effects imposed by the central nervous system. A discussion of the latter is included in Chapter 3; the purpose of the present discussion is to consider only the interaction between a muscle fiber and its nerve. A knowledge of the elementary physicochemical and cellular changes accompanying muscle contraction is essential if one is to truly understand the physiological significance of strength, speed, and other factors related to muscle contraction in man, and some of the techniques used to measure them.

Three processes are important in the contractile process: (1) the development of the *transmembrane potential* and the transmission of the nerve impulse, (2) the transfer of the nerve impulse to the fiber (*neuromuscular transmission*), and (3) the activation of the contractile process (*excitation-contraction coupling*). Only those physiological changes necessary for an understanding of exercise physiology will be discussed here. The student who is interested in studying this topic in depth should consult more advanced texts (see references *50, 200, 444*).

The Transmembrane Potential. Nerve and muscle cells are especially adapted to develop a transmembrane potential, an electrical charge across the cell membrane. This adaptation enables these cells to develop and transfer the nerve impulse that leads to muscle contraction.

The potential is developed in the nerve and muscle fibers as a result of the physical and chemical properties possessed by certain ions and by the cell membrane. The cell contains protein-bound *anions* (negatively charged ions) which cannot leave the cell, and therefore attract the *cations* (positively charged ions) of sodium (Na^+) and potassium (K^+) to the cell wall. The Na^+ ions are forced to the outside of the cell and the K^+ ions to the inside by the so-called sodium-potassium pump, the active transport system that forces this movement of Na^+ and K^+ with the concomitant expenditure of energy.

As a result of the active transport process, each kind of ion becomes unequally distributed across the cell membrane (Na^+ out, K^+ in), causing a *concentration gradient.* Because substances always tend to diffuse to the area of lowest concentration, Na^+ tries to move into the cell and K^+ tries to move out. K^+, however, tends to diffuse out 50 times more easily than Na^+ diffuses in because of physical characteristics of the ions and the cell membrane. Consequently, even though Na^+ and K^+ are pumped on a one-for-one ratio, more positive charges accumulate outside the cell than inside, creating an electrical potential. This potential can be measured, and has been found to be approximately −85 millivolts (1 millivolt is $1/1000$ volt), with the inside negative to the outside. Chlorine ions (Cl^-) remain outside the cell with Na^+ to maintain electrostatic balance, since anions within the cell cannot escape. The *resting potential* is thus created.[548]

The nerve or muscle membrane maintains the resting potential unless the potential is destroyed by a chemical, mechanical, or electrical stimulus. This process is called *depolarization.* When it occurs, an area of the membrane suddenly becomes permeable to Na^+, which rapidly diffuses into the cell, while K^+ diffuses out, causing a reversal of the transmembrane potential (the inside becomes positive to the outside).

The depolarization of one area causes adjacent areas to suddenly become permeable to Na^+, so there is a progressive depolarization along the length of the nerve or muscle fiber, leading to the transmission of the nerve impulse. The depolarization is soon followed by repolarization, as shown in Figure 8. The wave of depolarization is called the *action potential* and is an all-or-none process.

Neuromuscular Transmission. Neuromuscular transmission occurs at the end plate which consists of the nerve's terminal ending and the adjacent muscle fiber membrane (see Fig. 14). The material of the neuron is not continuous with the material of the muscle fiber, so some provision must be made for the transfer of the impulse. This is accomplished via a *transmitter substance* called *acetylcholine* which is found in packet-like vesicles located in the terminal ending. The action potential sweeps along the axon to the terminal ending. On arrival of the impulse, the packets of transmitter substance are first mobilized and then released onto the adjacent muscle fiber where its depolarization begins. The latter depolarization precedes *excitation-contraction* coupling.

Excitation-contraction Coupling. The currently accepted explana-

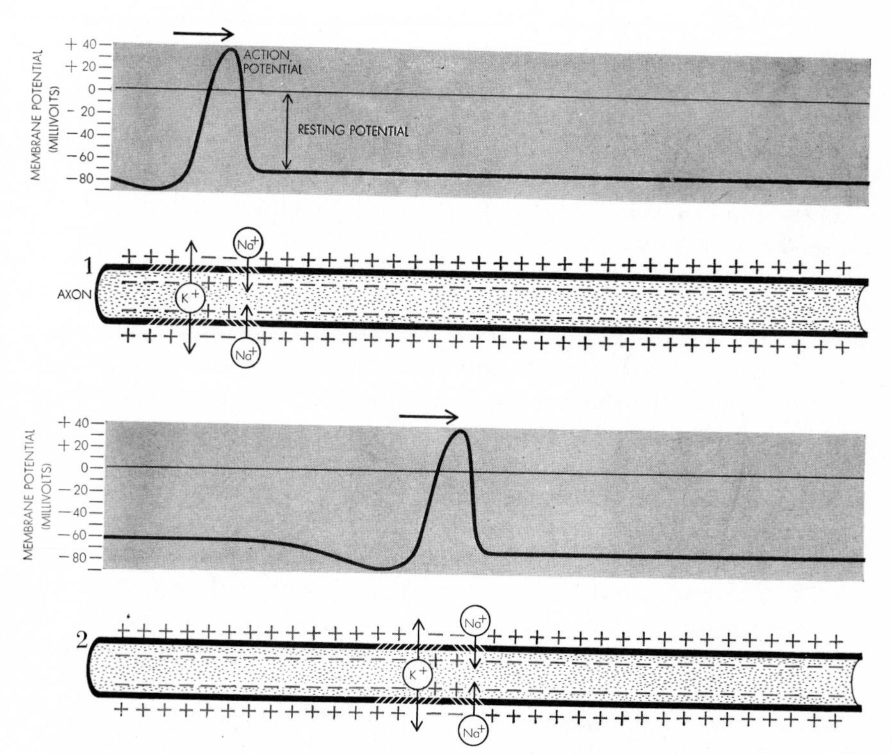

Figure 8. The process of depolarization and repolarization of the nerve fiber with the passage of the nerve impulse. (From How Cells Communicate, by B. Katz. Copyright © 1961 by Scientific American, Inc. All rights reserved.)

tion of muscle contraction is the *sliding filament theory*.[257] Essentially, the theory is that contractile force is developed by the myosin filaments pulling the actin filaments between them by means of the interfilamentous bridges (Figs. 2 and 3). As the actin filaments slide between the myosin filaments, the Z-lines are pulled toward the middle of the sarcomere, causing overall shortening. Contraction requires the energy which is provided by the breakdown of adenosine triphosphate (ATP) to adenosine diphosphate (ADP), inorganic phosphate, and energy. In the process, a portion of the myosin filament located in the interfilamentous bridge apparently acts as the enzyme precipitating the reaction.

The coupling between excitation and contraction is not completely understood, but it apparently depends on the rapid release of calcium ions (Ca^{++}) from the sarcoplasmic reticulum. The Ca^{++} is released by the wave of depolarization sweeping down the fiber after stimulation. The T-system, because it penetrates into the innermost areas of the fiber, permits the effect on Ca^{++} due to changes in the Na^{+} and K^{+} concentrations at the cell membrane to be transmitted rapidly to the depths of the contractile machinery, thereby facilitating the rapid response which is necessary for effective contraction. The Ca^{++} is recovered (*sequestered*) to the sarcoplasmic reticulum during fiber relaxation, and the ATP is restored to its resting levels.[548]

Chemistry of Muscle Contraction. All cells, including those of muscle, derive the energy needed for life from the universal source, ATP. This compound aids in the transportation of various substances through the cell membrane and in chemical synthesis within the cell, and provides energy for muscle contractions. ATP is formed in all cells from carbohydrates, fats, and proteins. The amount of ATP available at any moment is so small that it can maintain maximum muscle contraction for only one-half second. This means that some mechanism must be put in operation to cause an immediate resynthesis of ATP. The emergency mechanism is the action of phosphocreatine (PC), which combines with ADP to form ATP. This emergency mechanism, however, lasts for only seconds, until the PC supply is exhausted. After this has happened, ATP is synthesized by glycolysis, which is the breaking-down of muscle glycogen into pyruvic acid and then, if oxygen is not available, into lactic acid. This process extends the ability of the muscle to contract, until an accumulation of lactic acid stops it.

The aspect of muscular chemistry so far discussed is called the anaerobic phase, because no oxygen is required. For contractions to continue, oxygen is required, and this phase is called aerobic. When oxygen is present, glycogen is broken down only to pyruvic acid, which then enters into a chemical process called the Krebs cycle and is finally reduced to carbon dioxide and water. The energy released during this process is used for resynthesis of PC and ATP. During the oxidative process, 18 times more energy is released than during anaerobic glycolysis. However, not all the energy obtained anaerobically and aerobically is utilized. About 39 per cent is dissipated as heat. Anaerobic and aerobic processes may go on simultaneously.

The provision of energy for muscular activity is a complex process that will be discussed in more detail in Chapter 6. It is well to note at this time, however, that it is the breakdown and resynthesis of ATP that triggers into action the systemic body changes that accompany exercise.

MUSCULAR CONTRACTION

The terminology applied to various types of muscular contraction is well differentiated only philologically. In practice, there is some unavoidable deviation from the true meaning of the words. Muscle is said to contract *isometrically* when its length does not change. In order to record tension developed in isometric contraction, however, a slight shortening is allowed. Muscle shortening during contraction, while the load remains the same, is known as *isotonic* or *dynamic* or *phasic* contraction. Strictly speaking, there are no purely isotonic contractions in the body. This misnomer is probably used in order to avoid such a cumbersome term as hetero-tonic-metric contraction, which indicates a change in both length and tension. When a contracting muscle lengthens while resisting an external force, this state is referred to as an *eccentric* contraction; whereas, when a muscle shortens, it is called a *concentric* contraction.

Excised Muscle Experiments. Most available knowledge on the chemical and physical changes accompanying muscle contraction has come from experiments on excised muscles. It is expected that the reader will be familiar with experiments commonly used in elementary physiology courses. Pertinent basic equipment and experiments are presented here; if the student desires to review these extensively, elementary physiology books should be consulted.

The equipment used to study excised muscle contraction is diagramed in Figure 9. Sample recordings from kymograph records illustrate important phenomena. The *threshold* is the amount of stimulus necessary for minimal contraction and would involve the fibers of at least one motor unit. As the stimulus is increased, more fibers (as units) contract, eventually increasing the response to a maximum. A typical single contraction, or *twitch*, is recorded when the kymograph turns fast, and the stimulus is high enough to cause maximum contraction. The twitch is typified by a brief *latent period* between the application of stimulus and muscle response, a *shortening period*, when force is applied to the recording pen, followed by a longer *relaxation period*, when the

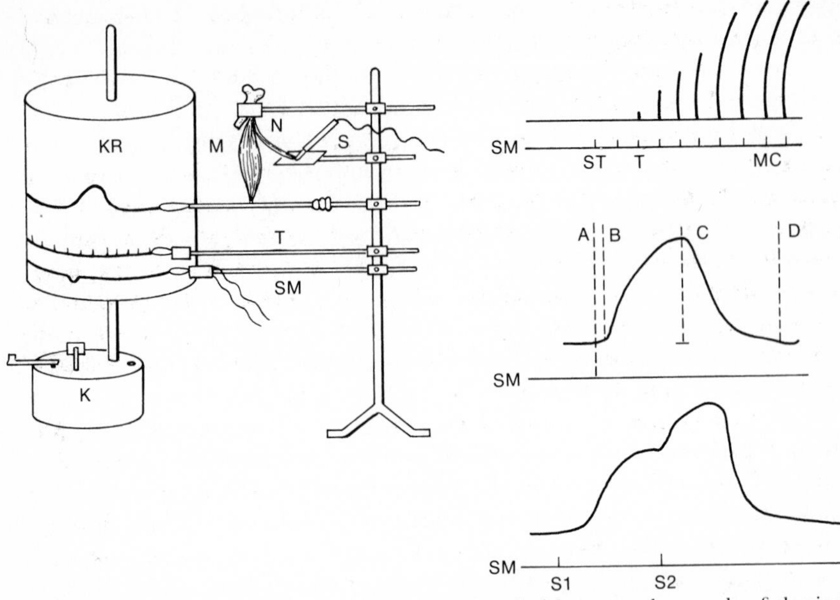

Figure 9. The equipment arrangement and typical kymograph records of classical excised muscle experiments. For the apparatus: *K*, kymograph; *KR*, kymograph record; *M*, muscle; *N*, nerve; *S*, stimulating electrode; *T*, time indicator; *SM*, stimulation marker. Top record showing recruitment of motor units: *SM*, stimulation mark (each mark represents a stimulus of higher voltage); *ST*, subthreshold stimulus; *T*, threshold stimulus; *MC*, maximum contraction. Note that this represents a very slow drum speed. Middle record showing simple muscle twitch: *SM*, stimulus mark; *A-B*, latent period; *B-C*, shortening; *C-D*, relaxation. Bottom record summation: *S1* and *S2* indicate the two maximum stimuli.

muscle returns to its resting length. If the muscle is restimulated before relaxation is complete, the recording for the second contraction will be higher than the first, a phenomenon called *summation.* The second stimulus must be applied after the refractory period, during which the muscle does not respond to further stimuli. The frequency of muscle stimulation determines the amount of summation.

Muscle Contraction in the Body. Excised muscle experiments provide information about the dynamics of contraction but they do not provide complete information about what happens *in vivo. Electromyography* is used to obtain an understanding of the latter. Electromyography is the recording of electrical changes occurring in muscles during contraction. The ionic changes accompanying the transmission of the nerve impulse and muscle contraction cause changes in the electrical potential which are transmitted away in waves via the adjacent fluids. These changes are picked up by *electrodes,* amplified, and recorded by appropriate equipment. Typical electromyograms are shown in Figure 10. Basmajian[44] has presented an excellent review of electromyographic technique and completed research. O'Connell and Gardner[395] have contributed brief, practical directions for the use of electromyography in kinesiological research.

The amount of force generated by a muscle is dependent upon the *summation* of the contractions of many motor units. Individual motor units can be consciously activated but produce either no movement or one that is barely perceptible.[44] The summation process is dependent upon the frequency of unit contraction (*temporal summation*) and the number of units active at one time (*spatial summation*). Some motor neurons are more sensitive than others and consequently pass impulses to their fibers at only minimal stimulus to the pool. As the amount of stimulation increases, less sensitive neurons become recruited, until maximum contraction occurs. Since motor unit *recruitment* is *asynchronous,* the contraction of the entire muscle is smooth.

Electromyographic studies have shown that individual motor units contract at frequencies of up to 40 per second, a rate that establishes tetany. This rate is apparently maintained in the more sensitive units as less sensitive ones are recruited to increase contractile force. Bigland and Lippold[56] found that the recruitment of units accounted for most of the force variation during submaximal contractions.. The electro-

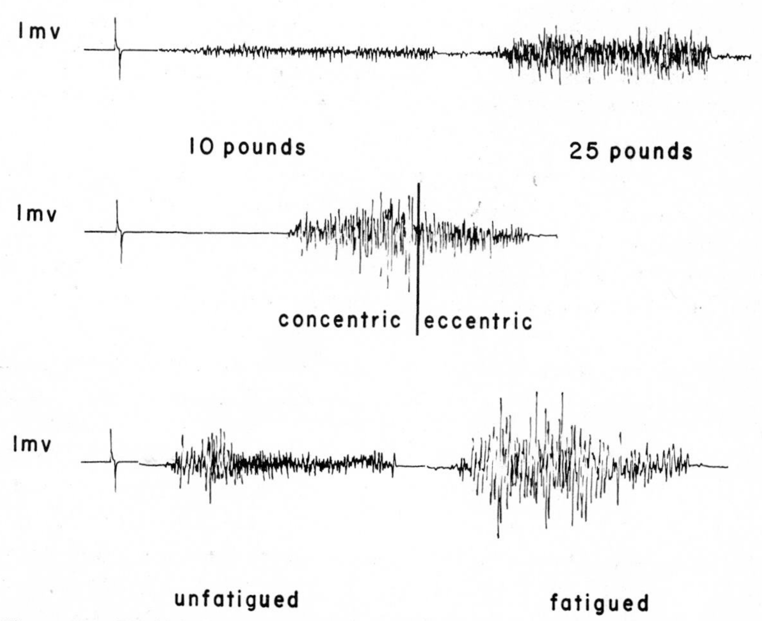

Figure 10. Typical electromyograms taken from the biceps brachii during forearm flexion and extension. Upper record taken while holding designated weights in hand, elbow flexed to 90 degrees. Middle record taken while performing concentric-eccentric contractions holding 25 pounds; bottom record shows same movements to fatigue.

myogram reflects both kinds of activity, that is, changes in frequency and number of stimuli.

Quantitative electrical changes recorded by electromyography have a high relationship with the amount of force developed.[56] Such a relationship for isotonic contractions is shown graphically in Figure 11. It is interesting to note that there is less electrical activity during the eccentric phase than during the concentric phase, when the same amount of force is developed. Electromyograms taken during concentric and eccentric contractions against a constant load are shown in Figure 10.

With continued submaximal contractions against a resisting load, there is a gradual increase in total electrical activity, while that of individual motor units becomes diminished but *synchronous* (each motor unit fires in a definite rhythm).[455] As motor units fire repetitively, some of the muscle fibers are apparently unable to respond, and more units are activated to exert the same amount of force. This phenomenon

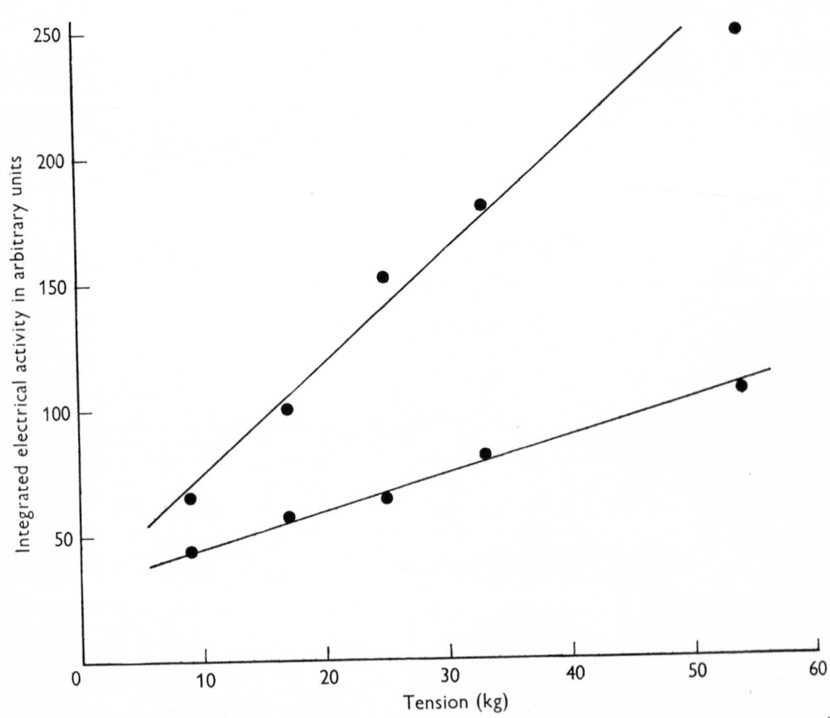

Figure 11. The relationship between integrated electrical activity during concentric (upper line) and eccentric (lower line) contractions. Records taken with surface electrodes from the calf muscles. (Bigland, B., and Lippold, O. C. J.: J. Physiol., *123*:214, 1954.)

occurs with the fatigue developed during activities such as push-ups, pull-ups, and repetitive movements with weights.

PHYSICAL PROPERTIES OF MUSCLE

Elasticity. Muscle possesses certain mechanical features as a result of its structure. The muscle can be envisioned as consisting of two elements: a *contractile element*, which is composed of the myofibrils, and an *elastic element*, composed of the tendons, the investing membranes, the sarcolemma, and the non-contractile materials of the muscle cell. The elastic component may be either in series or in parallel to the contractile mechanism.[239, 542, 543]

The series elastic element is of most concern to us here. Because of its presence, the contractile material must stretch the elastic material until the force stored in the elastic material is sufficient to overcome the inertia of the load. The student may illustrate the same phenomenon by attaching a rubber band to objects of different weights and lifting them. After the rubber band has been stretched enough to store sufficient energy, the objects will be lifted. Bouissett, Goubel, and Lestienne[61] have demonstrated that muscle contraction, as revealed by electromyography, precedes tension development. Peak tension is reached at the end of muscle tension, and then slowly deteriorates to resting values.

This arrangement does have value. Hill[239] noted that the elastic component serves as a buffer between the contractile element and its load when there are abrupt transitions from the passive to the active state. The elastic material smooths out rapid changes in tension and accumulates mechanical energy as contraction continues. The accumulated mechanical energy can thereafter be released, producing a movement velocity higher than the maximum rate of fiber shortening. The elasticity undoubtedly also serves to protect the contractile material by absorbing force in activities such as walking or jumping.

Cavagna, Saibene, and Margaria[90] have indicated that the elastic component may also serve as a storehouse of energy during rapid, rhythmical movements. They found that a frog muscle performed more positive work if it was stretched immediately before it was allowed to shorten; the less the interval between stretching and shortening, the greater the effect. During running, the muscles providing force are stretched immediately before applying forward thrust, with the elastic recoil providing some of the force for forward propulsion. The same authors estimated in an earlier study[91] that from 40 to 50 per cent of the energy expenditure during running was supplied from the recoil. Such findings indicate a basis for increased efficiency through the use of rhythmical movements.

Velocity and Tension During Contraction. The velocity of muscle shortening decreases as the load increases. The change in velocity with increased load is illustrated in Figure 12. The equation relating the two variables was derived by Hill and follows the form: $V = (P_0 - P)\,b/(P + a)$, where V is the velocity of shortening, P_0 is the maximum tension possible, and a and b are constants with the dimensions of force and velocity respectively.

When the intact muscle is stimulated for maximal rate of contraction, the number of motor units recruited is apparently the same, no matter whether the limb is unloaded or holding a heavy object.[540] In the latter case, however, stimulation would last a longer period of time. Because of the equality of stimulation for all loads, the force-velocity relationship must be explained on the basis of characteristics of the muscle tissue rather than of the central nervous system.

Fenn and co-workers[166] believe that the cause of the loss of velocity with increased load is chemical rather than physical, and that it may be explained as follows: Contraction of a muscle depends on energy liberated during certain chemical reactions in the muscle. These reactions are characterized by the constancy of the rate; thus the rate of liberation of energy necessary for contraction is also constant. Since less energy is required to lift a lighter load, it can be produced in a shorter time than a larger amount of energy required for lifting a heavier object, and, therefore, with a lighter load, the muscles will be able to contract faster.

Huxley[257] has also noted that the load-velocity relationship supports the sliding filament theory in that, during rapid contractions, there would be less time for interconnections to be formed between actin and myosin filaments and, therefore, less time for force to be developed.

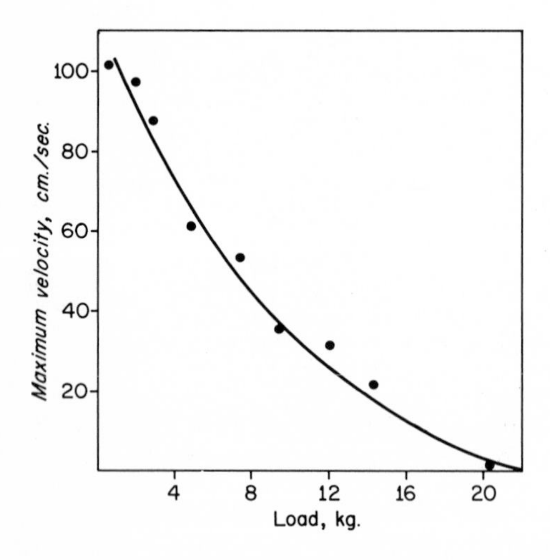

Figure 12. The relationship of load to maximum speed of shortening. (As adapted by Woodbury, J. W., Gordon, A. M., and Conrad, J. T.: Muscle. *In* Physiology and Biophysics, 19th ed. Edited by Ruch, T. C., and Patton, H. D. W. B. Saunders Co., 1965.)

The force-velocity relationship has application when one is considering the best load for optimal work and power performance. Wilkie[543] noted that the relationship between load and velocity is unknowingly observed by designers of tools and other equipment. Bicycles, for example, have gear ratios that take the best possible advantage of the load-velocity relationship. In bicycles with a number of available gear ratios, the cyclist can adjust to suit his situation.

QUESTIONS

1. Develop working definitions of the following terms: muscle fiber, myofibril, myofilament.
2. What is a sarcomere?
3. What are the mitochondria? What is their function?
4. Describe the way in which muscle fibers are bound together to eventually form a muscle.
5. What is the transmembrane potential of nerve and muscle cells? How is it developed?
6. What are the names of the "contractile" proteins of muscle?
7. What is muscle glycogen? How much is found in the muscle?
8. What is a motor unit? Motor pool?
9. What is the "sliding-filament theory" of muscle contraction?
10. What does "ATP" stand for? "PC"? Why are these substances so important?
11. Distinguish between the following adjectives describing muscle contraction: isotonic, isometric, concentric, eccentric.
12. Describe how motor unit contraction is varied in order to develop varying amounts of force by the muscle.
13. What is elasticity?
14. What function does the elasticity of muscle serve?
15. What is the relationship between velocity and tension? What is the physiological basis for this relationship? What practical functional value does it have?

Chapter Two

MUSCLE TRAINING AND RELATED PROBLEMS

Probably no other organ demonstrates as easily as the muscle the old Lamarckian slogan that "function makes an organ." If muscles are not used, they atrophy. If one wishes to develop them, one should use them. Children are more active than adults because exercise is needed for proper development, in general, and for muscles, in particular.

Although muscle training may affect the other organs, our discussion in this chapter will be limited to changes in muscles only.

Even though muscle training has been practiced since time immemorial, and obviously with remarkable success, one may be surprised to discover that even now there is no complete agreement as to the *best method* for muscle training. The usually accepted idea has been that one has to give all he has in order to get a maximum development. This chapter indicates that such a philosophy is not necessarily true. *It seems possible to get maximum gains without investing maximum effort.* The effects of muscle training are changes in structure, in strength, in endurance, and in speed.

CHANGES IN STRUCTURE OF MUSCLE

Regular and heavy muscular work tends to thicken and toughen the sarcolemma of the muscle fiber and to increase the amount of connective tissue within the muscle. As a result of the latter change, the meat of heavily worked animals is tougher and coarser than that of those that have lived inactive lives.

Muscle Size. The use of a muscle causes an increase in its size. This growth is, in some way, linked with the destruction of constituents of the muscle that takes place during strenuous muscular contraction. When nature replaces the lost materials, she overcompensates.

Examples of the law of overcompensation are seen in the production of antitoxin and other immunizing substances, in which case the body is subjected to the action of disease-producing toxins, or in the development of callus on the palm of the hand as the result of friction and pressure which remove the superficial layers of the skin. There is reason to believe that the number of fibers in a muscle is not increased by training. In every muscle there are latent or unused fibers and fibers that are small from lack of use. These develop in response to the increased demands made upon them. According to Petren and his coworkers,[409] some increase in size is due to an increase in the number of capillaries. Exercises of strength induce hypertrophy of the muscle fibers; exercises of endurance induce capillarization. The classic experiments on dogs indicate that about 7.7 per cent of the increase in cross section may be caused by hypertrophy of fibers.[385]

GAIN IN STRENGTH

The only way by which the strength of the muscles can be developed is by exercising them against gradually increasing resistance. For this purpose one can use springs, weights, or the weight of the body itself. Even though the same method of training is used, the rapidity and the ultimate degree of development in different persons will be different. Some of this variation may be explained and even predicted through examination of the anatomical characteristics of individuals. Usually, a person with small bones, or a tall, skinny individual, will reach the limit of development before a stocky man with large bones attains the limit. Anatomical variation only partially explains this. The complete explanation is still a mystery.

In this connection it may be worth while to recall the ancient story of Milo of Crotona, who was able to carry a four-year-old bull because he had practiced doing so from the time it was a calf. This story has been re-enacted on a radio program. A 17-year-old boy, weighing 149 pounds, started a daily feat of lifting a calf, which on the first day weighed 75 pounds. This continued for 201 days, and the boy had to give up the feat when the calf had gained 290 pounds; the boy had gained 3 pounds.

Systematic training may lead to tremendous development in muscular strength. L. Zhabotinski, Olympic weight-lifting champion in 1964, could press 413.5 pounds, snatch 369 pounds, and clean and jerk 479.5 pounds—an impressive total of 1262 pounds.

Reports coming from E. A. Müller's laboratory in Germany have made a great impact upon methods of muscular training.[388] These reports indicate that a single daily isometric contraction continued for six seconds and utilizing only two-thirds of maximum strength will give the best results in gaining muscular strength. At this amount of tension, capillaries of the muscles are compressed and oxygen supply becomes inadequate.[445] It seems that this oxygen deficit is an important factor in the acquisition of muscular strength.

It is hard to accept these reports, because they apparently contradict everyday experience. Just think about muscle-men working one to two hours a day at least three days a week in order to develop strength. Maybe they are just wasting their time. Maybe!

Analysis of a training session which may take from one to two hours shows that the time spent on actual lifting is not more than two to six minutes.[390] Moreover, the single lifts take only a few seconds; clean and jerk, 3.30 seconds; snatch, 3.48 seconds; and press, 4.12 seconds.

Müller and Hettinger[389] reported that repetition of contractions is not more effective than a single one. However, Asa,[10] in his ingenious experiments, obtained contradictory evidence. Asa used the abductor of the little finger, using one hand as experimental and the other as control. Isometric contractions repeated 20 times gave better results than a single contraction. But Asa substantiated Hettinger's report that isometric contractions give better results than isotonic ones. Asa's technique is already being used on patients, with remarkable results.

Isometric exercises are now widely used for conditioning and also as an adjunct in athletic training. It has been claimed that they have been responsible for weight-lifting records. However, it is difficult to find authenticated evidence for these claims.

Royce[446] reviewed a number of articles by Hettinger and Müller published between 1953 and 1963, and found some discrepancies regarding the effectiveness of isometric exercises as expressed in percentage of weekly increase over the initial strength. In 1953 it was 5 per cent, in 1958 it became 3.3 per cent, and in 1961 it dropped to 1.8 per cent. Since 1953, however, isometric training has mushroomed and even become "big business." Royce asks these questions: "Do top athletes *really* increase their static strength? Is a better athletic performance *really* related to such an increase?" And he suggests that it is now time to be "off the bandwagon and back to the drawing board."

Berger[51] found experimentally during three weekly exercise sessions for 12 weeks that the best results were obtained with three sets of six repetitions per minute. The training programs varied as follows: one, two, and three sets with 2, 6, and 10 repetitions.

BODY WEIGHT AND STRENGTH

Let B_W represent the body weight and W the total weight lifted. The total weight is the sum of the weights lifted in press, snatch, and clean and jerk. It is an established fact that strength is proportionate to the cross section of the muscle. For the sake of mathematical clarity, let us assume that a man, whose body weight is B_W, has the shape of a cube. The side of this cube is then equal to $\sqrt[3]{B_W}$, and the cross section is $(\sqrt[3]{B_W})^2$, or $B_W{}^{2/3}$. Therefore, we can write that W is proportionate to $(B_W)^{2/3}$, or log W is proportionate to ⅔ log B_W. If the weight is plotted against the body weight, a curve is obtained. If, however, the corresponding logarithms are plotted, a straight line results.

Following this reasoning, Lietzke[343] used 1955 world weight-lifting records and obtained a formula:

$$W = aB_W{}^{.6748}$$

Here, *a* is a constant equal to 28.7, and .6748 is very close to the theoretical exponent of ⅔. In order to obtain a straight line relation, Lietzke used a logarithmic formula:

$$\log W = 1.458 + .6748 \log B_W$$

After the publication of Lietzke's formula, several records were broken. To fit the new records, the senior author then calculated a new formula:

$$\log W = 1.4718 + .6748 \log B_W$$

Although the record of the 198-pound class and especially that of the 148-pound class were much below the new level, the prediction was made that they must soon improve.[293] The same year this prediction was made, a new record for the 148-pound class was established: 865.5 pounds, differing by about 1 pound from the predicted 866.6 pounds.

Since then more records have been broken. Examination of the records of 1963 indicated that a new prediction formula was in order. This was accomplished by solving an equation

$$X = \log W - .6748 \log B_W$$

where X is a constant for each record in every body weight class. The largest X found was 1.4978 for a body weight of 148 pounds. Therefore, this value of X was taken as the constant for a new prediction formula, which thus became:

$$\log W = 1.4978 + .6748 \log B_W$$

Line 2 in Figure 13 is representative of this formula. Since the record for 148-pound weight was used for construction of the formula, the

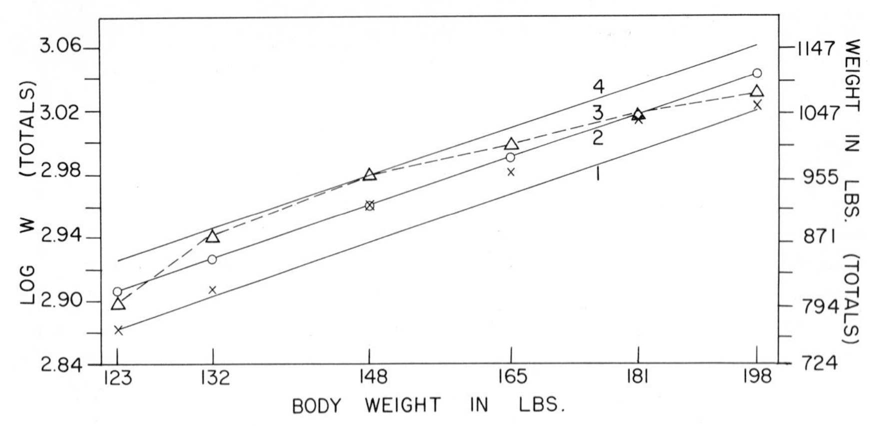

Figure 13. Relation between the body weight (B_w) and total weight lifted: (1) records predicted in 1958, log W = 1.4718 + .6748 log B_w; (2) records predicted on the basis of 1963 data, log W = 1.4978 + .6748 log B_w; (3) 1964 Olympic records; (4) records predicted after Olympic Games, log W = 1.5124 + .6748 log B_w. Body weight and totals in pounds were plotted according to logarithmic scale. *x*, official records up to January 15, 1964; *o*, records for different classes predicted before 1964 Olympic Games; △, Olympic records.

record for this class, 920 pounds, was already on the new level. The record for the 181-pound class had to improve by only one-half pound. The other classes had to lift from 18 to 59 pounds more (see Table 1).

Examination of Table 1 shows that, during the 1964 Olympic Games, all records but one (Class 181 pounds) were better than those given for 1963. Moreover, Classes 132 and 148 even surpassed the 1964 predictions, the first by 24.5 pounds and the second by 31 pounds. On the other hand, Class 123 failed by 23.5 pounds; Class 165 failed by 9 pounds; and Class 198 failed by 45 pounds. Although the

TABLE 1. *Comparison of Actual and Predicted Weight-Lifting Records in Pounds*

Body Weight Class	*I* 1958 Predicted	*II* World Records* 1963 Actual	*III* 1964 Predicted	*IV* 1964 Olympic Records	*V* Difference IV-III	*VI* 1965 Predicted
123.25	763	776.50	810	786.5 w	–23.5	838
132.25	801	832.00	850	874.5 w	24.5	879
148.75	867	920.00	920	951.0	31.0	951
165.25	930	969.75	988	979.0	– 9.0	1021
181.75	992	1052.50	1053	1045.0	– 8.0	1089
198.25	1052	1057.75	1117	1072.5 w	–45.0	1155

*World records for 1963 were official as of January 15, 1964 (Strength and Health, May, 1964). Olympic records were taken from the *New York Times*, October 26, 1964; those marked with a *w* are world records.

Olympic record for Class 181 is 8 pounds lower than the prediction, it really is not important because the 1963 record in this class differs from the predicted record only by one half a pound. The new records necessitate making a new prediction formula, which is

$$\log W = 1.5123 + .6748 \log B_w$$

The relation between the new prediction and the 1964 and 1958 predictions may be seen in Figure 13. In Table 1 may be seen the new predicted records for each body weight class.

One may ask a question: When will the absolute limits for records be reached? The authors have no answer to this question, but venture to say that the limit may be reached sooner by heavier classes than by lighter ones; and that the prediction line, instead of being straight, will curve downward as body weight increases. This supposition is based on the fact that lighter champion weight lifters are relatively stronger than heavier ones. While the first three classes can lift more than six pounds of weight per each pound of body weight, the heavier classes lift less than six pounds. To illustrate: for the class of 132 it is 6.61 pounds and for the 198 class it is 5.41 pounds.

The present discussion has purposely been limited to men weighing up to and including the class of 198 pounds. The heavyweight class has no body weight limitation and therefore a prediction will have to be made on an individual basis. It is expected that the record will be below the predictive straight line.

WEIGHT TRAINING AND ATHLETICS

Half a century ago, only wrestlers looked with favor upon weight lifting. Other athletes believed that weight lifting had detrimental effects and caused muscle-boundness. Three decades ago, even wrestlers began to look with suspicion upon lifting weights because it did not help in developing endurance.

During the past decade, there has been a radical change in the attitude toward weight lifting. Coaches and athletes are now using weight lifting as a training adjunct. To avoid the stigma of "weight lifting," a new term, *weight training*, has been coined. *Post hoc* and experimental evidence may be found in the literature.

One should remember that weight training is not a substitute for athletics, but merely an adjunct which can be conveniently used, especially off-season. Weight training may even be done without weights. A set of barbells or dumbbells is, however, more convenient because of its adaptability to various movements and the ease with which the load may be increased.

WHAT HAPPENS WHEN WEIGHT TRAINING STOPS?

Naturally, strength decreases. If training was slow and long, acquired strength remains longer than when it was gained rapidly.[388] In general, acquired strength may persist for a considerable time. Some of Müller's subjects retained 50 per cent after 40 weeks without training. It is also relatively simple to retain acquired strength by exercising once a week.

Many people have a tendency to become fat after discontinuing intensive training. This happens because their appetites remain the same, while the expenditure of energy is decreased.

IS TAPERING OFF NECESSARY?

There is a deep-seated notion among coaches and athletes that training should not be discontinued abruptly, but should be "tapered off." The authors do not know of any physiological reason to support this notion.

GAIN IN ENDURANCE

The gain in endurance is out of all proportion to the gain in size. This may be illustrated by experiments in which the flexor muscles of the second finger of the right hand were trained daily for one month by the Lombard type of ergograph, which records the number of contractions and the amount of work accomplished.[258] The number of contractions was increased by training from 273 to 918, and the distance the weight could be lifted from 375 to 1205 cm. The work was carried on to complete fatigue in each trial. Thus the number of contractions and the distance the weight was lifted were more than trebled by training. Casual observation is sufficient to prove that muscles do not make a similar gain in size. Such an increase in endurance, therefore, suggests that training improves the quality of contraction.

Many types of manual work requiring skill require also a great deal of endurance. For instance, a frail woman weighing 90 pounds, working in a pretzel shop, has been reported to twist forty-eight pretzels per minute. In an eight-hour work-day she could twist 20,000 pretzels.[393]

The increase in endurance may be associated with an actual chemical change by which (1) fuel is made more available, (2) fuel is stored in greater amount, or (3) oxygen is more abundant, owing to a more adequate circulation of blood through muscle. Since training for endurance causes an increase in the number of muscle capillaries, *1* and *3* are more important than *2*.

Maison and Broeker[356] demonstrated the inadequacy of anaerobic muscle training in developing endurance. In their experiment, a group of young men exercised isotonically for several months the extensor muscles of the fingers on a specially designed ergograph. The fingers of both hands were used, with the difference that the blood supply to the right hand was stopped by application of an inflated sphygmomanometer cuff around the upper arm. The endurance of the muscles with normal blood supply increased more rapidly and reached a higher level than that of muscles without blood supply. This experiment indicates that some important changes take place during exercise itself, and that an adequate blood supply is necessary for these changes. Maison and Broeker also showed experimentally that endurance for work with heavy loads is developed best by training with light loads.

Using a hand dynamometer, one may find that ischemia has a greater deleterious effect on isotonic contractions than on isometric ones.

CHEMICAL CHANGES

It has been shown that training muscles of rabbits for a period of five days suffices to bring about a definite increase in the phosphocreatine content. The increase is perceptible for about six days after the termination of training.[397] Embden and Habs,[67] by tetanizing single legs of rabbits while the opposite legs remained quiet, found after several weeks that the muscles of the worked legs invariably contained the larger content of glycogen—often two or three times more than those of the untrained legs. The quantity of non-nitrogenous substances regularly increased during training. The color of the trained muscles was darker than that of the untrained, indicating an increase in the amount of myoglobin and a more favorable condition for obtaining oxygen.

In adult dogs the amount of myoglobin in 100 gm. of muscle tissue was found to vary all the way from 400 mg. in a quiet house dog to 1000 mg. in an active, trained hunting dog.[535]

END PLATE TRANSMISSION

It has been suggested that the facility of transmission of the nerve impulse across the motor end plate in the muscle fiber is one of the gains of training.

This suggestion was substantiated when it was found that animals trained for speed running had an increase in the area of contact between the motor nerve endings and a muscle fiber. The ratio of the end plate area to the cross section of the muscle fiber also was increased.[323]

MORE COMPLETE USE OF ALL FIBERS

It may be assumed that, as end plate transmission is improved and idle fibers are activated, it becomes easier to use the muscle in its entirety. A muscle with idle fibers must be less strong than one in which all fibers can be called into action.

Gain in Speed. In speaking about speed, we should differentiate between movements requiring special skills and "natural" movements. To develop speed in the first type of movement is easy, while in the second it is very difficult.

Among movements requiring special skill we may mention piano playing, skating, swimming, and tennis. A beginner on a piano starts with "10 thumbs" and rapidly shows improvement; a person who could hardly stand on the skates or stay above water or hit a ball also shows progress in the speed of desired movements.

On the other hand, if a beginner happens to be a good sprinter, it is difficult to make him *run* faster. An advantage may be obtained by starting faster or by eliminating some errors of form, but it takes great effort to increase the speed of running. According to Fiodorov,[169] speed of muscular contractions will increase if the latent time of muscular relaxation becomes shorter.*

Tests conducted on the quadriceps femoris in beginners and well-trained swimmers (masters of the first class) showed that the latent period of relaxation of this muscle in well-trained men was much shorter than in beginners. Thus, speed in skill activities depends mostly on establishing conditioned reflexes, while speed in sprints depends on a sheer reduction of time needed for complete contraction and relaxation.

Speed in endurance events is a by-product of endurance.

CAPILLARIES

Petren and co-workers[409] studied the effect of running upon the changes in the muscles of guinea pigs. They found no difference in size

*On the electromyogram it is indicated as the time between the instant the signal for relaxation was given and the disappearance of the action potentials.

of the muscle fibers, but a great change in the number of capillaries, which, in the heart and gastrocnemius muscles, was 40 to 45 per cent more in the trained than in the control animals. Thus a gain in endurance is mainly a problem of improving transport of the blood in the muscles.

MUSCLE TONE

In physical education, one often hears the expression "muscle tone," which signifies that state in a muscle which gives it a quality of firmness. Such firmness can be ascertained by palpation and by the resistance offered by the muscle to passive movements. Elementary physiology used to define muscle tone as a state of constant contraction of a number of motor units in the muscles, and explained that these units "take turns" in their contractions so that, therefore, no appreciable sense of fatigue develops.

This physiological definition is only partly true, and fits only those conditions in which muscle reflexes are present. As we know, the contraction of a muscle, even a single motor unit, may be ascertained by the presence of an action potential. Therefore, muscle tone may be studied by recording action potentials.

The testing of muscles that are stimulated by reflexes reveals the presence of action potentials. But, when a muscle is completely relaxed, no action potential may be ascertained in an overwhelming number of tests. Exceptions are those muscles that cannot be relaxed at will, such as the laryngeal, the scaleni, and the abdominal muscles, which are stimulated with each inspiration. It is also difficult to relax completely the muscles of the face and the tongue.

Thus the consensus of thought is that, since no action potential can be elicited from relaxed muscles, their "tone" should be explained as a result of the physical property, elasticity. Nervous individuals may have definitely increased muscle tone, which can be proved by the presence of action current. One should not, however, jump to the conclusion that every nervous person who "knows" that he is tense cannot relax his muscles. Even a psychoneurotic person may exhibit perfect voluntary relaxation of his muscles, and a complete absence of action potentials.

EFFECT OF WARMING-UP

Effectiveness of muscular contractions depends on temperature. Lowering of the muscle temperature below normal decreases muscle contractility and capacity for work; during physical activity muscle temperature rises. These two observations, put together, have led to a belief that if muscles are warmed-up in some way before work, and especially before a competition, their performance will increase.

There are a number of physiological reasons why warming-up *should* be beneficial. Besides an increase in muscle temperature, blood circulation rate and volume are augmented, and so is pulmonary ventilation and oxygen transportation. The authors of this book were brought up to believe that warming-up is indispensable for good performance.

There are two types of warming-up: active and passive. The first type is most commonly used and is subdivided into two subtypes: formal and informal (also referred to as related and non-related). Formal warming-up involves the practicing of movements and skills which will be used later in competition. For example, shooting a billiard ball, hitting a tennis ball, or jumping hurdles, before taking part in contests involving these skills, belongs to this type. Informal or general warming-up usually consists of exercising the large muscles of the body in a manner not resembling their actual use in contest. Squats performed by a billiard man and calisthenics by a track man are examples. However, jogging done by a runner lies on the border between formal and informal warm-up. The true formal type obviously primarily affects the nervous system, while the informal type has as a purpose raising the level of activity of most organs and systems. To the passive type of warming-up belong showers, diathermy, and massage.

Just about 25 years ago no athlete had any doubt that warming-up was indispensable for good performance. Moreover, there was a good number of theoretical physiological reasons why warming-up *should* be beneficial. For this reason researches regarding warming-up were almost non-existent. Of the early ones may be mentioned experiments by Simonson and co-workers,[477] who reported a beneficial effect of warming-up on the 100-meter sprint. Then, in 1945, came a report from two Danish investigators, Asmussen and Bøje,[13] who found that while preliminary exercise, short-wave diathermy, and hot showers were beneficial for sprints and one-mile rides on a bicycle ergometer, massage had no effect whatsoever. In 1947, however, Schmid[457] of Czechoslovakia reported that all types of modalities used by Asmussen and Bøje, including massage, were beneficial for swimming 50 meters, running 100 meters, and riding a bicycle ergometer. Intrigued by this controversial effect of massage, Karpovich suggested to Creighton Hale, then a graduate student, that he conduct experiments concerning warming-up. The results were surprising: neither massage nor warming-up had any beneficial effect upon a sprint ride on an ergocycle or a 440-yard track run. As a matter of fact, the Springfield College "440" track record was broken by a man running in a "cold" condition.[204] Additional tests have supported these findings.[292, 296]

These reports regarding the ineffectiveness of warming-up led to a number of investigations in this country and abroad, only a few of which will be mentioned here. Some investigators showed a beneficial effect[83] and some showed no effect.[212] One investigator reported that just imagining warm-up improved athletic performance.[358]

Rochelle and co-workers[433] found that warm-up increases the distance of a baseball throw. When, instead of warm-up, subjects were given some money, performance was much better than in a "cold" state, but still was not as good as after warm-up.

Attention is called here to an old observation, published in 1941,[303] yet not appreciated by anyone, including the investigators themselves, until 1949. A group of excellently trained jail inmates was given a 15- to 20-minute period of warming up, during which they rode an ergocycle without their usual load. After this warming-up, they continued the ride with the usual load. Some of these men could not continue the ride for more than half an hour, rather than the usual two or more hours, and gave up in disgust in spite of the fact that they were paid by the hour and therefore lost money. In order to help them to overcome this critical half-hour barrier, the investigators had to stand by to encourage them. Even with this help, they could never reach their usual riding limit.

The last four references indicate the strong influence of psychological factors. To eliminate these factors, Massey and co-workers[367] hypnotized their subjects and told them to forget whether or not they had a warm-up. No beneficial effect of warm-up was observed on sprint rides on an ergocycle. As a matter of fact, "cold" subjects did better than "warmed" ones. Thus, the difference in experimental results may be explained by difference in the interaction of psychological factors on the part of subjects and sometimes on the part of investigators. At Springfield College, one complete study and a number of experiments had to be discarded because it was discovered that some subjects who were supposed to perform in "cold" state had warmed-up themselves surreptitiously.

This happened about 20 years ago, but there is no assurance that it has not happened more recently. It stands to reason that some honest subjects may be afraid to go all-out without warming-up, and therefore their performance will be adversely affected. This is probably the main reason for warming-up for all physical activities except those requiring a definite element of skill, such as gymnastics, tennis, billiards, or pitching a baseball.

After reading a report by Sedgwich and Whalen,[469] Karpovich wrote a comprehensive review of experiments with warm-up and prepared a detailed table showing the results. The summary is given in Table 2.

From this table, one can see that the sum of deleterious and zero

TABLE 2. *Effect of Various Types of Warm-up on Physical Performance*

	+	−	*0*
Formal	8	1	5
Informal	6	1	10
Passive	9	7	4
	23	9	19

+ = beneficial; − = deleterious; 0 = no effect.

effects was somewhat greater than the number of beneficial results: 28 versus 23.

The defenders of physiological reasons for a beneficial effect of warm-up say that this is because of "increased blood circulation." Increased circulation per se is not sufficient. Asmussen and Bøje[13] found that if blood circulation is increased by exercising legs, it will not affect work capacity of the arm muscles, although circulation through the arms is increased.

There is one very important factor frequently mentioned by believers in warm-up: They claim that warming-up reduces the incidence of athletic injuries, even though no experimental or clinical proof of this claim is available.

MUSCLE SORENESS AFTER EXERCISE

After prolonged or intensive work, one experiences not only local fatigue but also muscle soreness. This soreness, however, may not develop for several hours after work, and may continue for several days. If exercise is resumed the day after the soreness sets in, it will cause pain which, however, disappears in a few minutes. When exercise is ended, soreness reappears. It has been observed that mild exercises seem to reduce the duration of the soreness.

Hough[253] explained the soreness by suggesting that a number of muscle fibers become ruptured. There is no doubt that violent exercise may cause extensive muscle rupture, but the term "violent" can hardly be applied to most activity which results in muscle soreness. *Since no experimental evidence has been presented to substantiate Hough's explanation, the authors of this book do not subscribe to this theory.*

A plausible explanation of soreness is that prolonged, vigorous work results in an excessive accumulation of metabolites which causes an increased osmotic pressure inside and outside of muscle fibers. Thus retained, excess water causes edema and pressure on sensory nerves.

Asmussen[12] feels that this theory is not convincing either. He used two simple but ingenious experimental procedures to study soreness in 16 female subjects. The subjects stepped on a 50-centimeter-high stool, always using the same leg for ascent and the other for descent, keeping the same speed for both movements. In the other experiment the extensors of one forearm lifted a five- to seven-kilogram weight and the flexors of the other forearm lowered it. Both tests continued until subjects complained of fatigue. In both experiments, the lifting of the body or of the weight constituted positive work, while lowering produced negative work. It was noted that fatigue developed in the muscles doing positive work. The examination of subjects on the following days revealed that positive work caused soreness in only two subjects after step-up and in five after pulling the weight up. On the other hand, negative work caused soreness in 15 subjects after step down and in 12 after lowering weight.

Since the metabolism of negative work is five to seven times smaller than the metabolism of positive work, soreness cannot be explained as a result of excess metabolites. The soreness is evidently caused by a mechanical pull exerted by muscle fibers on intramuscular connective tissue. One may wonder why lighter, negative work might cause more soreness than heavier, positive work. It has been shown that, during negative work, muscle fibers lengthen and increase their tension, and that, therefore, the number of participating fibers decreases. The result is a greater pull by each fiber in connective tissue. It is postulated that this excessive pull traumatizes connective tissue, and local edema gradually develops, causing pain. During positive work, the fibers shorten and their tension decreases, compelling a greater number of fibers to take part in contraction and thus reducing the pull exerted by individual fibers on connective tissue.

MUSCLE CRAMPS

Every athlete is familiar with muscle cramps. They are spontaneous, sustained, painful contractions of one or several muscles.

The intimate mechanism of cramps is not known. The statements and explanations given in the following paragraphs are the results of observations made by one of the authors, mostly on himself, when, at various periods of life, he has happened to be affected by cramps.

Sometimes there is an advanced sensation giving warning of the onset of a cramp. In this instance, the involved muscle may be relaxed at will, forestalling development of the real cramp. Usually, however, cramps occur without warning and without any obvious cause: for example, while their victim is lying quietly in bed reading. Sometimes cramps develop only after intensive and prolonged use of certain muscles, and are brought on by even slight contraction of tired muscles; and sometimes they may be produced at will by vigorous contraction of fresh muscle. The latter are most easily produced in the calf muscles.

Because treatment of cramps sheds some light on their origin, this is mentioned before the discussion of their causes. For all cramps that come without warning, the best remedy is stretching the affected muscles by contracting the antagonistic muscles, or extending the involved muscles by some other means.

A firm kneading of cramped muscles seems to be of help on occasions. The writers who have recommended *gentle* massage have probably never had cramps themselves.

As to the cause of cramps, an inescapable conclusion is that they are either of nervous or neuromuscular origin. When a cramp is preceded by a warning sensation, it is possible that the cramp is caused by a lowered threshold of irritability of motor nerves, resulting in a sudden increase in the frequency of nerve impulses going to the muscle. In some mysterious way, the discharge of impulses can be inhibited at will. It also may be inhibited by contraction of antagonistic muscles—a process no less mysterious than that of inhibition at will. In cramps of neuromuscular origin, the threshold of irritability of motor units is also lowered. Some of the units may contract beyond their usual limits. As laboratory experiments have shown, an isolated muscle, allowed to shorten too much, will remain shortened for a long period of time unless helped back to normal length by stretching. Probably this happens during a cramp.

One more possibility may be considered. Muscle contraction occurs because the chemical, acetylcholine, is produced by the nerve endings at the motor end plates. Normally this substance is quickly destroyed by cholinesterase. But, may it not be possible that there is local lack in esterase, and, therefore, that contraction persists until acetylcholine is destroyed by an additional supply of esterase?

Abdominal Cramps. Frequently one hears of the danger of "stomach" cramps contracted while swimming. Investigation made by Lanoue[336] showed that, of 30,000 swimmers who passed through his hands, none had experienced abdominal cramps. Evidently such cramps occur very seldom. One of us (P.V.K.) once experienced a cramp in his left rectus abdominis while fishing. It occurred when he bent forward trying to get something from under the bench on which he sat. The pain was considerable, and since it was augmented by breathing, he hyperextended his trunk and held his breath. Eventually the cramp disappeared. There is no doubt that cramps of this severity could be fatal if they occurred while one was swimming. Even a good swimmer, if far from shore, might become incapacitated.

MENSTRUATION AND MUSCLE STRENGTH

Early dynamometric studies[120] indicated that women's strength suddenly decreases a few days before menstruation begins, and continues at a lower level throughout the menstrual period.

The relation between athletic performance and the menstrual cycle presents a more complicated picture. In a study[145] of 111 athletic women participating in field and track events, it was shown that 55 per cent suffered no decrease in efficiency, the performance of some of them even being increased on days of bleeding. The other 45 per cent showed a decrease in performance either during menstruation or immediately before the onset of the flow.

The question is often asked: Is participation in athletics during menstruation harmful? There is no evidence to prove that it is. Some menstruating girls faint during severe exertion, of course; but severe exertion also causes fainting in some men.

In speaking about possible harm, it is logical to assume that the period immediately preceding the menstrual flow is most critical, and that a woman should be especially protected during that period. Yet, at that time, some women have a compelling urge to undertake tasks requiring physical effort such as rearranging furniture or cleaning a cellar or garage. The results of such activity appear to be beneficial. Also, there seems to be sufficient evidence that other exercises alleviate dysmenorrhea. Excusing a menstruating girl from classes requiring mild physical activities is not warranted.

In general, more information is needed about the relations between athletics and menstruation. Iwata of Japan examined 418 athletic girls[269] and found that menstrual irregularities among them were more prevalent than among average girls. Whether this was a coincidence or a result of athletics is impossible to tell.

IRREGULAR CHROMOSOME ARRANGEMENTS

The suspicion that some girl-champions were not real girls compelled the International Amateur Athletic Federation to require medical inspection. As a result of this rule, some candidates refused to be examined, giving all kinds of non-convincing reasons. Almost all girls passed inspection without any difficulty. However, besides a mere inspection, a chromosome test of cells scraped from the inside of the cheek was also introduced. In a normal woman, there are 22 pairs of non-sex chromosomes and a pair of sex chromosomes, called XX. (The pair of sex chromosomes in a male is called XY.) Examination of a Polish girl-athlete, who had been brought up as a girl and believed that she was a girl, revealed an abnormal combination of sex chromosomes.

Although the examiners did not make their findings public, an educated guess is that some sex chromosomes were XYY combinations and the others just a single X. Besides tragic Y-victims of biology, there were also frauds: men who masqueraded as women and became champion athletes among women.

QUESTIONS

1. How can you apply the Lamarckian postulate to muscular development?
2. What has been reported regarding the use of isometric contractions in muscle training?
3. What intensity of exercise is needed to increase strength? To increase endurance?
4. What is the relation between body weight and weight lifted?
5. Who is relatively stronger, a 148-pound or a 181-pound champion?
6. Approximately how much of the total time of a weight-lifting session is spent in actual lifting?
7. What is the relation between weight training and athletics?
8. What happens to the muscles when training stops? Is tapering off necessary?
9. How often should one exercise to retain acquired strength?
10. Is ischemia helpful in the development of strength, or in the development of endurance?
11. What chemical and structural changes occur during training?
12. What effect has training upon the capillaries of the muscles?
13. Why does the muscle soreness felt at the beginning of training later disappear?
14. What is muscle tone? Define it and explain its nature.
15. Discuss the causes of muscle soreness.
16. What effect does menstruation have upon muscular strength and athletic performance?
17. Are physical exercises harmful during menstruation?
18. What is formal and what is general warming-up? Discuss the importance of each in attaining maximum physical performance.

Chapter Three

NERVOUS CONTROL OF MUSCULAR CONTRACTION

It was shown in Chapter 1 that the amount of muscle contraction depends on the number of motor units recruited from the motor neuron pool. Every pool receives both excitatory and inhibitory stimuli, the final amount of contraction being determined by the interaction between the two. If the excitatory effects on the pool are great enough, contraction occurs; but if there are enough inhibitory stimuli to counteract the excitatory effect, no nerve impulse passes and contraction does not take place.

It is the purpose of the present discussion to identify some of the excitatory and inhibitory factors and to provide an overview of the control of muscle contraction. The sensation of movement, or kinesthesia, will also be discussed. The nervous system is so complex that it is impossible to provide a complete discussion of it in an introductory text. The student who is interested in studying its processes in more detail is referred to bibliographical references which have been used extensively in the preparation of this chapter: *200, 447, 500.*

ORGANIZATION OF THE NERVOUS SYSTEM

The nervous system illustrates perhaps better than any other the relationship between structure and function. It is, therefore, advantageous to review its major components before beginning a discussion of how it controls movement.

The *neuron,* or nerve cell, is the structural unit of the nervous system. It is especially adapted to develop a *membrane potential* and transmit a *nerve impulse,* a process described in Chapter 1. Neurons differ widely in appearance and function. The essential parts of a motor neuron are shown in Figure 14. The dominant features are the cell body and its fibrous structures, the axons and dendrites. The latter structures connect neurons to one another, sometimes becoming quite long. Some of these fibers have a *myelin sheath* which acts to speed the transmission of impulses. Such fibers are said to be *myelinated.*

Every neuron receives a vast number of *terminal endings* from the axons of other neurons on its *soma* (body) and its *dendrites.* These junctions between the neurons are called *synapses,* and serve to transmit nerve impulses from one neuron to another; the impulse always goes from the terminal endings of the axon of the transmitting neuron to the dendrites and soma of the receptor neuron.

The nerve impulse is passed across the synapse by the release of a transmitter substance from the terminal endings of the axons. Some endings release substances that are *excitatory* in that they tend to depolarize the receptor neuron, which causes an impulse to be passed on via the axon of the receptor neuron. Others release substances that are *inhibitory* and tend to hyperpolarize the receptor neuron, making it resistant to the passage of an impulse. The release of the transmitter substance at only a few endings will not induce an impulse in the receptor neuron; there must be a summation from several endings. Eccles[149] has presented a very clear description of synaptic function.

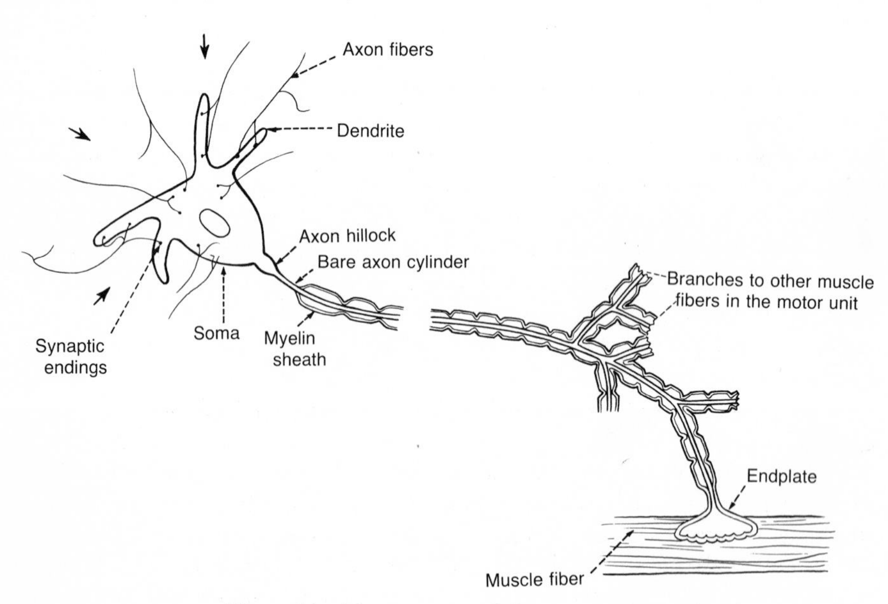

Figure 14. The structure of the motor neuron.

Anatomically, the nervous system can be divided into the *central* and *peripheral* divisions; the former consists of the brain and spinal cord and the latter consists of the nerve fibers which arise from them and pass to the various parts of the body. Functionally, we can identify the *somatic* and *autonomic* systems. The somatic system is the topic of the present discussion due to its role in control of muscle contraction. The autonomic system, which consists of the *sympathetic* and *parasympathetic* divisions, acts as a physiological regulator. Its functions are discussed where applicable.

The brain is made up of three major subdivisions, which are shown in Figure 15: the *forebrain* (prosencephalon), the *midbrain* (mesencephalon), and the *hindbrain* (rhombencephalon). Each of the subdivisions has important structures which play specific roles in the control of muscle contraction. The activities of the various parts are coordinated by interconnecting *tracts*, which are collections of neuronal fibers.

Forebrain. The most prominent structures of the forebrain are the cerebral hemispheres, the development of which overshadows all other structures. The cortex of the hemispheres is made up of neurons of various types. Its surface can be divided into the functional areas shown in Figure 15*A*. The grooves in the surface are called *sulci*, and the raised areas, *gyri*. The *motor* and *premotor* areas, which are located anterior to the central sulcus, give rise to nervous stimuli which produce voluntary muscle contraction for very precise movements. The *sensorimotor* (or *somesthetic*) area is located posterior to the central sulcus. It is the area responsible for the simple perception of touch, pressure, heat, and cold, as well as the recognition of spatial relationships related to touch and joint movements.

The different areas of the cerebral cortex do not act independently of each other or of the lower brain centers. *Commissural tracts* pass between the hemispheres, while *association tracts* pass between segments of the same hemisphere. *Projection tracts* link the cerebral cortex with lower areas of the central nervous system.

The *basal ganglia* form a structural unit located deep in the forebrain. This structural unit is made up of the *lentiform nucleus*, the *caudate nucleus*, and the *amygdala*. The *amygdala* is felt to play a role in the control of behavior, while the other nuclei are important in the control of muscle contraction. The lentiform nucleus has two structures: the *putamen* and the *globus pallidus* (Fig. 15*C*). The putamen and caudate nucleus together form the *striate body*. This structure apparently initiates and regulates gross, intentional movements, while the *globus pallidus* provides background muscle tone to stabilize body segments during precise movements initiated in the cerebral cortex.

Other forebrain structures of importance are the *thalamus* and the *hypothalamus*. The thalamus acts as a switchboard for the brain in that it receives afferent stimuli from the body as well as from other areas of

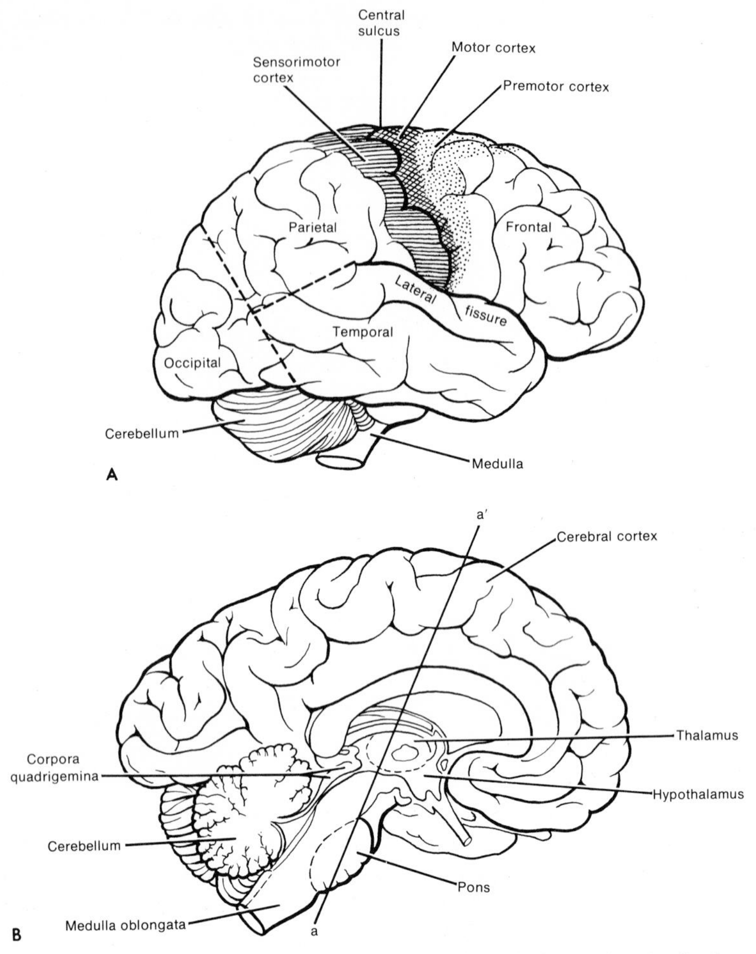

Figure 15. The major structures of the brain. *A*, lateral view; *B*, longitudinal section; *C*, cross-section. The line a-a[1] in *B* indicates the plane of the cross-section shown in *C*. (*A* and *B* are adapted from Francis, C. C.: Introduction to Human Anatomy, 5th ed., The C. V. Mosby Co., 1958. *C* is adapted from Lockhart, R. D., Hamilton, G. F., and Fyfe, F. W.: Anatomy of the Human Body. J. B. Lippincott Co., 1959.)

(Illustration continues on facing page.)

the brain and routes the information to appropriate cerebral areas. The hypothalamus contains many regulatory centers and apparently plays an important role in coordinating neural and endocrine functions. There are also collections of neurons in the lower thalamic area, forming *subthalamic nuclei*, structures which are involved in the control of muscle contraction.

Midbrain. The midbrain is a small area (Fig. 15). Its dominant external features are the superior and inferior colliculi, which appear as large, rounded prominences on its dorsal surface. Together they are

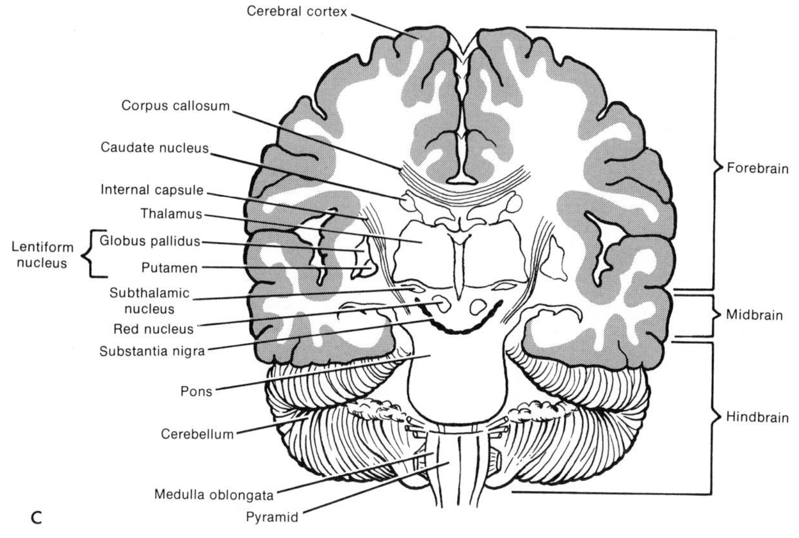

Figure 15. *Continued.*

called the *corpora quadrigemina.* There are many fiber pathways, which interconnect various areas of the central nervous system, passing through the midbrain.

Centers of the midbrain which are important in the control of muscle contraction are the *red nucleus,* the *superior colliculi,* and the *substantia nigra.* A cross-sectional view of the midbrain illustrating some of the centers is shown in Figure 16. The red nucleus receives impulses from the caudate nucleus, putamen, and cerebellum, and apparently plays a role in the regulation of movement through its effect on tonic contraction of muscle. Its ablation (destruction) leads to involuntary movements and rigidity. The function of the substantia nigra is not clearly understood, but its ablation leads to effects similar to those for the red nucleus. The superior colliculi are centers for the control of eye, head, trunk, and limb movements in response to visual stimuli, while the inferior colliculi play a similar role relative to auditory stimuli.

Hindbrain. The hindbrain has three important subdivisions: the *pons,* the *medulla oblongata,* and the *cerebellum* (Fig. 15). Together, the midbrain, pons, and medulla are sometimes referred to as the *brain stem* and their functions are interrelated.

Several nerve fiber tracts pass through the pons. The *pontine nuclei* act as relay stations between fibers from the cerebral cortex and the cerebellum. Nuclei in the pons also play an important role in the reflex fixation of the eyes, which enables us to track a moving object.

The medulla oblongata is a very primitive control center forming the lowermost portion of the brain. Several fiber tracts are found there,

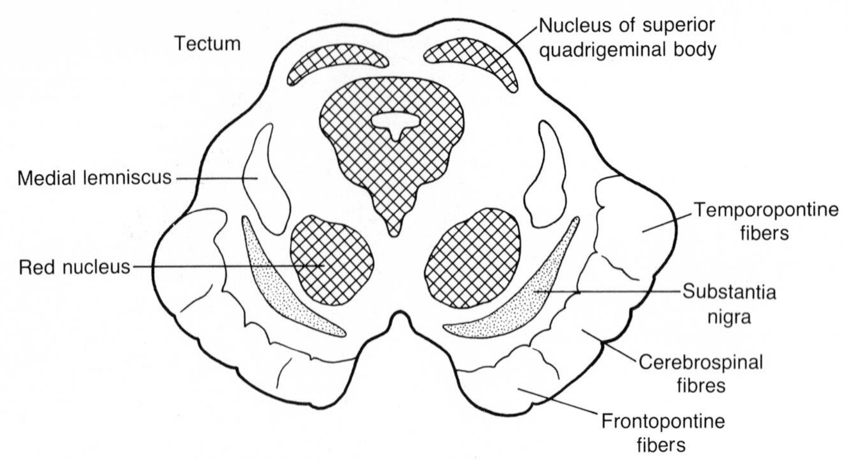

Figure 16. A cross-section of the midbrain showing nuclei of importance in the control of muscle contraction. (Adapted from Bell, G. H., Davidson, J. N., and Scarborough, H.: Textbook of Physiology and Biochemistry, 7th ed. Williams and Wilkins Co., 1968.)

one of which passes from the cerebral cortex to the motor neurons located in the spinal cord, forming the *pyramids.* The *nucleus cuneatus* and the *nucleus gracilis,* which are shown in Figure 20, act as important sensory relay stations.

The cerebellum, which cannot itself initiate movement, is an important center for coordinating movements originating in other parts of the central nervous system. The ablation of its parts leads to different degrees of *ataxia,* the inability to coordinate muscle movements. Its cortex can be divided into three areas: the *anterior palleocerebellum,* the *posterior palleocerebellum,* and bilateral projections, called the *neocerebellum.* The bilateral projections form what is usually referred to as the cerebral hemispheres. The three areas are sometimes referred to relative to their main connections with other parts of the nervous system: *spinal cerebellum* for the anterior portion, *cerebral cerebellum* for the neocerebellum, and *vestibular cerebellum* for the posterior portion.

The cerebellum has direct or indirect interconnections with all of the centers that play a role in controlling muscle contraction (Fig. 17). Information is transmitted to the three parts of the cerebellar cortex from the peripheral body parts and the brain stem via the *inferior peduncles,* and from the forebrain via the *middle peduncle.* Fibers pass from the three areas of the cerebellar cortex to the *dentate nucleus,* from which fibers then pass to the red nucleus and other upper portions of the brain stem and thalamus via the *superior peduncle.* Other fibers pass from the vestibular cerebellum to the *fastigial nuclei,* which then sends fibers to the lower portions of the brain stem.

The *vestibular nucleus* plays an important role in reflex control of

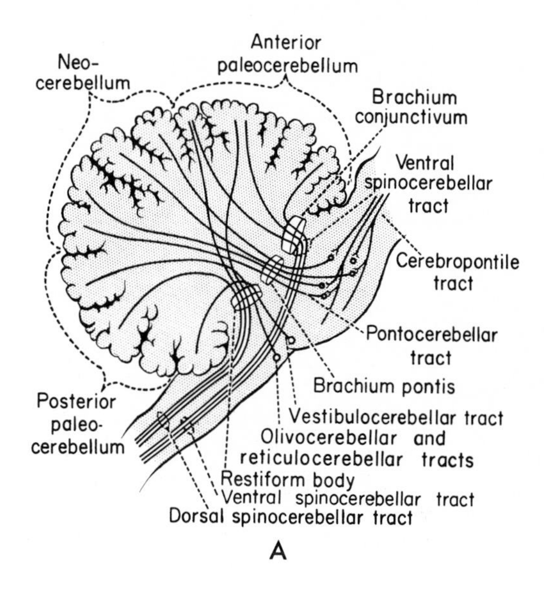

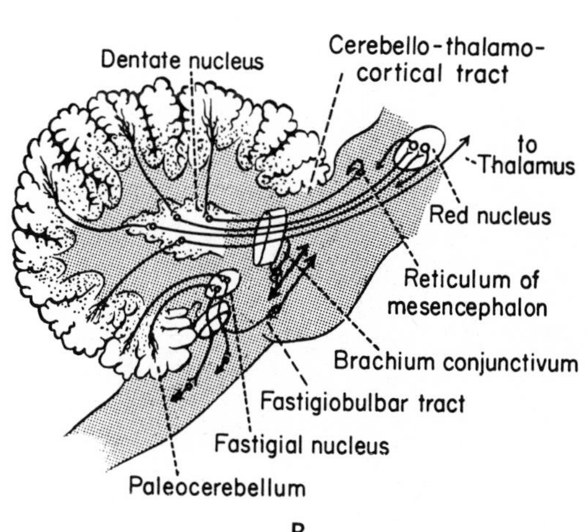

Figure 17. The principle afferent and efferent tracts of the cerebellum. *A* indicates afferent tracts; *B,* efferent. (Guyton, A. C.: Textbook of Medical Physiology, 3rd ed. W. B. Saunders Co., 1966.)

body equilibrium in that it receives most of the nerve fibers from the *vestibular* or *labrynthine structures* of the inner ear—the *semicircular canals* and the *utricle* (Fig. 18). The semicircular canals are sensitive to angular acceleration, while the utricle senses the position of the head in space and linear acceleration.

The *reticular formation* is a loosely defined mass of neurons scattered throughout the brain stem. Its approximate distribution is shown in Figure 18. It receives fibers from other centers in the brain and sends tracts through the spinal cord, providing either excitatory or inhibitory stimuli to motor neuron pools. It also directs fibers upward as part of the *reticular activating system* which keeps us awake and directs our attention to specific facts of our environment.

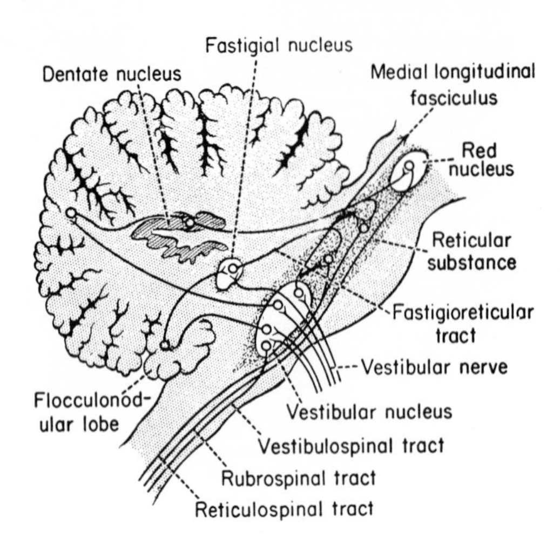

figure 18. The connections between the vestibular nucleus and other parts of the brain. Note also the locations of other nuclei and the reticular substance (Guyton, A. C.: Textbook of Medical Physiology, 3rd ed. W. B. Saunders Co., 1966.)

Spinal Cord. The spinal cord, which is shown in cross-section in Figure 19, also plays a role in the control of muscle contraction. The gray column contains unmyelinated motor neurons and fibers. The neurons form anatomically distinct motor pools within the anterior horn of the gray matter. A motor pool includes all the neurons innervating a muscle. Two zones can be identified within each pool. The middle of the pool is called the *threshold* or *liminal* zone. It is richly innervated and its neurons apparently discharge impulses leading to contraction with only minimal stimuli. The outer areas of the pool make up the *subthreshold* or *subliminal* zone, which is not as richly innervated. This is sometimes called the *facilitated zone,* because stimuli which activate threshold zone neurons will only increase the excitability of subthreshold neurons.

The white columns contain tracts of fibers providing communication between the brain and spinal cord or between different segments of the cord. These tracts have been named according to their points of origin and termination. Their distribution within the spinal cord is shown in Figure 19, and a summary of their functional significance is shown in Table 3. Ascending tracts, those going from the spinal cord to the brain, transmit sensory information from the body, while descending tracts transmit either excitatory or inhibitory stimuli to the motor neuron pools.

The spinal cord acts as a reflex center for the control of muscle contraction. These reflexes are innately set during the anatomical development of interneuronal circuits and are always active as possible modifiers of human movement.

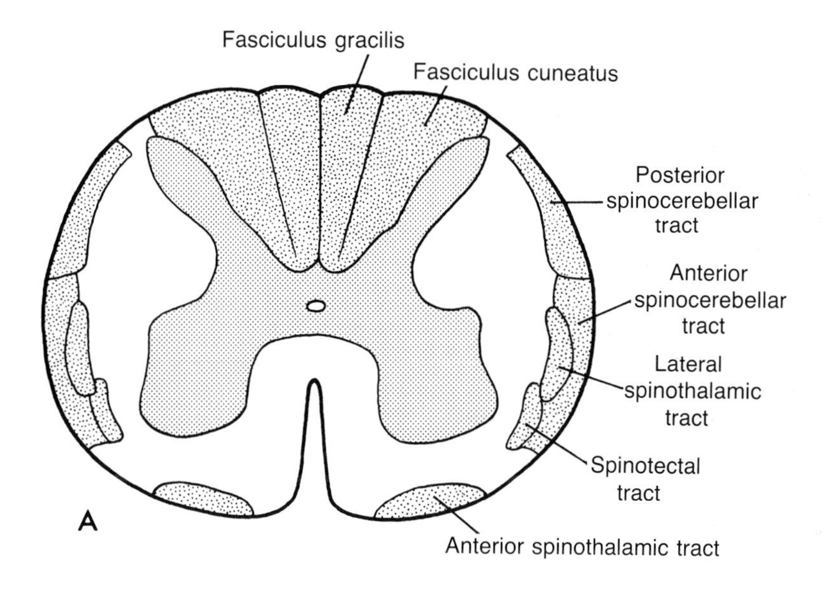

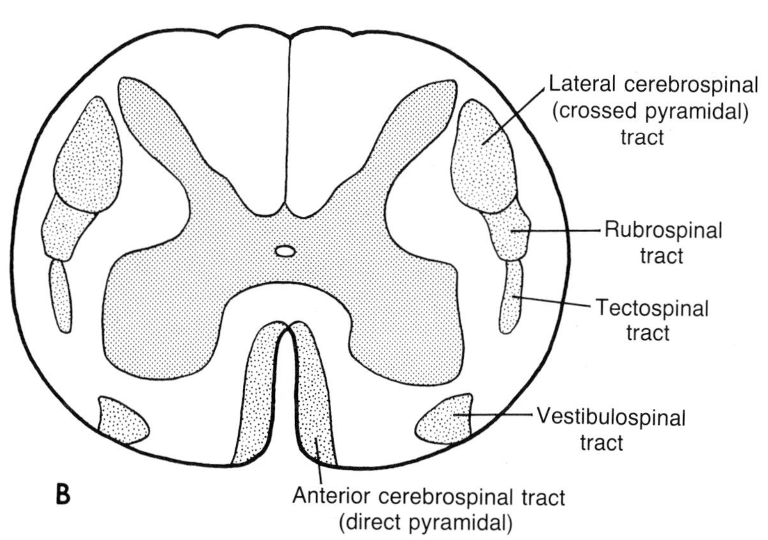

Figure 19. Cross-section views of the spinal cord showing ascending or sensory tracts (*A*) and descending or motor tracts (*B*). (Adapted from Bell, G. H., Davidson, J.N., and Scarborough, H.: Textbook of Physiology and Biochemistry, 7th ed. Williams and Wilkins Co., 1968.)

TABLE 3. *Important Tracts of the Spinal Cord and Their Functional Significance**

Tract	Type	Function
Ventral spinothalamic	Afferent	Light touch
Spino-olivary	Afferent	Reflex proprioception
Ventral corticospinal	Efferent	Voluntary motion
Vestibulospinal	Efferent	Balance reflexes
Tectospinal	Efferent	Audio-visual reflexes
Reticulospinal	Efferent	Muscle tone
Spinocerebellar (dorsal and ventral)	Afferent	Reflex proprioception
Lateral spinothalamic	Afferent	Pain, temperature
Spinotectal	Afferent	Reflex
Lateral corticospiral	Efferent	Voluntary motion
Rubrospinal	Efferent	Muscle tone, synergy
Olivospinal	Efferent	Reflex
Fasciculus gracilis and Fasciculus cuneatus	Afferent	Vibration, passive motion, joint discrimination

(*Adapted from Chusid, J. G., and McDonald, J. J.: Correlative Neuroanatomy and Functional Neurology, 12th ed. Lange Medical Publications, 1964.)

THE SENSATION OF MOVEMENT

To be perceived, sensory impulses must reach the sensorimotor cortex. The tracts carrying such impulses, which are shown in Figure 20, all have three orders of neurons. The *first order neurons* are located in the spinal ganglia of the dorsal roots of the spinal cord. Their afferent fibers have endings of various kinds which are sensitive to the different physical or chemical factors that give rise to the stimuli. Their efferent fibers follow one of two patterns. Some synapse in the cord with a *second order neuron,* which sends a fiber contralaterally to ascend via either the *ventral* or *dorsal spinothalamic tracts.* Other first order neurons send their efferent fibers up the ipsilateral *fasciculus cuneatus* or the *fasciculus gracilis,* where they terminate on second order neurons in the medulla in nuclei of the same names. The second order neurons send fibers to the contralateral portion of the thalamus via the *medial* or *lateral lemnisci.* The *third order neurons,* which are found in the thalamus, transfer the stimuli to the appropriate area of the sensorimotor cortex. The sensory areas in the cortex, where the sensations are perceived, are quite well defined, as is shown in Figure 21.

Kinesthesia is the sense of position and of movement of the joints.[438] The kinesthetic receptors are located in the ligaments and the joint capsules, and are of three types (Fig. 20). There are Golgi and spray-type endings which are sensitive to the position of the joint. Pressure sensitive pacinian-type receptors, located about the joints, apparently play a role in the detection of the rate of movement. These fibers ascend in the fasciculi cuneatus and gracilis. The fiber portions

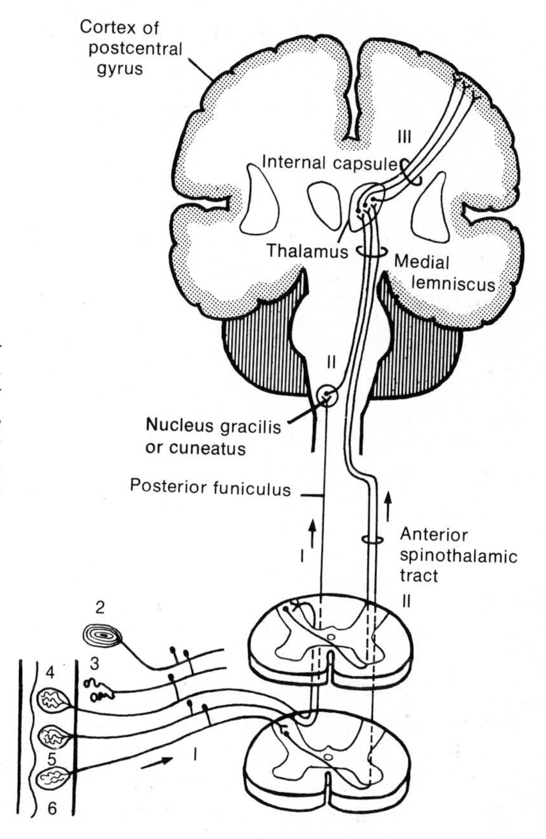

Figure 20. Sensory pathways to the sensorimotor cortex. Roman numerals indicate order of neuron. Arabic numerals: *2*, Pacinian corpuscle; *3*, Joint ending; *4*, *5*, and *6*, Pressure sensors. (Adapted from Gardner, E.: Fundamentals of Neurology, 5th ed. W. B. Saunders Co., 1968.)

of some of these neurons must extend almost the entire length of the body: for example, receptors in the ankle joint.

Touch and pressure sensations are also important in the performance of physical skills. This is especially true in contact sports (such as wrestling), where the competitor must be aware of the position and movement of an unseen opponent. The processing of such sensations is similar to that for kinesthesia. Finer touch and pressure impulses are transmitted in the dorsal spinal tracts with the kinesthetic stimuli. Grosser sensations, however, are transmitted via the spinothalamic tracts. The latter tracts apparently also transmit stimuli to the brain stem which play a role in the reflex control of muscle contraction.

Other tracts are also listed in Table 3. Of these, the most important for the present discussion are the spinocerebellar tracts (Fig. 19). Stimuli arising from the muscle spindles and the Golgi tendon organs are transmitted via this double route to the cerebellum to keep it informed of the state of muscle contraction. These tracts ascend ipsilaterally and

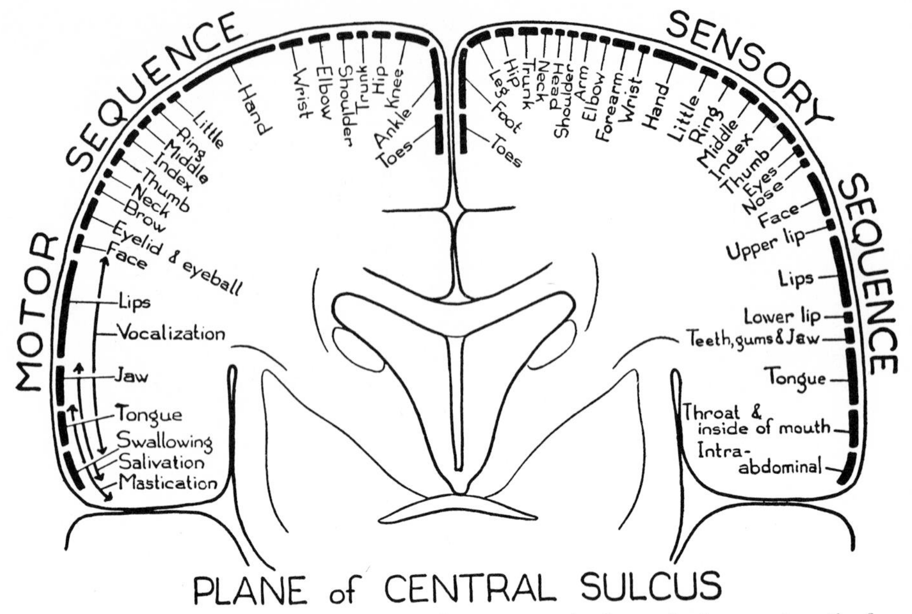

Figure 21. Distribution of sensory and motor areas in the cerebral cortex. Length of bars indicates the general extent of cortical areas devoted to each structure. See also Figure 15. (Rasmussen and Penfield: Fed. Proc. *6*:452, 1947.)

transmit stimuli very rapidly. Their nerve impulses do not reach the sensorimotor cortex and, therefore, do not play a role in kinesthesia.

The vestibular apparatus, which is part of the membranous labyrinth, is also an important source of sensory information for the control of movement. The vestibular apparatus consists of the *utricle* and *semicircular canals*, both of which are shown in Figure 22. The utricle is important in sensing the position of the head relative to the pull of gravity. Nerve impulses arise from the displacement of the *hair cells*; these bend in the direction in which the head is tilted due to the displacement of the *otoconia* embedded in the *gelatinous layer*. The utricle is also important in sensing linear acceleration. As the body moves in any direction, the otoconia lag behind due to their inertia, displacing the hair cells. Reflex leaning such as one experiences riding in an automobile is dependent on this stimulus.

The three semicircular canals of one labyrinth are also shown in Figure 22. Their function is to detect rotary movement. The canals are arranged in the skull so that, through the interpretation of stimuli from both labyrinths, movement in all planes can be detected. The *crista ampullaris* forms the functional sensory ending. The *cristae* are located in the bulbous *ampullae* found at the end of the canals. A fluid called *endolymph* circulates through the canals and utricle. The crista

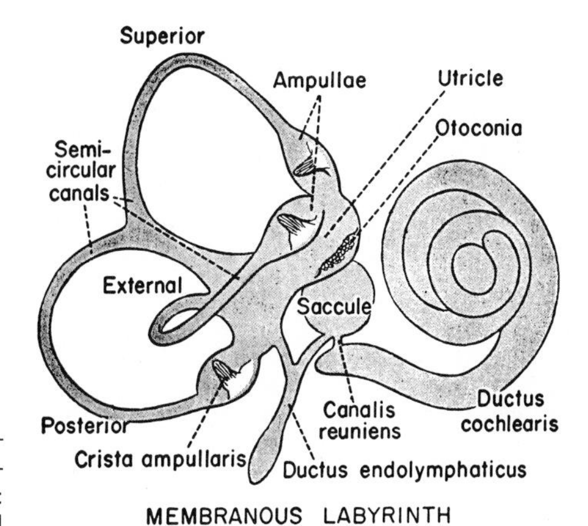

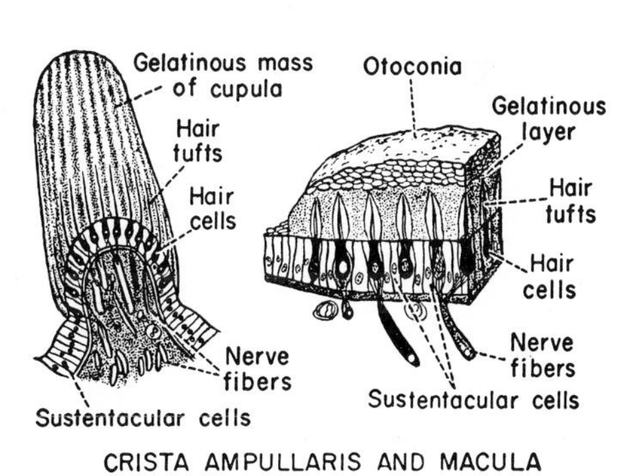

Figure 22. The membranous labyrinth and its specialized sensory structures. (As adapted for Guyton, A. C.: Textbook of Medical Physiology, 3rd ed. W. B. Saunders Co., 1966.)

ampullae act as swinging doors at the junctions between the ampullae and utricle. When the head is rotated, the endolymph tends to lag behind because of inertia; it thereby bends the cristae varying amounts, which varies the nerve impulses sent to the sensory cortex.

The labyrinth structures send some nerve fibers to the cerebellum, but the major distribution is to the vestibular nuclei (Fig. 18). The integration of these stimuli with those from stretch receptors located in the neck which sense the position of the body with respect to the head are important in reflex control of postural body tone. Some fibers pass to the opposite thalamus, which transmits nerve impulses to the temporal lobe where the sensory cortex for both auditory and equilibrium sensations is found.

NEURAL INTEGRATION OF MUSCLE CONTRACTION

The stimuli for precise, voluntary movements originate in the cerebral cortex. There are, however, a number of involuntary actions, reflexes, constantly originating at subconscious levels. A *reflex* may be defined as an involuntary act involving sensory reception, transmission within the central nervous system, and resultant activity of a muscle or gland. We must undoubtedly learn to inhibit some innate, protective reflexes in order to perform some sports skills.

The purpose of this discussion is to note some of the more important neural controls of muscle contraction and the integration of neural stimuli. This integration involves the simultaneous interaction of all levels of the central nervous system. We have already reviewed the physiological anatomy, and the student who is still unfamiliar with the terminology would do well to refer frequently to that discussion.

Spinal Level Reflexes

The typical spinal reflex includes a *sensory (afferent) neuron,* either an *excitatory* or *inhibitory internuncial neuron,* and a *motor (efferent) neuron.* There is also some kind of a *reverberating circuit* that helps prolong muscle contraction even after a sensory stimulus is removed.

Several known spinal reflex patterns are illustrated in Figure 23. The *nociceptive reflex,* which is also termed either a *flexor* or a *withdrawal*

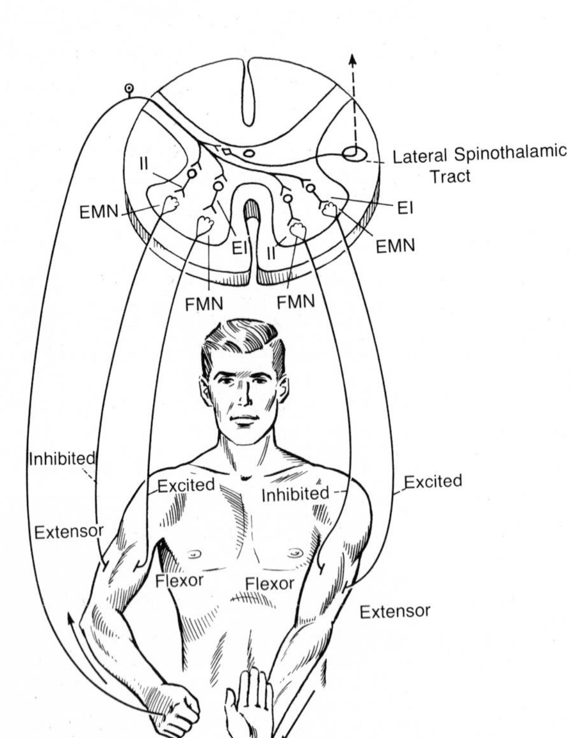

Figure 23. Summarization of spinal level reflexes. Note that the intent is to show the nature of the interconnections and not the actual neurons involved.

reflex, is protective in that it permits an instantaneous removal of a body part from a painful stimulus. In this reflex, the receptor neuron acts on an excitatory interneuron (*EI*), which depolarizes the motor pools of the appropriate flexor muscles (*FMN*), resulting in muscle contraction. A *reverberating circuit,* which is not shown, causes contraction to continue even after the transmission from the receptor has stopped due to the imparted flexor movement. The sensory neuron will also form synapses on second order neurons, which conduct the impulse to the thalamus via spinothalamic tracts and subsequently to the sensory cortex; this is the reason the pain is felt after the completion of the reflex act.

Other phenomena associated with the nociceptive reflex are also illustrated. Reciprocal inhibition is demonstrated by the effects of the stimulus on the antagonist extensor motor neurons (*EMN*). An inhibitory interneuron (*II*) acts to prevent the extensor muscles from contracting, thereby assuring relaxation and removal of interference during the completion of the movement. The process of reciprocal inhibition is also in effect during voluntary movements. Another reflex pattern is the *crossed extensor reflex.* The sensory impulse is passed to the contralateral motor neuron pools, where there is an excitation of extensor muscles with an inhibition of the flexors.

Stretch-Stress Reflexes. The stretch reflex is important for the maintenance of posture and control of muscle contraction. The presence of the stretch receptor or spindle in the muscle has already been noted in Chapter 1. A detailed review of research on the muscle spindle has been completed by Mathews[368] and the interested student would enjoy studying it. The stretch receptor is another example of the way structure determines function. For this reason, it is well for the student to know the structural organization of the stretch receptor before attempting to understand how it works.

The stretch receptor is composed of a rather fascinating combination of nerve endings and muscle fibers (Fig. 24). These receptors are embedded in the connective tissue of the muscles, in parallel with the regular muscle fibers previously discussed. However, they are much smaller, and are bound into bundles of three to ten *intrafusal* fibers, the confining structure being the *capsule.* The intrafusal fibers receive innervation from the *gamma* (γ) *efferent* nerve fibers, which arise from the *gamma motor neuron* pools in the spinal cord. Both the γ-efferent fibers and neurons are smaller than the *alpha* (α) *efferent* fibers and neurons, which supply innervation to the regular muscle fibers. Since small fibers transmit impulses more slowly than large fibers, the γ-system is the slower of the two.

Another interesting feature of the intrafusal fibers is their lack of contractile material in the *equatorial zone* (Fig. 24). Because of this, if the ends of the receptor are fixed in place, contraction of the intrafusal

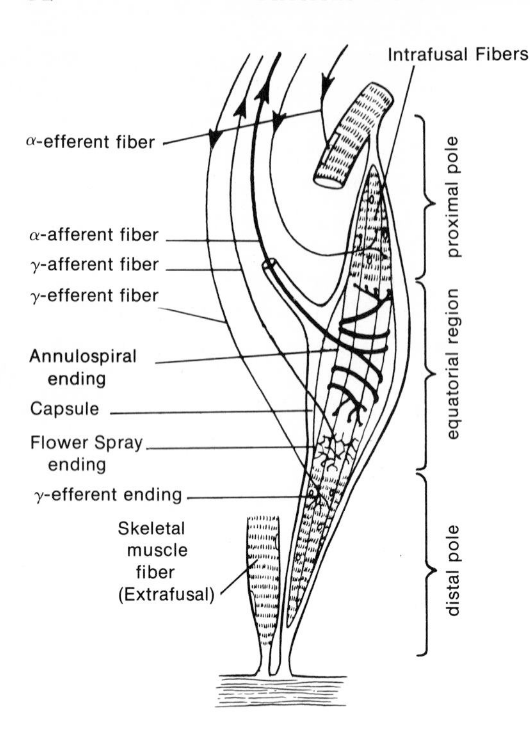

Figure 24. The stretch receptor showing essential structures. (Adapted from Barker, D.: Quart. J. Micr. Sc. *89*:143, 1948; and Gardner, E.: Fundamentals of Neurology, 5th ed. W. B. Saunders Co., 1968.)

fibers will tend to shorten the *polar regions* and stretch the equatorial region. If the receptor ends are not fixed, however, contraction leads to overall reduction in receptor length.

Two kinds of specialized sensory endings are found within the capsule in the equatorial zone. The *annulospiral* ending, which is stretch sensitive, occupies the major portion of the capsular area. It gives rise to the large, rapid-conducting α-afferent nerve. As the receptor is stretched, the frequency of impulses transmitted over this nerve increases. The annulospiral ending is also called the *primary ending.* The other receptor, which is called the *flower spray* or *secondary ending,* gives rise to the γ-afferent fiber. This ending is also sensitive to stretch.

A simplified diagram of spinal synaptic connections is shown in Figure 25. The α-afferent ending makes a direct synapse with the motor-neuron pool of the α-efferent neurons. This is the basis for the *stretch reflex.* As an entire muscle is stretched, the annulospiral endings are also stretched, increasing excitation of the α-efferent neurons, which leads to reflex contraction of the muscle. In reality, a group of synergistic muscles (called the *myotatic unit*) contract and, with concomitant inhibition of antagonists, cause an integrated, involuntary movement. This reflex is felt to be important in sustaining posture against the force of gravity, which tends to cause joint flexion.

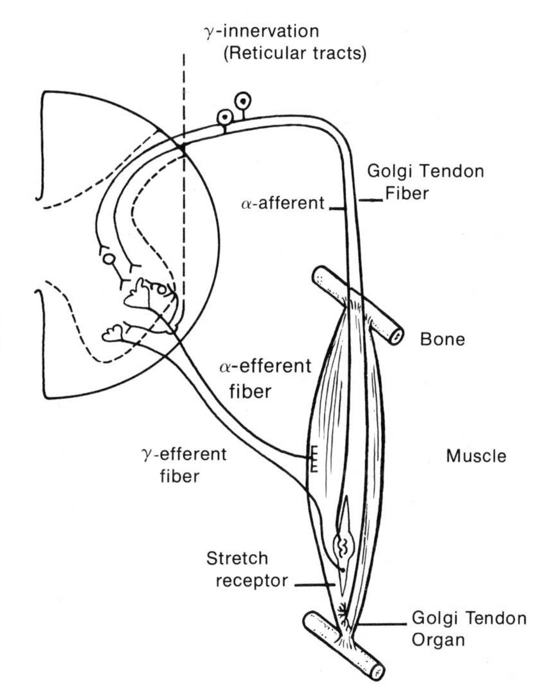

Figure 25. The spinal synapses of the stretch and stress reflexes.

It should also be noted that the stretch reflex is the only known monosynaptic (one-synapse) reflex. Because of this, and the high conduction rate of the involved nerve fibers, the response to stretch is very rapid. The familiar *patellar tendon jerk,* or *quadriceps reflex,* is a manifestation of the stretch reflex, and similar reactions can be induced at other joints. In these reflexes, sudden stretching of the tendon stimulates several receptors at once, leading to an exaggerated contraction not found under normal physiological conditions, where force is applied only gradually.

The *stress* or *tendon reflex,* which is also shown in Figure 25, operates in opposition to the stretch reflex. The Golgi tendon organ (see Chapter 1) acts as the sensory receptor. It, like the stretch receptor, responds to stretch on the muscle. The tendon organ, however, is arranged in series with the muscle fibers rather than in parallel, so it is sensitive to force applied to the muscle as a whole rather than only to the elongation of the fibers. As tension on the tendon increases, either by external force or from muscle contraction, these receptors increase their rate of discharge. They influence the α-motor neuron pool via an inhibitory interneuron and hence tend to cause muscle relaxation.

The stretch-stress complex is an example of a physiological servocontrol system. In a servocontrol, the output of a system (which in this case is contractile force) is reverted to the system for its control. The contractile force due to stimulation of the stretch receptors is actually controlled in three ways: As the muscle shortens, stretch is removed, thereby reducing spindle discharge; stress receptors send inhibitory stimuli to the motor neuron pool; and a reverberating circuit, called the *Renshaw Loop*, causes self-inhibition by the motor neuron.

The student must remember that the inhibitory interneurons do not act exclusively as a reaction to the stretch reflex. They are also active during all voluntary contractions. The tendon reflex, especially, is probably important in protecting the muscle fibers from excessive outside load. It has been postulated that the reduction of such inhibitory effects is an important factor in increasing contractile force as a result of strength training.[384]

Up to this point we have ignored the role of the intrafusal fibers. Their most apparent function is to regulate the length of the stretch receptor to assure that it remains sensitive at all muscle lengths. As the muscle shortens, the stretch receptor will become slack and thereby lose its ability to respond to stretch. Contraction of the intrafusal fibers will adjust it to the total length of the muscle, thereby keeping the sensitivity of the annulospiral endings relatively constant at all times.

The nervous system apparently uses this length-regulation mechanism to control static limb position. We voluntarily move a limb to a desired position and at the same time set the length of the stretch receptor (for example, flexing the elbow to 90 degrees). Thereafter, we adjust to constant forces, such as gravity or small unexpected additions of force, through the stretch reflex.[207] The significance of such a mechanism can be illustrated by visualizing what would happen if a weight were dropped in the hand, using our example of the flexed elbow. By the time the sensations of pressure and movement were transmitted to the brain through the three synapses, and the necessary cerebral connections made, the limb would surely have extended a considerable distance due to the added weight. The stretch reflex is, however, monosynaptic, and its response is almost instantaneous, providing immediate supplementary muscle contraction to maintain the limb in the desired position.

The γ-motor system can also induce the stretch reflex by stretching the annulospiral ending. The latter process results from the shortening of the contractile polar regions with concomitant elongation of the equatorial region. Intrafusal fiber contraction is controlled by the anterior lobe of the cerebellum[194] rather than by the motor cortex. Apparently, the reticular tracts are the route used to the γ-neuron pool.

By inducing the stretch reflex in such a manner, the intrafusal fibers also act as part of a *servomechanism*. In a servomechanism, small forces are used to control larger forces. In the nervous system, small amounts of force are applied to the stretch receptors via the intrafusal fibers as described above. Large force is developed when the elongated annulospiral endings activate the extrafusal fibers via the stretch reflex.

This process is supplemental to direct stimulation of the α-motor neuron pools from the motor cortex. The reflex apparently helps activate the neurons in the subthreshold zones, thereby involving the entire motor pool as much as possible during the exertion of great amounts of force. It has been shown that the anterior cerebellum must be intact for this mechanism to operate.[194]

Supraspinal Controls

Supraspinal controls are the result of the interactions of the various brain nuclei. The nuclei having most significance for the control of muscle contraction have already been described. The purpose of the following discussion is to show some of their main effects on muscle contraction.

Two main pathways are utilized to effect voluntary muscle contraction: the *pyramidal tracts* and the *extrapyramidal tracts.* The pyramidal tracts originate in the motor cortex, while the extrapyramidal tracts have origin in other brain centers. The activity of both pathways, however, is influenced by the variety of sensory information fed back into the brain from the body parts. The cerebellum acts as a comparator of the stimuli sent to the muscles and the sensory information returned; it then makes adjustments in the overall contractile pattern by increasing or decreasing, as necessary, the activities of the other centers.

The Pyramidal System. The essential features of the pyramidal (corticospinal) tracts are shown in Figure 26. Most of the fibers in the pyramidal tracts originate in the motor area (see Fig. 15), but some also arise from the pre-motor area. The motor area contains giant Betz cells, which account for 30 per cent of the fibers in the tracts.[50] The motor cortex, like the sensorimotor cortex, has a fairly definite representation of the various body segments, the distribution of which is shown in Figure 21. Other smaller cortical cells also contribute to the total fiber count.

The pyramidal tracts, the pathways of which are shown in Figure 26, are the longest motor tracts. Their origin is in the motor cortex and their termination is on interneurons, which, in turn, excite the lower motor neurons in the spinal cord. As the fibers pass downward, collateral branches form synapses which inform other brain centers of motor cortex activity. Stimuli are sent to the cerebral cortex itself, the striate body, the hindbrain nuclei, the reticular substance, and the cerebellum.

Most of the pyramidal fibers cross to the opposite side in the medulla forming a structure called the *decussation of the pyramids.* The name of the tract is derived from its pyramid-like shape in the medulla. From the medulla, these fibers descend in the spinal cord as the *lateral*

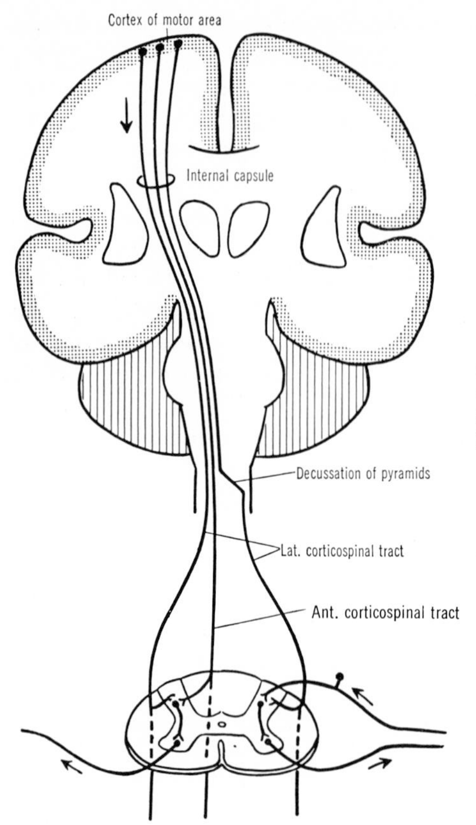

Figure 26. Essential features of the innervation of motor pools by the cerebral cortex. (Gardner, E.: Fundamentals of Neurology, 5th ed., W. B. Saunders Co., 1968.)

corticospinal tract, sometimes called the *crossed pyramidal tract.* About 20 per cent of the fibers do not cross at the pyramids, but descend in the ipsilateral ventral white column of the spinal cord as the *anterior corticospinal tract* before crossing to the opposite side in the cord. This tract is sometimes called the *direct pyramidal* or *direct corticospinal tract* (see Fig. 19).

The Extrapyramidal System. The term extrapyramidal is somewhat nebulous in that it is used to describe all of the influences on the lower motor neurons that are not effected through the pyramidal system. The extrapyramidal system is activated during voluntary contractions and involves the integration of the activity of several subcortical nuclei. The most involved final common pathway by which these effects are transmitted to the motor pools is the reticulospinal tract, but other descending tracts are also involved. It should be noted that, while the pyramidal system is only excitatory in its effect, the extrapyramidal system may be either excitatory or inhibitory. This system may only facilitate the motor neurons rather than excite them to action.

Control of Muscle Contraction. The various neural structures provided for the control of muscle contractions have been previously described. We have already noted that the cerebellum is important in integrating the activities of the various centers. The role of the cerebellum is illustrated in Figure 27.

Information of the motor cortex stimulation of spinal motor pools is passed to the cerebellum via the pontocerebellar tracts, informing it of the desired rate and force of muscle contraction. As the muscles contract, proprioceptive stimuli are sent from them back to the cerebellum over the spinocerebellar tracts, informing it of the actual state of contraction. The cerebellum can then compare the two sources of information and either increase or decrease the activity of various

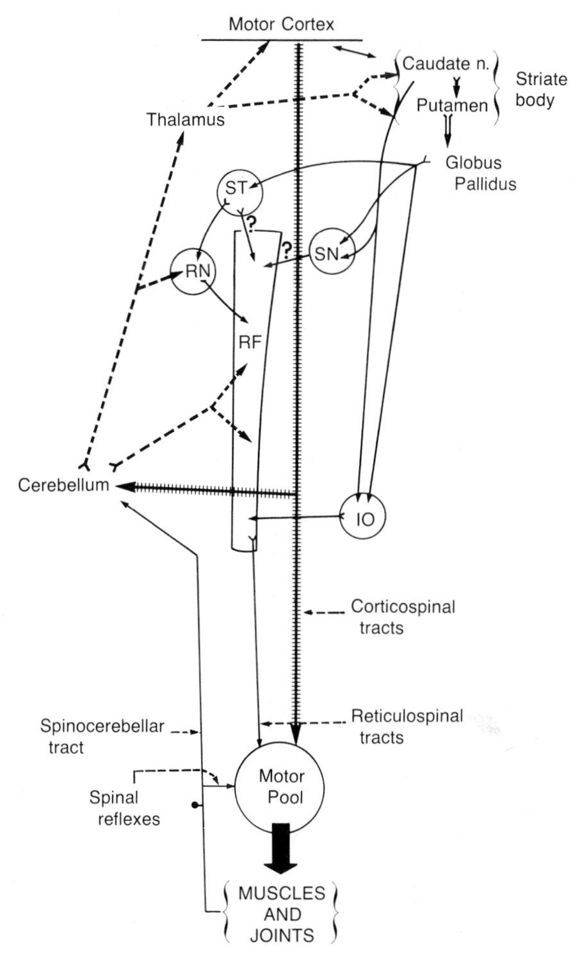

Figure 27. Summary of effects on the motor neuron pool. *RN*, Red nucleus; *ST*, Subthalamic nucleus; *SN*, Substantia nigra; *IO*, Inferior olive; *RF*, Reticular formation. (Adapted from Guyton, A. C.: Textbook of Medical Physiology, 3rd ed. W. B. Saunders Co., 1963.)

nuclei in order to adjust the rate and force of contraction to the intended amount. This is, briefly, the way in which the cerebellum acts as a comparator in a servomechanism.

The pattern of control for the extrapyramidal system is similar. The striate body is activated, with the motor cortex, stimulating the motor neuron pools in order to bring into action the necessary gross movements and static fixations that position and stabilize the limbs for the more precise movements of the cerebral cortex. Again, the cerebral cortex compares and acts to make the necessary adjustments through various nuclei.

The diagram presents a highly simplified description. Not only are the cerebellar controls active, but the various spinal cord and hindbrain reflexes previously described constantly act on the neuron pools. It should be noted especially that visual reflexes are very important in the control of muscle contraction. The vestibular apparatus also affects the cerebellum directly. The inhibitory and excitatory effects of these various centers culminate at the motor neuron pool and, as was noted at the beginning of this discussion, the amount of muscle contraction at a given time depends upon their summation.

Control of Highly Skilled Movements

The previous discussion presents an adequate description for the control of postural and slow voluntary movements. The rapidity with which some highly skilled acts are performed, however, makes inadequate a complicated feedback control system which requires transmission over a number of synapses. When striking a ball, one does not have time to make slow, deliberate adjustments.

The control of such skills is apparently complete within the cortex, and theories about how it is effected include some sort of self-contained, patterned sequence of controls over the discharge of nerve impulses causing the movement. Paillard[396] suggests that the learning of such an act requires a restructuring of existing movement combinations into new, workable units which become the new act. Guyton[200] refers to the development of sensory "engrams" which are called forth to effect the desired skill. Henry and Rogers[226] conceptualize a "memory drum" which calls forth the proper sequence and rate of muscle contraction on appropriate command. These patterns may be summoned into action by other areas of the brain.

It should also be noted that such patterns are complete and, once the sequence of contraction is started, we can do little to change it. Hence, we sometimes experience the frustration of knowing we are doing something wrong while performing the skill, but not being able to change the course of events during the movement. The sensory feedback, however, does enable us to modify subsequent performances.

It is not the purpose of this text to include a discussion of motor learning. However, it should be noted that these patterned sequences are supposedly developed by practice, which provides sensory input permitting us to modify subsequent acts and thereby, hopefully, improve our performance.

Skillful acts are typified by highly coordinated muscle contractions. Electromyographic analysis of shot-putters revealed that the highly skilled, in comparison to the unskilled, could better increase contraction at decisive moments, develop more force, utilize fewer muscle fibers to perform a submaximal activity, relax antagonistic muscles better during decisive movements, and better employ supporting body structures.[170] Kamon and Gormley[277] used electromyography to demonstrate improved integration of muscular contraction as schoolboys acquired skill in performing the single knee circle mount on the horizontal bar.

QUESTIONS

1. Develop a glossary with concise definitions of terms that are used to describe the structure and function of the central nervous system.
2. Develop, in tabular form, a list of the nuclei of the brain that affect muscle contraction, and make brief, descriptive statements of their function.
3. How are nerve impulses transmitted from neuron to neuron and from neuron to muscle?
4. The spinal cord acts as a conductor of information and as an integrator of muscle contraction. Explain.
5. The stretch receptor plays an important role in kinesthesia: true or false? Explain.
6. Compare the roles of stimuli transmitted in the dorsal and lateral sensory tracts as they relate to kinesthesia.
7. Describe the function of the semicircular canals.
8. How is the stretch receptor used to provide supplementary stimulus during maximum muscle contractions?
9. The statement was made in the text that some innate reflexes must undoubtedly be inhibited to perform some sports skills. Formulate two possible examples.
10. How would the stress reflex serve a protective role relative to the muscle fibers?
11. The final amount of muscle contraction depends on the summation of excitatory and inhibitory stimuli imposed on the motor neuron pool. Identify as many sources of each kind as you can from the discussion in the text.
12. What are the structural adaptations of the intrafusal muscle fibers that permit them to serve their function? (Not the sensory endings within them.)
13. What is a servomechanism? Give illustrations of how the nervous system acts in this manner.
14. Why is information transmitted from the muscles inadequate for the control of rapid, highly skilled movements?
15. What is the general nature of the theories that describe the way that rapid, highly skilled movements are controlled?

Chapter Four

PHENOMENA ASSOCIATED WITH NERVE CONTROL

Overall neuromuscular coordination is of primary concern to anyone responsible for teaching physical skills. It is only logical that measurement of different phases of this process should provide clues to the performance capacity of accomplished or potential athletes. Consequently, many studies on reflex time, reaction time, and performance time have been completed. Some investigators have also studied the role of innate reflex patterns during performance, while others have concerned themselves with higher central nervous system controls.

There is a considerable amount of overlap here between exercise physiology and physiological psychology. Attaining the level of performance to which we aspire in athletics is dependent upon the realization of both our physiological and psychological potentials. Consequently, it is usually impossible (or at least unwise) to ignore one factor in favor of the other. It is our purpose here to describe selected neuromuscular phenomena, the knowledge of which might help the athlete perform better or the teacher have a better understanding of his or her students.

THE GUIDING ROLE OF THE HEAD

Sherrington[472] observed that when the head of a decerebrate animal is turned to one side, muscle tone on that side increases. This is a normal postural reflex that is dependent on the tonic neck and labyrinthine reflexes and helps the animal maintain its equilibrium. When a cat is dropped with its feet pointing upward, the movement of the head initiates the muscular contraction that infallibly causes the animal to land on its feet. The head thus plays a guiding role.

Humans react similarly. Hellebrandt and co-workers[221] showed that head position and movement have a direct effect on the positions of the limbs. Head rotation causes the arm on the side to which the chin is turned to abduct at the shoulder joint while the opposite arm adducts. When the upper extremities are placed in a quadrupedal position, ventroflexion of the head causes the arms to relax and the legs to extend while dorsiflexion causes the opposite. Limb movements also appear to have a similar but reciprocal relationship to the head, especially movements of the shoulder girdle and shoulder joint.[525] There is, apparently, a highly developed coordination of head and shoulder movements.

The body also shows predictive reflex patterning of movement in response to stressful exercise. Performance of exhaustive exercise progressively leads first to head rotation, and eventually to positioning of other body segments and strong contractions of muscles not directly involved in the exercise.[218, 222, 526] These movements facilitate the motor pools of the muscles directly involved in the exercise, thereby recruiting a maximum number of motor units. The stimuli activating the reflex patterns apparently arise from proprioceptors in the active limbs.

Teachers frequently recommend that participants in weight training programs keep their body parts in a rigid position in order to "isolate and exercise only one muscle group." The experimental evidence reviewed above would seem to make such a practice inadvisable. Performance of any exercise requires the integrated activity of several components of the entire neuromuscular apparatus. Rigid posturing would seem to prevent the normal physiological adjustment that the body must make to adapt itself to load and work.

In a normal person, the head plays a guiding role, even if the individual is blindfolded. If, however, the utricles and saccules of the labyrinths are destroyed, a blindfolded person loses his sense of orientation in space and will not be able to maintain equilibrium when a platform on which he is standing is tilted. Deaf-mutes, who frequently have damaged labyrinths, have to depend on their sight to maintain equilibrium. For this reason, deaf-mutes should not be allowed to swim after dark.

REACTION AND REFLEX TIME

Reaction time is the time elapsing between the moment of application of a stimulus and the moment of response. Usually the term is applied to reactions requiring a conscious response. In a purely unconscious response, the time is referred to as "reflex time." It is, however, often impossible to differentiate *reaction time* from a *conditioned reflex time*. In sports and games, in which movements of a participant are conditioned by signals, by movements of opponents, or by motion of the ball, reaction time is of great importance. A sprinter who can start faster than other contestants; a baseball catcher who can react faster to the change in the direction of the motion of the ball; a ping pong player who is always in the right place at the right time—all have a definite advantage over slower-reacting men.

The principle of testing reaction time is very simple. The testing device consists of an electric circuit with two switches, one to be operated by the tester and the other by the subject. Before a test begins, the subject's switch is closed and that of the tester is opened. To start the test, the operator closes his switch. This immediately starts a timing device, and gives the required stimulus to the subject. The timing device usually consists of an electrically operated stopwatch measuring time in 1/100 or 1/1000 of a second. The stimulus may be auditory, by means of a buzzer; visual, by means of an electric bulb; or tactile, by means of an electric shock. Upon the perceiving of the stimulus the subject opens his switch, and the length of time needed to respond, or his reaction time, is automatically registered, because either the stopwatch stops or the signal marker makes a mark on the kymograph.

It is also possible to test the stretch reflex, using either the quadriceps or triceps surae reflex. The former is tested by striking the

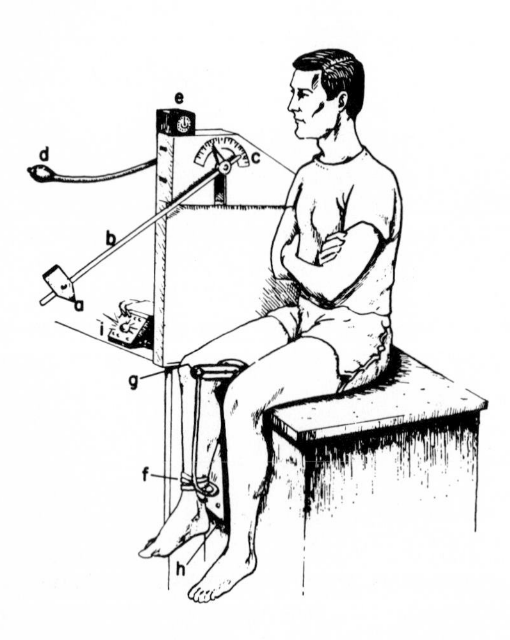

Figure 28. A device for testing the patellar tendon reflex. *a*, hammer with built-in switch which starts the clock upon striking patellar tendon; *b*, shaft; *c*, protractor; *d*, hammer release; *e*, electric timer; *f*, metal swing; *g*, ball-bearing hinge; *h*, contact plate; *i*, light indicating contact between metal plate and swing. (Tipton, C. M., and Karpovich, P. V.: J. Appl. Physiol. *21*:15, 1966.)

patellar tendon and the latter by striking the Achilles tendon. A simple device for testing the patellar tendon reflex is shown in Figure 28.[512] A hammer mounted on a swinging arm is allowed to strike the patellar tendon. The reflex time is the amount of time elapsing between the strike of the hammer and the extension of the leg. The result of this test can be made more exact by using a sensitive photoelectric cell to detect the slightest leg movement[511] or an electromyograph to determine the actual moment of muscle contraction. Typical average values for the patellar tendon reflex are 78.5 milliseconds[512] and 70.79 milliseconds.[333]

The amplitude and velocity of limb movement during the reflex and the return of the limb to its resting position (relaxation) can be measured by an elgon (short for electrogoniometer) attached over the

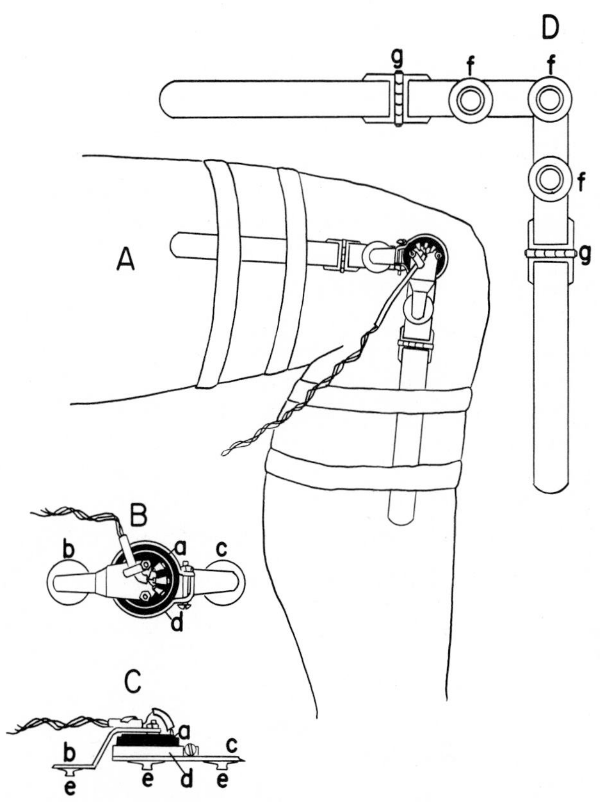

Figure 29. *A,* Electrogoniometer (elgon) attached to a chassis, which is strapped to the leg. *B,* Top view of the elgon; a, potentiometer; b and c, snap buttons; d, a clamp for attaching the arms to potentiometer. *C,* Side view of the elgon; a, potentiometer; b and c, arms; d, clamp; e, studs of snap buttons. *D,* Chassis; f, sockets for the elgon studs; g, hinges.

knee or ankle. The elgon, which is shown in Figure 29, is a goniometer in which a potentiometer has been substituted for a protractor, and it is connected with a battery and direct current oscillograph.[298] Joint movement causes rotary movement in the elgon, while resistance to electric current changes proportionately to the angle of the joint. Therefore, the record on the oscillograph can be read in degrees of joint motion. Prior to testing, the elgon is calibrated so that displacement of the oscillograph record can be translated into degrees of movement. The device can also be used in the study of locomotion during athletic movements.[187, 318, 429, 481]

Numerous studies have shown that athletes have a faster reaction time than nonathletes.[309, 533] There also seems to be some relation between reaction and reflex times. For instance, sprinters have shorter reaction and shorter patellar reflex times than do long-distance runners.[339]

Exhausting exercise apparently increases the stretch reflex time. Tipton and Karpovich[512] found the reflex time decreased after one minute of bicycle ergometer riding but increased 15.7 and 16.3 per cent after rides to five minutes and exhaustion respectively. Similar results were obtained with ipsilateral leg extensions with a 25-pound weight to exhaustion.

In experimenting with reaction time, one should remember that reaction time has diurnal variations, the best time usually being obtainable in the afternoon. The condition of the subject should also be taken into consideration, because fatigue, as a rule, slows down reaction time. Age also should be considered, because reaction time is slower in younger children and gradually improves with age, reaching its maximum at college age.

A comparison of hand and total body reaction times for different age groups is shown in Table 4. Henry and Rogers[226] have shown that simple reaction time becomes longer when subsequent movement is increased in complexity. In a comparison between mongoloid and normal children and young adults, Kramer[327] found differences in favor of the normals in reaction time, performance time, and grip strength, but found no difference between the groups in patellar reflex time. Since the reflex is a spinal level phenomenon, it was not necessary to involve any cognitive processes.

Strength training has been shown to have a beneficial effect on the stretch reflex time. Francis and Tipton[178] have shown a 5 per cent decrease in the patellar tendon reflex time with strength increases. Reid[427] found a decrease of 7 per cent in Achilles tendon reflex time with a 27 per cent increase in plantar flexion strength. The mechanism responsible for these effects has not been identified. It is of interest to the researcher and might be a result of an improvement in one or

TABLE 4. Age and Reaction Time in Fractions of a Second

Age	*No. of Cases*	*Hand*		*Body**		*r*
		MEAN	S.D.	MEAN	S.D.	
14	69	.2754	.0436	.8173	.0780	.34
15	62	.2716	.0358	.8263	.0795	.33
16	58	.2631	.0325	.7973	.0855	.43
17	59	.2593	.0279	.7948	.0785	.24
All subjects	248	.2675	.0349	.7986	.0782	
University subjects	200	.2469	.0212	.7824	.0829	

*The subject made one step and opened switch by hand, in response to an auditory stimulus.

r = coefficient of correlation between hand and body reaction times. (From Atwell and Elbel: Res. Quart. Amer. Ass. Health Phys. Educ. 19:22, 1948.)

more factors: stretch receptor sensitivity, afferent or efferent conduction rate, synaptic transmission, or muscle function.

Physiological and Psychological Limits

Both physiological and psychological factors play a role in determining the limits to performance at a given time. The *psychological limits* would be revealed by performance when muscle contractions are voluntary, while the physiological limits would be revealed by performance when muscle contraction was independent of voluntary exertion.

Ikai, Yabe, and Ischii[266] found that the maximal strength of the thumb adductor was 30 per cent higher when muscle contraction was induced electrically than when it was induced voluntarily. More recently, Ikai and Yabe[265] reported on the physiological and psychological limits to muscle endurance after having subjects perform thumb adductor exercises against a resistance until they were forced to stop due to fatigue. The adductor muscles were then stimulated electrically via their nerve until they no longer responded. Subjects were also submitted to a muscle endurance training program. The difference between the limits, as shown in Figure 30, remained quite constant even as endurance improved.

The difference between the physiological and psychological limits apparently acts as a reserve which can be used when needed. Different ways by which the athlete may more nearly approach his physiological limits are discussed in the remainder of this chapter.

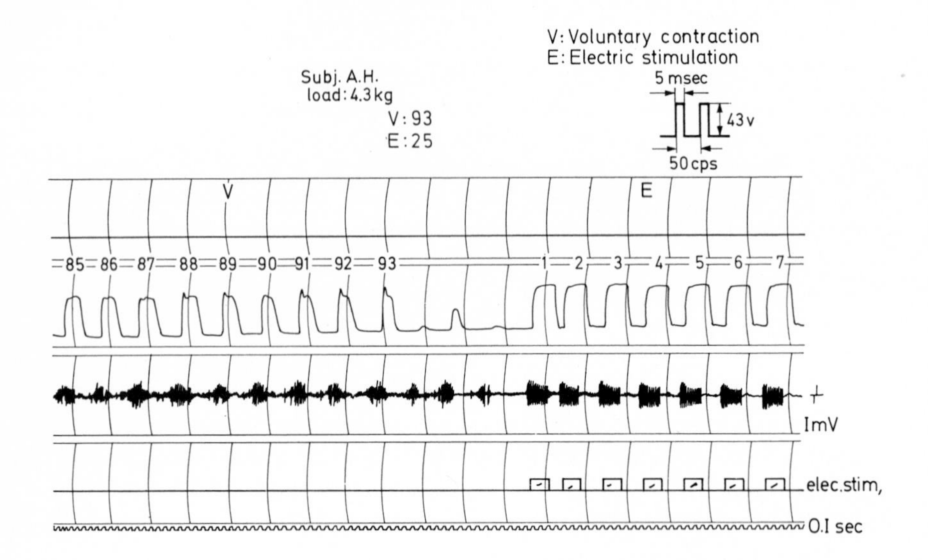

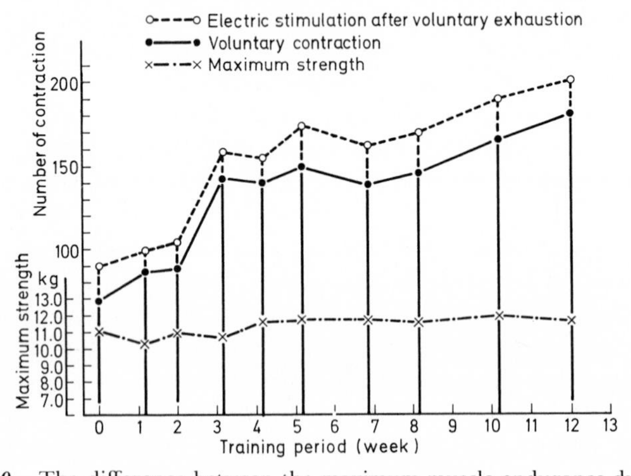

Figure 30. The difference between the maximum muscle endurance due to voluntary contraction and electrical stimulation, and the effects of endurance training. The thumb adductor muscles were stimulated via their nerve when voluntary contractions could no longer be continued. The top part of the figure shows a recording taken during experiments: upper trace, movement of the thumb; middle trace, electromyogram; lower trace, electrical stimulus. (Ikai, M., and Yabe, K.: Int. Z. Angew. Physiol. *28*:55, 1969.)

Dynamogenic Effect of Muscular Contractions

Every athlete has experienced a feeling of helplessness when, near the finish line, he has been outdistanced by an opponent. He remembers a sensation of extreme heaviness in his feet, and his inability to move them any faster. Something besides mere desire was needed to make him run faster. Some runners used to carry, in each hand, a piece of cork loosely secured to the hand by string. Approaching the finish line, they would squeeze the corks as hard as they could. Some carried handkerchiefs for the same purpose. This practice seems to have been discontinued; yet, physiologically, it is based on a sound principle.

A better effect, however, may be obtained by vigorous movements of the arms. This will cause a momentary revival of the legs. One should remember, however, that this vigorous arm movement should be reserved as a last resource. Numerous laboratory experiments conducted by one of the authors[283] showed that, when some muscle groups have been exhausted by continuous work on an ergograph, they can be revived immediately if hitherto idle muscles are brought into the action. Imagine the author's embarrassment when later he found that a similar discovery had been made in 1903 by the father of Russian physiology, Setchenov.

During the past three decades, many investigators in the U.S.S.R. have substantiated Setchenov's findings. Moreover, they have concluded that recovery from acute muscular fatigue may be reached faster if, instead of taking a complete rest, the subject proceeds to exercise using the muscles which had not been previously tired. This has led to development of "active rest," which is now widely used in the U.S.S.R.

Ergographic investigations by Kozlowski[325] showed that the effect of active rest was noticeable even when the blood circulation in the tired limb was stopped by a tourniquet; however, the result was not as good as when circulation was not interrupted. The beneficial effect of active rest was also observed on muscles which were fatigued by stimulation with electric current. Active rest is also used in other countries but usually almost as a ritual, such as walking after running. If one asks an athlete or a coach why one should walk after running, one may get a very simple answer: "It helps." Yet, in 1937, the Harvard Fatigue Laboratory investigators discovered that the rate of lactic acid removal after an exhaustive race was greater if the subject slowed down to a moderate pace instead of completely stopping.[391] It has been reported that massage of the untired limb also beneficially affected recovery of the tired limb.[325] However, when the skin was anesthetized, massage produced no effect. "Active rest" evidently affects tired muscles mainly through the central nervous system, and is related to the dynamogenic effect described in this chapter.

Krestovnikoff and his co-workers[331] made observations on changes in the strength of the hand grip during exercises with barbells. An alternate lifting by each arm had a beneficial effect on the resting arm. A more effective result was obtained after squatting exercises with a barbell. The strength of the right hand increased, on occasions, by 18.7 pounds, and the left grip by as much as 17.05 pounds.*

Cross Training

The dynamogenic effect has been utilized in "cross education" or cross training. Hellebrandt and others[216, 217, 219] have reported that if, for example, one leg is immobilized by a cast, exercising the other leg will increase the strength of the immobilized one. Hodgkins[246A] demonstrated that exercising one leg with an 18 pound boot, three times a week for three and a half weeks, increased endurance of that leg 966 per cent, and the endurance of the non-exercised leg by 275 per cent. This cross effect is possible, however, only if the motor nerves leading to the inactive muscles are functioning. Severing these nerves precludes any cross training. It has been shown that contraction of a muscle may cause the appearance of the action potentials in the identical muscle of the other side.[475] It is probable that the effect is more noticeable in those muscles which are frequently involved in a synergetic action. When, for instance, a biceps contracts against resistance, the back muscles will have action potentials present because their tension is increased by reflex. If, however, an abductor of the small finger of one hand is contracted, no reciprocal stimulation of the identical muscle of the other hand is observed.

Hettinger and Müller[235] and Asa,[10] using isometric exercises, could not observe any cross training effect.

Müller thinks that the difference in results of cross training obtained by different investigators has depended on the difference in methods of training used. Positive results were obtained with isotonic contraction, whereas negative results were obtained with isometric contraction. Movements involved in isotonic contraction require more complex nervous regulation than those in isometric.

Ergogenic Effects of Cheering and Music

In armies, when soldiers have to march long distances, singing is effective. Each company has at least one man who will sing a simple

*Experiments conducted at Springfield College could not detect that these squatting exercises had a beneficial effect upon the strength of the hand grip.

rhythmic ditty, and the company will take part in the chorus. Musical instruments and bands are also used. Musicians sometimes are allowed to ride, so that they will not be too tired to play effectively while the others have to march.

Krestovnikoff and his co-workers[334] studied the effect of music upon oxygen consumption during calisthenics. In most cases, the consumption increased slightly, but participants reported that it was easier to exercise with music than without it. It must be concluded, therefore, that suitable music acts as an additional reinforcing stimulus which compels participants to work more energetically. Yet, in spite of a slight increase in energy expenditure, work seems to become easier.

Laboratory experiments have also shown that cheering has ergogenic effects. The work output of men riding a bicycle ergometer, or working on an arm ergograph, always increased when cheering was used.[283]

In actual competitions, cheering may be a powerful but also a dangerous weapon. If an athlete, stimulated by cheering, starts his final spurt too soon, he may exhaust himself prematurely and have to slow down before the finish. Fortunately, many runners, because of concentration on the race or because of fatigue, become deaf to all cheers, and follow their predetermined pace.

Ikai and Steinhaus[264] found that firing a gun behind a man pulling a cable tensiometer with flexors of the elbow increased the force of the pull. The subject was instructed to exert a maximal pull whenever the second sweep hand of an electric clock placed before him crossed the 1 o'clock position on the dial. During each 30-minute session the operator standing directly behind the unwarned subject occasionally fired a gun 2, 4, 6, 8, or 10 seconds before the pull was to be expected. The effect was greater when a shot preceded the pull. When they coincided, the force of the pull often decreased.

Ergogenic Effect of Excitement

It is a matter of common observation that, while a state of excitement may impair the skill and accuracy of performance, it may greatly augment the muscular strength of an individual. A frail woman has been known to lift and move heavy objects when the life of her child was in danger. It may take several attendants to restrain an excited and disturbed patient who, under ordinary conditions, is weaker than one attendant.

It may be of interest to cite an unbelievable feat of strength which occurred during World War II.[509] An Air Force officer, with a parachute on and a crash ax in one hand, was knocking the bombs loose from a disabled plane, in order to make a safe landing. Suddenly he started to fall; but, fortunately, he caught hold of the bomb rack with his free hand and slowly pulled himself up. After he did so, he realized that his other hand was still holding the ax. This story is interesting for several reasons. When one thinks how seldom a one-arm pull-up is demonstrated, even by members of varsity gym teams, one will fully appreciate the difficulty of doing the same dressed in flying togs, with a parachute on one's back and an ax in one hand. This story also illustrates impairment of rationality because of fear. Under normal conditions the flyer would have dropped the ax and used both hands to pull himself up. Yet he clutched the ax, and did not think of using the hand which held it. In some individuals, fear may cause a complete inhibition of voluntary movements. They become speechless and "freeze" in place, instead of running away or defending themselves.

Effect of Hypnosis and Suggestion

Many claims have been made regarding extraordinary feats of physical strength and endurance exhibited when one is in the hypnotic state. Probably the most impressive feat consists of placing a hypnotized person on two chairs so that the body weight is supported by the neck and ankles alone. On the other hand, reports that speed records have been broken by runners in an hypnotic state have not been substantiated.

Laboratory experiments have shown that, under well-controlled conditions, even a mild degree of hypnosis causes an increase in strength and endurance.[160, 256, 440] These findings were recently again substantiated.[261] Tests conducted on both men and women have shown that the strength of the grip and the arm flexors, and the ability to hang from a bar by the hands, have been increased. A hypnotized subject who has been told that the feeling of discomfort and pain should be disregarded and that he or she can do better than before usually does so.

Coaches use another powerful tool—suggestion. A strong suggestion may bring about a mild degree of hypnosis. The effectiveness of suggestion depends on the personality of a coach. Identical ideas coming from two different coaches may have diametrically opposed effects. A skilled coach possessing poise and a strong, commanding personality will restore with a few words self-confidence and a desire to do the utmost; while an insecure and excitable coach will only unnerve his men by a long and superfluous pep talk.

Athletes may put themselves in an almost hypnotic state by means of self-suggestion. They become at first obsessed and then possessed by a desire to win. This condition compels them to train more thoroughly; and, if they have the ability, they may eventually reach their goal.

QUESTIONS

1. What is the *leading* role *of the head*?
2. Why is it dangerous for the deaf-mute to swim after dark?
3. What is the effect of anticipation of activity upon physiological functions, and how is it achieved?
4. Explain development of athletic "form."
5. Explain the physiological meaning of "skilled hands."
6. Describe the principle of measurement of reaction time.
7. What is reaction time?
8. What is the relation between age and reaction time?
9. What is the relation between reaction time and success in athletics?
10. Discuss the dynamogenic effect of muscular contractions.
11. Discuss cross training.
12. Discuss the dynamogenic effect of music and cheering.
13. Discuss the effect of hypnotism on performance.
14. What is the difference between reflex time and reaction time?
15. Return to Chapter 3 and trace, within the limitations of the information supplied there, the neural paths followed for both the reaction to a stimulus of pressure placed against the leg and the reflex response to tapping the patellar tendon.
16. What is meant by the terms "physiological limitation" and "psychological limitation"?

Chapter Five

THE PROVISION OF ENERGY FOR MUSCULAR WORK

The energy used for muscular work comes from food. All three classes of nutrients—carbohydrates, fats, and proteins—can provide this energy. The process by which this energy is made available for muscular work is complex, but it gives us a basis for understanding many of our physiological adjustments to exercise.

THE RELEASE OF ENERGY IN THE CELL

It was noted in Chapter 1 that the energy-rich compound, adenosine triphosphate (ATP), serves as the direct source of energy for muscle contraction. ATP is not stored in significant quantities and the available supply can support muscle contraction for only a fraction of a second. Consequently, ATP must constantly be resynthesized in the cell after its breakdown to ADP, if contraction is to continue.

The energy for resynthesizing ATP comes from the foodstuffs we ingest. The process takes place in three phases, as shown in Figure 31. The first phase is digestion, where complex molecules are broken down into simpler ones which can pass through the intestinal wall into the blood stream. The second phase consists of several series of reactions that eventually yield one of three common end products: *acetyl coenzyme A*, *α-ketoglutaric acid* and *oxaloacetic acid*. These end products then enter the Krebs cycle where they are further degraded into carbon dioxide and water.

It is important to note the role of the biological catalysts, which are called *enzymes*, in these reactions. If we were to ignite a quantity of

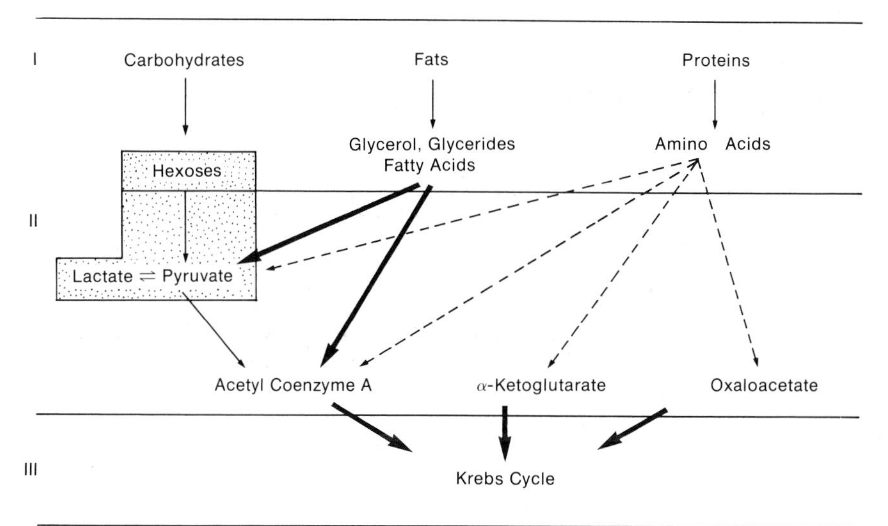

Figure 31. An outline of the routes followed by foodstuffs in the production of energy. Shaded area, anaerobic glycolysis. (Adapted from Krebs, H. A., and Kornberg, H. L.: Ergebnisse Der Physiol. *49*:212, 1957.)

sugar, it would burn freely, changing to carbon dioxide and water and releasing its energy only as heat. The enzymes enable the cells to capture the energy for the synthesis of ATP.

Figure 32 shows an overview of some of the biochemical reactions that help us understand the bodily response to exercise. Despite its formidable appearance, it does not show all of the reactions or products formed. It shows, primarily, the way in which glucose is broken down and the specific points at which the phase III products shown in Figure 31 enter into the Krebs cycle.

The breakdown of glucose is important for several reasons: most carbohydrates are metabolized as glucose; it is the main source of energy during exercise; and it can yield a limited amount of ATP when oxygen supplies are inadequate.

In the first step, glucose is changed to glucose 6-phosphate. It is interesting to note that ATP is the source of energy for this endergonic reaction, even though the ultimate purpose is to produce more ATP. The glucose 6-phosphate (*2*, Fig. 31) is an important intermediary in that *glycogen* may be formed from it. Glycogen is stored in the liver and muscles to be called forth when needed. Molecules of glycogen are composed of long chains of glucose molecules.

The glycogen stores are mobilized during exercise, when energy requirements are high. The enzyme phosphorylase catalyzes the break-

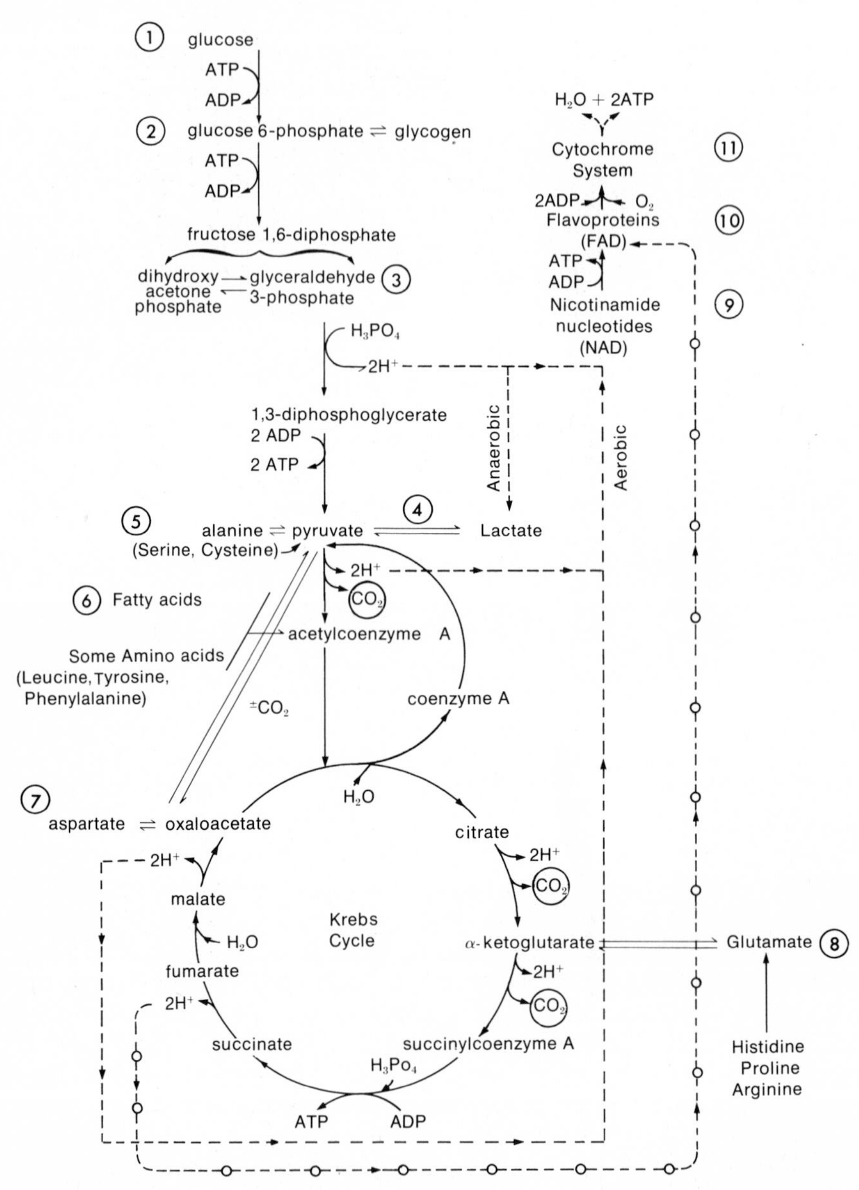

Figure 32. A summary of important chemical reactions that provide energy for physical activity. (Adapted from Bell, G. H., Davidson, J. N., and Scarborough, H.: Textbook of Physiology and Biochemistry, 7th ed. Williams and Wilkins Co., 1968.)

down of glycogen back to glucose 6-phosphate, which can then be subjected to the successive reactions, continuing through the Krebs cycle. The exact mechanism by which the phosphorylase activity is triggered is not known, but it has been shown to be activated by the hormones epinephrine and norepinephrine in the muscles and liver, and by glucagon in the liver. Anoxia (lack of oxygen) has also been shown to influence this activity.[189]

In the liver, the glucose 6-phosphate can be reconverted to glucose, which is then free to diffuse into the blood and be carried to the active muscles. The enzyme glucose 6-phosphatase is necessary for this reaction, but the muscles do not contain it. Consequently, once the glycogen is formed within the muscle cell, it cannot be released back into the blood stream and made available to other cells.

Further reactions (*2* through *3*) cause the original 6-carbon glucose to be broken into two 3-carbon molecules, one of which, phosphoglyceraldehyde, is eventually transformed to pyruvate (*3* through *4*). During this transformation, two ADP molecules are taken into the reactions and are changed to ATP which can be used for muscle contraction or other physiological processes needing energy. As the two molecules of ATP are formed, two hydrogen ions are released.

The reaction can continue as long as the nicotinamide adenine dinucleotide (NAD, *9*, Fig. 31) is available to remove the hydrogen ions by forming NAD2H. If oxygen is available, the hydrogen ions are transferred from the NAD2H to the *flavoproteins* and then to the *cytochrome system*, to be combined with hydrogen to form water and more ATP. However, if there is insufficient oxygen available, the pyruvate assumes the role of removing the hydrogen ions from the NAD2H by combining with them to form lactic acid, a reaction catalyzed by the enzyme *lactic dehydrogenase* (*4*, Fig. 31).

The process of forming ATP by the breakdown of glucose, with the formation of lactic acid, under anoxic conditions is referred to as *anaerobic glycolysis* (comprised in steps *1* through *4* of Figure 31). This process is important in that it provides a temporary emergency source of energy when the body is deprived of oxygen, such as during immersion or upon unexpected exposure to a smoke-filled room. It also provides the main source of energy for activities, such as running sprints, when the rate of energy expenditure is so great that the cardiovascular and respiratory systems have neither the time nor the capacity to make an adequate increase in oxygen delivery.

The reactions included in the breakdown of glucose to pyruvate occur in the cell protoplasm. Pyruvate then combines with coenzyme A to form acetyl coenzyme A, which enters the Krebs cycle, a series of reactions which go on inside the mitochondria. Here, further degradation occurs, with carbon dioxide molecules, hydrogen ions, and a unit of ATP being formed. The carbon dioxide is removed via the blood stream and lungs, while the hydrogen ions are combined with molecular oxygen to form water and ATP via the *electron transport chain.*

The electron transport chain is complex, but some clarification is appropriate due to its importance in biological oxidations. It is represented by numbers *9*, *10*, and *11* in Figure 31. The substance broken down is called the *substrate.* As the substrate is systematically degraded in the presence of enzymes, hydrogen ions are removed generally by the NAD. The NAD is then relieved of the hydrogen ion by the flavo-adenine dinucleotide (FAD), which in turn passes it to the cytochrome system. As the ion is passed from the NAD to the FAD, there is a loss of energy which is used to form more ATP. In one instance in Figure 31, the FAD acts directly on the substrate. The hydrogen ions are then transferred through the cytochrome system, which has the ability to combine them with oxygen to form more ATP and water.

Continued activity on the part of the electron transport system depends upon an adequate supply of oxygen. Hence, these reactions are *aerobic.* If oxygen is not available, the hydrogen ions cannot be removed from the NAD. The capacity to provide energy for muscular contraction via this route is thereby dependent on the amount of oxygen available.

In addition to glucose, Figure 31 also shows where other products of phase II of metabolism enter the final aerobic process. Fats are previously broken down into glycerol, which enters as glyceraldehyde 3-phosphate (*3*, Fig. 31), and fatty acids, which enter as acetyl coenzyme A (*6*, Fig. 31).

The fatty acids undergo a series of reactions which are referred to as *beta-oxidation.* These reactions yield acetyl coenzyme A, which also enters into the cycle. The β-oxidation process is aerobic, and it produces a considerable amount of ATP before entering the cycle. Continued metabolism of the fatty acids relies on continued metabolism of carbohydrate to provide the oxaloacetate (*4* and *7*, Fig. 31) for continued reception of the acetyl coenzyme A into the cycle.

The amino acids enter these metabolic pathways after the oxidative removal of chemical group NH_3 which is their identifying characteristic. This process is called *deamination*, and takes place primarily in the liver. Some of the amino acids are stored as glycogen after this process is completed.

Aerobic and Anaerobic Energy Sources During Exercise

Exercise may be referred to as either aerobic or anaerobic, depending upon which kind of metabolic pathway serves as the primary source of energy. Aerobic exercise is carried on at an intensity within the capacity of the body to continue aerobic reactions. In anaerobic exercise, we use energy at a rate beyond our capacity to deliver oxygen and use it in the cells.

Another route by which energy may be provided anaerobically is through the substance phosphocreatine (PC). Creatine (C) stores and releases high energy bonds according to the reaction:

$$C + ATP \rightleftharpoons PC + ADP$$

The phosphocreatine is formed when other demands for ATP are not high, and the reaction is reversed when the high energy bonds are needed for physiological activity.

It should be noted that the duration of aerobic reactions is limited only by the availability of the foodstuffs made available to the cells. During anaerobic glycolysis, the lactic acid tends to inhibit muscle contraction, and its accumulation to a concentration of 0.3 to 0.4 gm. per cent in the muscle, or 0.032 to 0.140 gm. per cent in the blood, eventually leads to the cessation of exercise. This level is seldom, if ever, reached. A level of 0.2 gm. per cent has been observed in a very well-trained athlete.[135]

Examination of Figure 31 also reveals that the total energy yield via anaerobic glycolysis is small. Only 2 moles (gram-molecular weights) of ATP are formed for every mole of glucose metabolized to pyruvate and then to lactic acid. If the hydrogen ions are removed via the electron transport chain so that lactic acid is not formed, 6 additional moles of ATP are provided, and the pyruvate goes into the Krebs cycle, where 30 more moles of ATP are synthesized. The total energy yield for the system, utilizing both aerobic and anaerobic routes, is 38 moles.

Overall performance capacity depends upon the optimum utilization of the aerobic and anaerobic phases of energy metabolism. Energy for short, sprint-type activities, such as the 100-yard dash, can be derived exclusively via anaerobic routes. Extended activities, such as distance running, must have energy provided primarily via aerobic routes because of their higher efficiency and because of the necessity of preventing the lactic acid accumulation of anaerobic glycolysis. The relative roles played by the two systems depend upon the amount of oxygen supplied to the cell. The use of oxygen is, therefore, very important, and is discussed separately in Chapter 6.

CELL FUNCTION AND TRAINING

Participation in strenuous training programs has long been known to increase the capacity to perform aerobic exercise. Recent studies of biochemical changes in the cell have shown that this increased capacity can be at least partially caused by changes within the cell. Böhmer[60] found decreases in the amounts of several cell enzymes as a result of limb immobilization in orthopedic patients. Holloszy[247] found a twofold increase in enzymes essential to aerobic reactions in rat muscle as a result of strenuous exercise.

The cellular structure has also been shown to change with training. Gollnick and King[188] showed that mitochondria in trained rat muscle

Figure 33. Electron micrographs taken from human vastus lateralis muscle before and after exhausting bicycle ergometer exercise. Structures identified: Z lines (*Z*), I bands (*I*), A bands (*A*), Mitochondria (*M*), Glycogen particles (small dots around *G*), and Sarcotubules (*ST*).

were more numerous than in untrained muscle and also appeared to be larger and have more densely packed cristae. The cristae are mitochondrial structures that purportedly contain the oxidative enzymes. Chemical analyses performed by Holloszy and Oscai[248] have also indicated changes in the cristae as well as changes in the number and size of mitochondria.

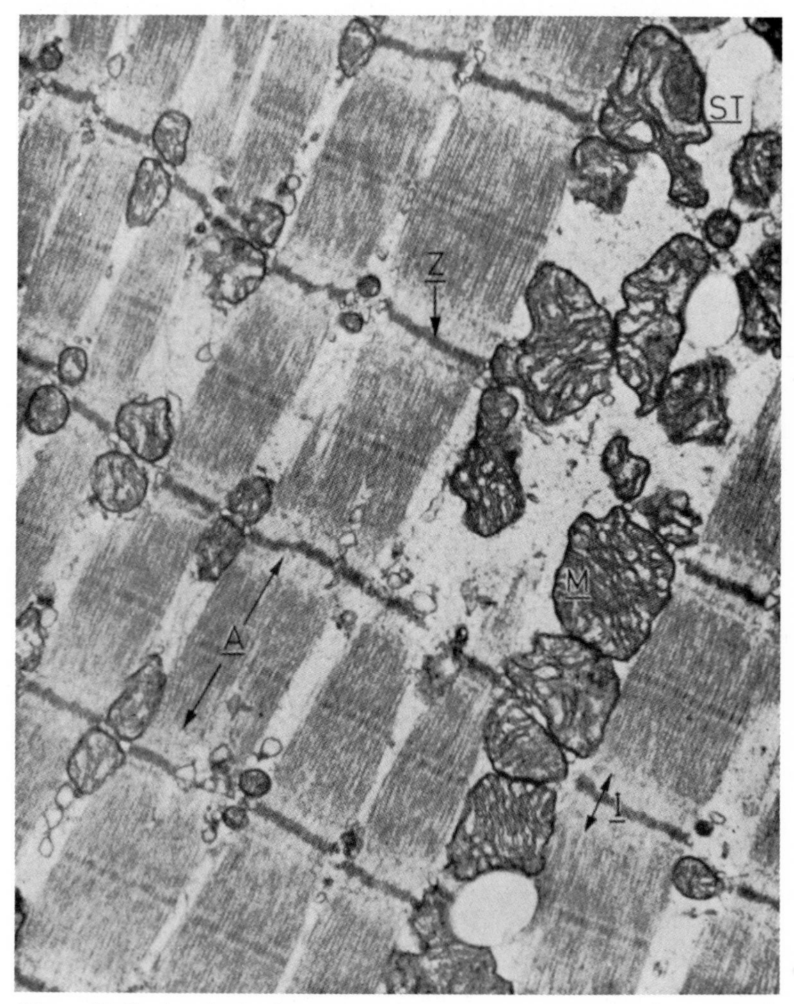

***Figure 33** Continued.* In the picture of the post-exercise sample above, note the absence of glycogen. (Gollnick, P. D., Ianuzzo, C. D., Williams, C., and Hill, R. T.: Int. Z. Angew. Physiol. *27*:257, 1969.)

There may also be a disruption of cellular function with exhaustion. Mitochondria from skeletal rat muscle became swollen and internally disorganized during exhaustive exercise.[188] This apparently interferes with the cell's ability to use oxygen in its metabolic reaction.

PROTEIN AS A SOURCE OF ENERGY DURING EXERCISE

There was a time when muscular work was supposed to be performed at the expense of energy derived from proteins in the protoplasm of the muscle fiber. This was later refuted, and it was shown that, under ordinary circumstances, muscular work is not performed primarily at the expense of nitrogenous products, for increased work does not call forth any noteworthy increase in the metabolism of protein. Most of the energy for muscular contraction, therefore, must be derived from the non-nitrogenous compounds.

Commercial advertising, extolling the energy-giving qualities of protein, merely represents an unscrupulous exploitation of public ignorance. The amount of protein indispensable for a normal output of energy is usually present in the diet. Although an athlete may need additional protein during intensive training, he particularly needs fat and especially carbohydrate, because fat and carbohydrate are the sources of energy. For this reason, it is suggested that his diet have 183 gm. protein, 134 gm. fat, and 732 gm. carbohydrate, with corresponding values in large calories of 750, 1246, and 3001, the total being about 5000.

RESPIRATORY QUOTIENT

Under ordinary circumstances, athletes and hard-working people depend on carbohydrates and fats as the source of muscular energy. It is possible, by laboratory tests, to determine how much of each of these ingredients is used during muscular activity. The determination is based on measurements of the amounts of carbon dioxide given off and oxygen consumed during activity.

It has been found that, when carbohydrate is oxidized, the volumes of oxygen used and carbon dioxide produced are equal. On the other hand, when fat is oxidized, more oxygen is used than carbon dioxide produced.

The ratio of carbon dioxide produced to oxygen consumed, $\frac{CO_2}{O_2}$, is called the *respiratory quotient* (R.Q.). Fats and carbohydrates, because of differences in their chemical structure, have different R.Q.'s. This knowledge may be utilized in determining the amounts of these substances used during muscular work.

Explanation of the reason for differences in respiratory quotients is simple. When carbohydrate is oxidized, since the proportion of hydrogen and oxygen in the molecule is the same as that of water, extra oxygen is needed only for the oxidation of carbon. Consequently, for every molecule of oxygen used, a molecule of carbon dioxide is released.

$$C_6H_{12}O_6 + 6O_2 = 6CO_2 + 6H_2O$$

$$\text{Hence the R.Q.} = \frac{6\ (\text{Volumes } CO_2)}{6\ (\text{Volumes } O_2)} = 1.0$$

When fat is oxidized, more oxygen is used than carbon dioxide is given off. This is because the amount of oxygen present in fat is not sufficient for the oxidation of its hydrogen, and therefore oxygen from the inspired air must be used for this process as well as for the oxidation of carbon. Oxidation of a typical fat proceeds as follows:

$$2C_{51}H_{98}O_6 + 145O_2 \rightarrow 102CO_2 + 98H_2O$$

$$\text{and R.Q.} = \frac{102\ (\text{Volumes } CO_2)}{145\ (\text{Volumes } O_2)} = 0.7$$

The respiratory quotient for proteins has been found to be about 0.8. The amount of protein that is oxidized during exertion is usually so small, as compared with carbohydrate and fat, that it can be disregarded. Yet, if desired, its amount can be determined over any extended period of time from the nitrogen in the urine.

In experiments of short duration, since the metabolism of proteins may be disregarded, the variations in respiratory quotient may be used to compute the relative carbohydrate and fat metabolism. The following calculation illustrates how the relative amounts of fat and carbohydrate used for an exercise may be determined if the respiratory quotient is known.

Let us suppose that, as a result of a certain exercise, 10 liters of oxygen were used and 9 liters of carbon dioxide were given off, making the respiratory quotient equal to 0.9.

It has been established that 1 gm. of glycogen in oxidation takes up 0.828 liter of oxygen and gives off 0.828 liter of carbon dioxide. Likewise, 1 gm. of fat takes up 1.989 liters of oxygen and gives off 1.419 liters of carbon dioxide. If we assume that during exercise x

grams of glycogen and y grams of fat were used, we can write two equations, which are easily solved:

$$\begin{array}{rl} 10\text{L. } (O_2) & = 1.989\,y + 0.828\,x \\ -\ 9\text{L. } (CO_2) & = 1.419\,y + 0.828\,x \\ \hline 1\text{L.} & = 0.570\,y \end{array}$$

$$\text{Therefore } y = \frac{1}{0.570} = 1.75 \text{ gm. of fat}$$

Substitution of 1.75 for y in either of the two equations will give $x =$ 7.87 gm. of carbohydrate.

EFFECT OF EXERCISE ON THE RESPIRATORY QUOTIENT

If exercise is of short duration or long, but not exhaustive, the respiratory quotient rises. If at rest it is 0.85, during work it may be between 0.90 and 0.97. After exercise it may fall to below 0.90.[108] If expired air is collected for a minute after an intensive exercise, such as

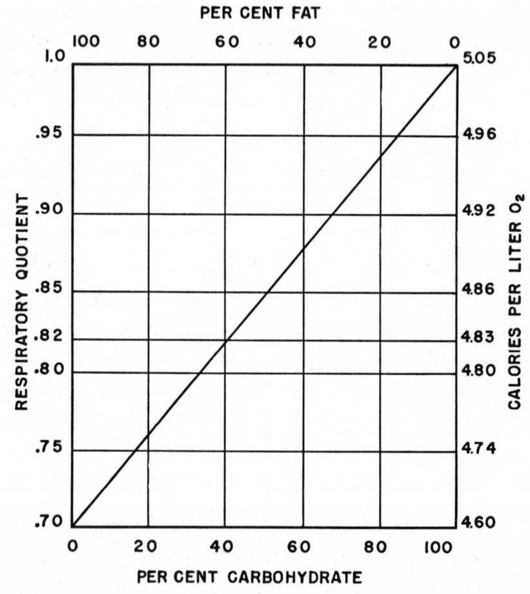

Figure 34. Relation between the respiratory quotient and the percentage of energy (calories) obtained from carbohydrate and fat, and also the number of calories obtained per liter of oxygen. Example: When R.Q. is 0.82, per cent of energy derived from carbohydrate is 40 and that from fat is 60, and 4.83 calories are obtained per liter of oxygen.

a 100-yard run, the R.Q. may be as high as 1.5. This is a spurious R.Q., resulting from overbreathing after exercise. Because of violent postexercise breathing, more carbon dioxide is removed than is produced. During prolonged and exhaustive work, the R.Q. goes steadily down toward the 0.70 value, indicating a steady increase in dependence on fats.

As has been previously mentioned, in an excised muscle the fuel for work is carbohydrate, and the R.Q. is 1.0. If, however, the muscle is not separated from the animal's body, the O_2 and CO_2 contents in R.Q. is found not to be unity but less—the same as for the entire working animal[244, 245]—indicating that fat is used as a fuel for work.

FAT AS A SOURCE OF ENERGY

It can be said that carbohydrates alone are seldom, if ever, used for muscular work. A series of studies made at the Harvard Fatigue Laboratory has furnished a striking illustration of this fact. In one, a fasting dog did twenty-seven hours of work on a treadmill in which the total work performed was ten times as great as could be accounted for by the glycogen reserves of its body.[133] In a similar experiment on an athlete, the respiratory quotient fell steadily from 0.83 to 0.75; that is, toward the respiratory quotient of fat. It was found that, even with an abundance of carbohydrate, some fat is used in exercise; and, as the carbohydrate reserve diminishes, the proportion of energy derived from fat may increase from 8 to 77 per cent.

In the light of these findings, it is no longer necessary to consider carbohydrate the sole available foodstuff for muscle. Muscles utilize either carbohydrate or fat, or both. Observations made by Dill and his co-workers[151] on Harvard football players indicated that, in spite of an abundance of carbohydrates in their diet, the players derived 44 per cent of their daily energy from fats.[143] Carlson and Pernow[84] found that, during work of medium intensity, lipids (unesterified fatty acids) serve as a fuel. During maximal work, however, lipids are not used as extensively and energy is derived primarily from carbohydrates. Hermanson, Hultman and Saltin[230] found that, when the relative work load of subjects was increased from 29 to 78 per cent of their capacity to maintain aerobic metabolism, the energy provided by fats changed from 2.5 to 3.6 kilocalories per minute, while that provided by carbohydrates increased from 3.4 to 12.3.

Krogh and Lindhard[332] found, however, that fats, as a source of energy, are 10 per cent less economical than carbohydrates. This has been substantiated by Henschel, Taylor and Keys,[227] who found that starving volunteer subjects deriving 50 to over 90 per cent of the energy from fat had 7 per cent less efficiency than before starvation. In order to do the same amount of work, athletes will need more oxygen on a fat diet than on a carbohydrate diet. Consequently, there will be an additional strain imposed on their respiration.

SUGAR AND ENDURANCE

Dill's experiments[133] dramatically demonstrated the value of sugar as a source of energy. A hungry dog running on a treadmill was exhausted in 4.5 hours; yet the same dog, when given sugar during the run, continued it for 17 hours. The speed and the degree of inclination of the treadmill were the same during both experiments. The depletion of glycogen stores during prolonged, strenuous exercise has been shown in man by taking biopsies of working muscles before and after exercise. There is a close relationship between the amount of glycogen used and carbohydrate burned, as determined from exercise R.Q.'s, indicating that the glycogen stored in the muscles is the main limitation to work of this kind.[230] An example of the depletion of glycogen from muscle is shown in Figure 33.

On the other hand, there is always a sufficient store of fuel in the body for strenuous muscular activity of short duration. Haldi and Wynn[202] found no difference in the speed of 100-yard swimmers whether their last meal before swimming was high or low in carbohydrate, a result which was to be expected.

All over the world the man engaged in hard physical work lives mainly on a carbohydrate diet, be it rice, bread, or potatoes. This in itself may be considered as sufficient proof that carbohydrates furnish energy for muscular work. In this connection, observations made by Karpovich and Pestrecov[303] supply a pertinent illustration. The inmates of a county jail worked on bicycle ergometers five days a week for several months, each day trying to ride as long as they could, because they were paid by the hour. As their riding time increased, so did their appetites. Since the only food available in any amount beyond their regular allowance was bread, they ate bread. One inmate, particularly outstanding for endurance, could ride over six hours without stopping, doing 7150 foot-pounds of work per minute (2,575,000 foot-pounds in six hours), equivalent to climbing 3.3 miles straight upward. The amount of bread he ate was 12 to 14 slices at breakfast, 14 to 19 slices at lunch and 23 to 25 slices at supper. His respiratory quotient was 0.97, indicating that 90.4 per cent of his energy for riding was derived from carbohydrates—in this case, bread.

There seems to be general agreement that, in strenuous but short exertion, carbohydrates furnish the main source of fuel. This, however, does not justify the taking of sugar in any form by athletes immediately before a short race. Such a practice cannot increase the rate of energy production when the supply of fuel is otherwise adequate, just as an extra gallon of gasoline added to an almost full tank in your car will not make it go faster during a short ride. The beneficial effects reported after taking sugar before short races should be interpreted as psychological. A rabbit's foot would probably have the same effect.

THE GLYCOGEN OF THE MUSCLE AND THE LIVER

The amount of glycogen stored in the body can be calculated from the respiratory quotient and the amount of energy used by a man who worked to complete exhaustion. The quantity of glycogen varies, being sometimes close to one pound and a half. Of this amount muscles may contain over one pound.

It should be emphasized that liver glycogen is unquestionably merely a store of reserve food, like starch in plants. It is known that, when the blood is loaded with sugar, the product of digested carbohydrates, a considerable portion of the sugar is quickly laid down in the liver as glycogen. Later, when the amount of sugar in the blood is reduced by work, the supply is made up at the expense of the glycogen in the liver. As much as 7 ounces of glycogen may be stored in the liver.

The relation between liver and muscle glycogen has been shown by experiments. The livers of untrained rats, after forced running on a treadmill, were almost free from glycogen, while in trained rats the same exercise did not materially reduce the glycogen content.[524]

Whether the blood sugar level falls below normal during exercise depends on the amount of glycogen in the liver and upon the energy demand of activity. If work is undertaken during excitement, the sugar content of the blood will rise above normal at the beginning of, or sometimes immediately preceding, the exercise. This anticipatory meeting of energy requirements is caused by an action of epinephrine on the liver, which causes glycogen to be converted into glucose and released into the blood stream. What happens to the blood sugar content during exercise depends on the duration and severity of activity.

The glycogen stores of the muscles can be increased after depletion by utilizing an appropriate diet. Bergström and others[52] have shown via muscle biopsies that the amount of glycogen stored following exhausting exercise is greater when using a carbohydrate-rich diet rather than either a protein-and-fat or a regular diet. The increased storage is accompanied by a subsequent increased work capacity. The implication to the athlete is to reduce the intensity of training a few days prior to competition and increase the amount of carbohydrate in the diet.

ATHLETE'S DIET

One often hears about "training tables" or special athletes' diets that are considered to be indispensable for best performance. On closer examination of these special diets one is impressed by three factors:

1. There are many varieties of athletes' diets; therefore, one cannot speak about the "athlete's diet."
2. Essentially, all diets on which athletes can train successfully without losing or gaining weight are alike in their basic nutritive elements, and represent nothing more than normal diets designed for hard-working people.
3. The fancy notions differentiating diets from each other are futile, but usually they are harmless.

In this connection, an investigation conducted by Bohm[59] is of special interest. He questioned many coaches, trainers, and athletes from various countries who participated in the 1936 Olympic Games; athletes who participated in the 1938 British Empire Games; and many coaches and athletes in this country. He came to the conclusion that, in spite of great variation in the diets of athletes, the common tendency is to eat, in moderation, a balanced diet consisting of plain, wholesome foods. Bohm also says: "Common sense in training methods has supplanted ones governed by superstition and ignorance." If he had added the word "generally," it would have represented the true situation better, since we still witness superstition and ignorance in the choice of food.

A short, comprehensive review of the literature pertaining to athletes' diets may be found in reference 268.

MEAT-EATERS VERSUS VEGETARIANS

In general, it is futile to argue which of these two diets is the better. One can hardly find an athlete who is pure vegetarian; and, on the other hand, meat-eaters also use vegetables. The so-called vegetarians ordinarily eat milk, eggs and butter. Some of them eat fish!

Inclusion or exclusion of meat is a matter of custom and taste. An Argentinian cowboy may consume from 4.4 to 6.6 pounds of meat daily;[254] a vegetarian from India may never taste meat in all his life; yet both of them are able to work hard and stay healthy. Meat is not indispensable. The only advantage of meat (or fish) over vegetables is that it is a good source of essential amino acids indispensable for life. Vegetarians, of course, can get along without meat if they eat a variety of vegetables to assure an adequate supply of the essential amino acids. A meat diet, however, is more palatable to most people, with the exception of convinced vegetarians.

It is beyond the scope of this book to discuss the fundamentals of dietetics, and the reader who wishes to pursue this matter should consult books on nutrition and diet.

THE PRE-GAME MEAL

As a rule, food should not be eaten for three to four hours before a contest. The reason can easily be found if one attempts to swim, wrestle, or run on a full stomach. We must admit, however, that some athletes have made their records after heavy meals eaten close to the time of competition. Asprey and co-workers[20, 21, 22] have shown that eating very light meals (cereal, milk, and toast) up to one-half hour before activity has no adverse effect on running distances, under experimental conditions, which range from the 440-yard dash to the 2-mile run.

Some coaches go to extremes and require their athletes to go without food for six to seven hours before a contest. Neither physiology nor common sense can offer anything in defense of this practice. A good coach should see that food on the day of a contest is more palatable than ever and thoroughly digestible. For this reason, so-called "greasy" foods should be avoided because they delay emptying of the stomach. The practice of giving broiled steaks has two good reasons: steaks taste good, even to nervous athletes, and are easily digested.

The use of a liquid pre-game meal has come into vogue in recent years. These substances are readily available under various trademarks for weight control. They generally have a high protein and vitamin content and range from approximately 250 to 300 kilocalories per serving. The rationale for their use is that they are readily digested, thereby preventing the muscle cramps and pre-game vomiting that might follow a heavier, only partially digested meal.[439] Research has indicated that they have no adverse effect on performance,[449, 539] and their use as a pre-game meal is not contraindicated if the athlete finds them palatable. Claims of any special powers for liquids in terms of muscle-building or performance capacity must be attributed to salesmanship rather than physiology. Their greatest beneficiary might be the athletic department budget, since the cost of a liquid meal is certainly less than that of the traditional pre-game steak.

EFFECT OF LACK OF FOOD AND FASTING UPON WORK

An underfed man cannot sustain hard work as long as a well fed man. Quantitative measures of this relation were made in Germany during and after World War II.[329] When miners who worked as "cutters" received about 2800 calories daily, each man produced 7 tons of coal daily. Assuming that the basal metabolism was 1600 calories, we see that they received 1200 *work* calories, and that the cost per ton was 170 calories. When rations were increased by 400 calories per day, work output increased to 9.6 tons per man, and the cost per ton went down to 167 calories. This, however, was only an apparent cost, because, in six weeks, the miners lost an average of 2.6 pounds in body weight, probably because the men worked too hard to show appreciation of the increased rations. Then 400 calories more were added to the diet. The output rose to 10 tons per day, and body weight slowly returned to normal.

Haggard and Greenberg[201] have advocated between-meal feeding as a means of increasing industrial production. Other investigators have not been able to notice any effect of either mid-morning or mid-afternoon meals upon the efficiency of industrial workers.[203] It is possible that it is not the extra meal, but just the rest itself, combined with appreciation for both, which has been responsible for increased efficiency. In the German experiment referred to previously, a greater work output was also obtained by distributing hard-to-get cigarettes among the workers.

Tuttle and his associates[519] reported that the work output of girls is smaller on days when they have had no breakfast. Unpublished observations made on male students at Springfield College showed that breakfast made no difference in the students' work output. Some of them even broke their records on a bicycle ergometer on breakfastless days.

An explanation may be offered that Springfield men are accustomed to going without breakfast, while Iowa girls are not. It has been shown[499] that, after repetitive periods of fasting, people can tolerate fasting better. The blood sugar level becomes higher, and motor speed and coordination deteriorate less.

Tuttle[516] experimented with wrestlers weighing from 145 to 217 pounds. A loss of 5 per cent of body weight had no deleterious effect on strength or vital body functions.

QUESTIONS

1. What are the three phases through which foodstuffs pass to release their energy for physiological work?
2. What are enzymes? Why are they important?
3. Differentiate between aerobic metabolism and anaerobic metabolism. When might you use each during physical activity?
4. What cellular changes occur with training that could increase our ability to do aerobic and anaerobic exercise? (See also Chapter 2.)
5. What experimental evidence do we have on the use of carbohydrates and fats as sources of energy during physical activity?

6. Are proteins important as a source of energy for muscular work?
7. What is a respiratory quotient?
8. Give the values of R.Q. for protein, carbohydrate, and fat.
9. Can the R.Q. be greater than 1.0 immediately after a vigorous sprint? Why?
10. During a five-minute exercise, 9 liters of O_2 were used and 8.1 liters of CO_2 were produced. Find the R.Q. and the amount (in grams) of carbohydrate and fat used. (Consult section on respiratory quotient.)
11. What per cent of energy is obtained from carbohydrate and fat if the R.Q. is 0.75? 0.82? 0.95? (Use Fig. 34.)
12. Why is fat less economical as a source of energy than sugar?
13. Can the taking of sugar improve performance in sprint runs or in marathon runs?
14. How can we calculate the amount of glycogen in the body?
15. What is the best athlete's diet?
16. Can a pure vegetarian become a champion athlete?
17. When should one eat before an athletic contest? Would you advise an athlete to eat a heavy meal an hour before participating in a contest? Give reasons for your reply and any instructions or advice you would give as to the time to eat or type of foods to include in the meal.
18. What measure should be taken to assure an ample supply of glycogen in the body?
19. Is there any relation between the food supply and the amount of work done?
20. What is the effect of breakfast upon work capacity in the morning?

Chapter Six

THE ROLE OF OXYGEN IN PHYSICAL EXERTION

THE DEMAND FOR OXYGEN

An adequate supply of oxygen is necessary for normal life and activity. It is used by all the cells for oxidative processes in the metabolic changes from which energy is derived. Whenever more energy is required, metabolism is increased, and hence the need for oxygen is also increased.

It is important to realize that man practically lives a hand-to-mouth existence as far as his oxygen supply is concerned. There is, however, a certain amount of oxygen present in the body, only a small part of which can be used in an emergency. This is the oxygen found in the blood and in the lungs, the total amount being between 1800 and 2250 cc. An additional 40 to 400 cc. may be present in combination with myoglobin.

When at rest, the body requires from 200 to 300 cc. of oxygen each minute. In vigorous exertion this need may be increased more than 20 times. Since muscles constitute about 40 per cent of the body weight, their consumption of oxygen may increase at least 50 times.

If exercise is moderate and uniform, the oxygen intake rises gradually and then, in a minute or two, levels off and remains at this level for the duration of the exercise (Fig. 35). Since the other bodily functions, such as respiration, heart rate, and lactic acid production, also maintain a steady level, this state is called the "steady state." During this state, the oxygen intake is equal to the oxygen expenditure.

Oxygen intake, during exercise, can be increased up to the maximum rate of oxygen consumption, which is referred to as the *maximum oxygen intake* or *aerobic power*. It is usually expressed either as an absolute value in liters per minute or as a relative value in milliliters per kilogram per minute. The latter is, in most instances, a better indicator

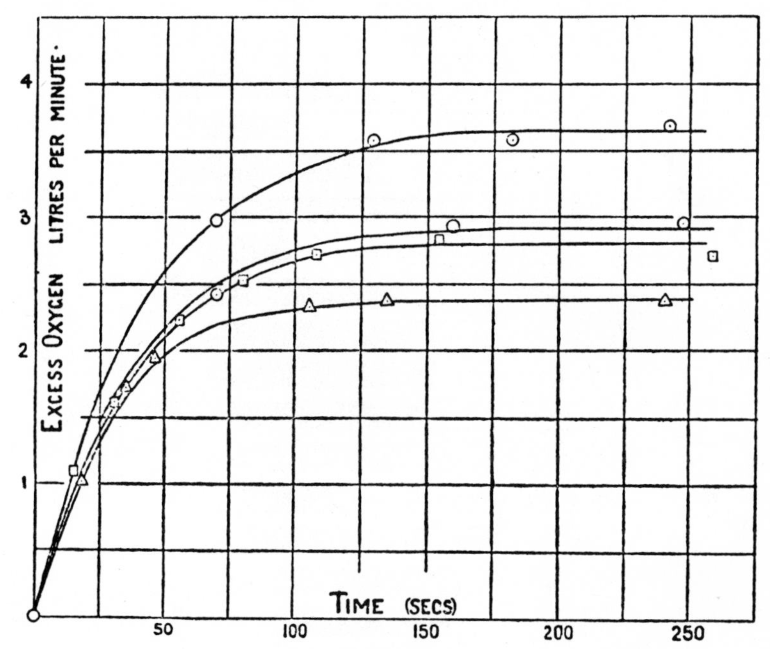

Figure 35. The true and apparent steady states. The lower three curves represent the true steady state. The upper curve is an apparent steady state, because activity can be maintained only until the maximum oxygen debt is reached. (Hill: Muscular Activity. Williams and Wilkins Co., 1926.)

of performance capacity, since a large person may have a high intake merely by virtue of his greater muscle mass. The procedures for measuring the maximum oxygen intake are described elsewhere (see Chapter 7).

The oxygen intake during a given exercise depends on the intensity of work and the size of the muscle groups involved. Well trained athletes are typified by high maximum oxygen intakes, the highest values being recorded for those who participate in endurance activities. Figure 36 shows values obtained by Saltin and Åstrand[450] from exceptional athletes as well as from a college-age population. The highest value relative to body weight was 85.1 milliliters per kilogram per minute in a cross-country skier, while the highest absolute value, 6.17 liters per minute, was found in a participant in orienteering. (The latter event, which is not common in the United States, requires the competitor to find his own way in the least possible time through rough terrain, using a map and a compass.)

As long as the energy requirements can be met by aerobic reactions, and a steady state can be maintained, the intensity of the exercise is said to be within the range of a "normal load." The greatest normal

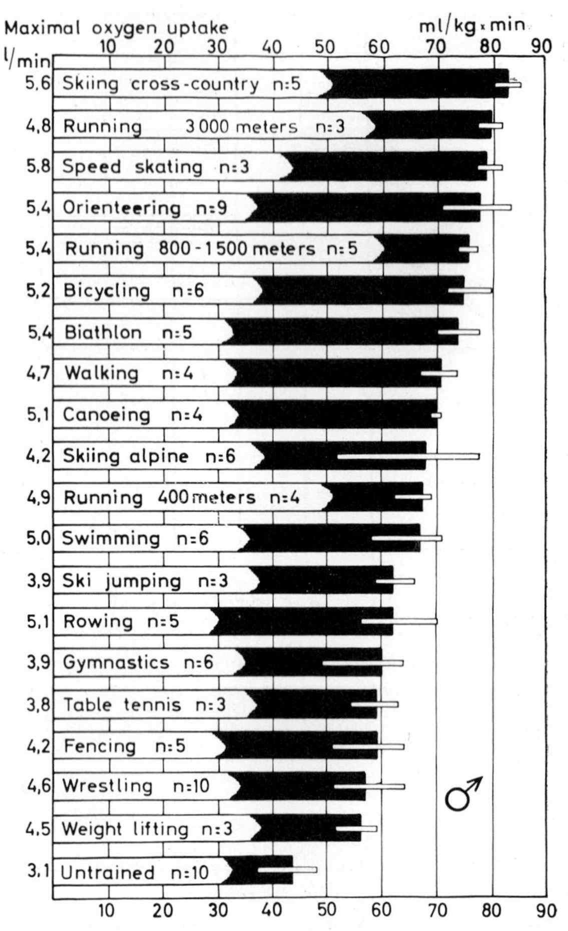

Figure 36. Maximum oxygen intakes of a normal college population and exceptional athletes from the male Swedish National Team in different sports events. (Åstrand, P.-O., and Saltin, B.: J. Appl. Physiol. *23*:353, 1967.)

load is called "crest load"; anything beyond that load is called "over load." The magnitude of crest load depends on individual differences and on states of training.

OXYGEN DEBT

If the intensity of exercise continues to rise, and the aerobic capacity is exceeded, it is obvious that additional work has to depend entirely on anaerobic chemical processes in the muscles. The amount of additional work will be limited by the degree of body tolerance for the accumulation of the products of anaerobic decomposition, chiefly lactic acid (see Chapter 5). After the cessation of totally or partially anaerobic work, oxygen consumption remains increased for some time, until the oxidation of the accumulated products of the anaerobic reactions has been completed and the muscle has been "recharged."

The rest period immediately after exercise is called the recovery period; the subject repays his oxygen debt at this time. The extent of the debt is determined by measuring the total amount of oxygen consumed during the period of recovery, and subtracting from it the amount of oxygen which would have been consumed during the same period if the subject had remained at rest. It has been supposed that the total amount of oxygen needed for an exercise is equal to the amount of oxygen taken in during the exercise itself, in excess over the resting level, plus the oxygen debt. A. V. Hill called this sum the *oxygen requirement* for exercise.

Because of oxygen debt, it is possible for a man to do a muscular exercise that requires far more oxygen than can conceivably be supplied during the period of exercise itself. In the severest forms of exercise, something over 22 liters of oxygen may be required to provide the energy used in one minute. This is an impossible accomplishment for the respiratory and circulatory systems, which even in a well developed athlete may supply only between four and five liters in a minute. By contracting an oxygen debt, the demand of 22 liters a minute can be met. The largest oxygen debt reported was 22.8 liters, after a 10,000-meter race.[331] This is a rather unusually high figure. More common values are those reported by Hermanson,[229] which are shown in Figure 37. Training apparently leads to an increased capacity to develop an oxygen debt.

The name "debt," however, has its shortcomings. Only too frequently, students forget the real meaning of oxygen debt and think that oxygen *has been borrowed* from somewhere in the body. As has been stated previously, even under most favorable conditions, the organism contains not more than 2.25 liters of oxygen; moreover, it is physiologically impossible to utilize it all. One cannot borrow 10 to 15 liters from 2.25 liters. Therefore, oxygen debt is actually a payment on a deficit incurred during anaerobic work, and the lactic acid and other by-products of work serve as promissory notes assuring prompt payment.

The oxygen deficit is incurred during the initial stage of activity before a steady state is reached, and therefore can be estimated by subtracting the amount of oxygen actually used during the initial stage from the amount which would have been used if the steady state were reached instantaneously. This sounds simple in theory but is not so in practice.[96] In very strenuous activities, especially in those of short duration, no steady state is obtainable.[534] If the activity lasts long enough, the maximum oxygen intake possible minus the oxygen con-

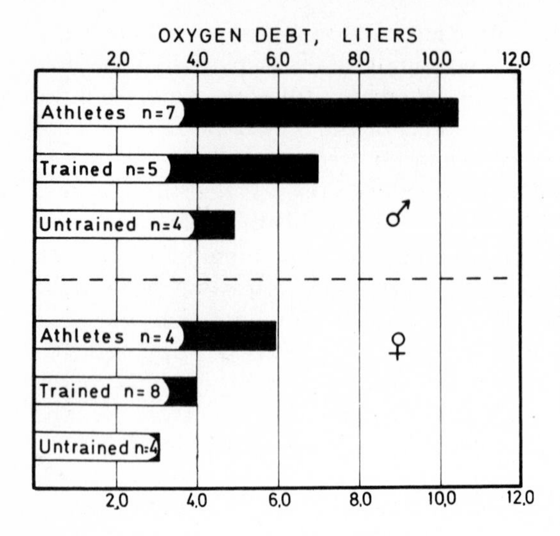

Figure 37. Common values for oxygen debt. (Hermanson, L.: Medicine and Science in Sports *1*:32, 1969.)

sumed will determine the extent of the deficit. If activity is of short duration, the oxygen deficit cannot be determined.

A. V. Hill assumed that the ratio of oxygen deficit to oxygen repayment (or debt) was 1:1. This seemingly logical and simple relationship has not been accepted by some investigators, who have found this ratio to be approximately 1:2.[349] Recent work suggests that the debt exceeds the deficit only when a steady state is not reached, but when the latter is attained, there is no further increase in debt during exercise, and Hill's assumption is applicable.[534]

THE RECOVERY PROCESS AFTER EXERCISE

Even after mild exercise, recovery is necessary, owing to the "lag" in adjustment of the organism to new demands for oxygen during exercise. The supply runs temporarily short of the demand. With an increase in the severity of exercise, there is also an increase in the amount of oxygen debt (Fig. 38. This figure was based on an assumption that oxygen debt equals oxygen deficit, and therefore the deficit was not measured).

Formerly, it was believed that oxygen debt depended exclusively on the excess of lactic acid production. This belief led to the statement that lactic acid is the security given for the payment of oxygen debt. Today we know that the theory is only partly correct. Margaria and coworkers[364] have shown that, in a good athlete, no extra lactic acid appears in the blood during or after exercise involving an oxygen

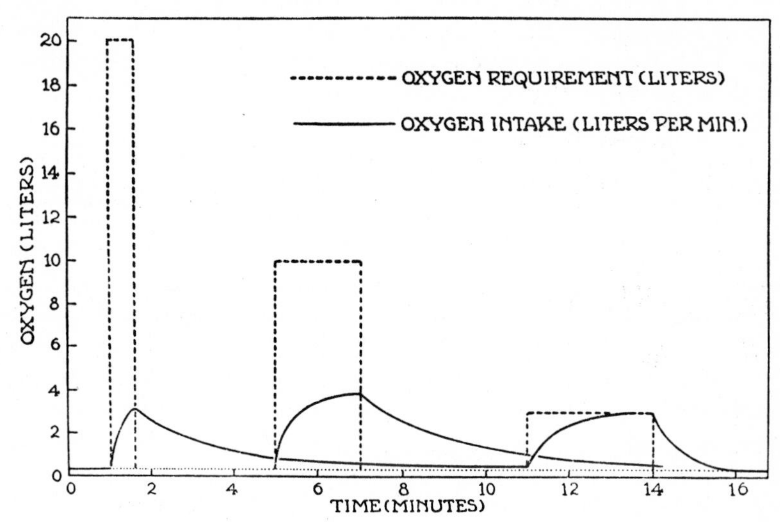

Figure 38. The oxygen requirement, oxygen intake, oxygen deficit and oxygen debt during three periods of exercise of different intensities. The rectangles represent oxygen requirements for each exercise; their areas are equal to the corresponding sums of oxygen intake, during both the exercises and the recovery periods. The parts of the rectangles above the curve represent oxygen deficit and are equal to the excess oxygen consumption during recovery, or oxygen debt. (Fulton: Howell's Physiology. W. B. Saunders Company, 1946.)

consumption of less than 2.5 liters. In an untrained person, this quantity will be decidedly less. When exercise requires a larger amount of oxygen than this, lactic acid accumulates in the blood to the extent of 7 gm. for each liter of additional oxygen debt. Yet after strenuous work, the lactic acid debt may be equal to 70 per cent of the total oxygen debt. The amount of lactic acid in blood becomes less when a person is less fit, is old, or exercises at high altitude.[135]

Knuttgen,[320] using four intensities of work (300, 700, 1000, and 1600 kgm./min.) on an ergocycle, found that oxygen debt was contracted during each intensity. However, excess lactate did not appear during the two lower intensities. Only when the oxygen consumption was 1.5 liters was there a rapid rise of the lactate. Lactic acid does not usually accumulate in large quantities until the maximum oxygen intake is approached or exceeded. However, some lactic acid may accumulate during the adjustment period at the beginning of exercise; but it is removed if a steady state is reached and exercise is continued long enough.[449A]

Margaria and co-workers,[365] in 1963, carried out additional experiments and found that no excess lactic acid is produced if the energy cost is below 220 cal./kg./min. for nonathletes, and higher for athletes. In nonathletes weighing 70 kg., this will amount to an energy cost of

15,400 cal./min., or 15,400 ÷ 4.86 = 3.17 L./min. This is a higher figure than Margaria found before,[364] and much higher than that reported by Knuttgen.

Thus oxygen debt consists of two parts: *alactacid* and *lactacid.* The alactacid debt is paid 30 times faster than the lactacid debt. It takes only about three minutes to pay the alactacid portion. While the identity of the substances oxidized is not known, the hypothesis is advanced that the energy used in the resynthesis of phosphocreatine and the replacement of oxygen lost from hemoglobin account for the alactacid part.

The question of the exact relation between oxygen debt and oxygen repayment after exercise is a problem. The efficiency of anaerobic chemistry is much less than that of aerobic, probably just 40 to 50 per cent. Thus, interest will have to be paid, and some investigators reduce measured oxygen debt values accordingly. Whip, Seard, and Wasserman[534] have found, however, that anaerobic work is as efficient as aerobic work if the exercise can be continued until a steady state is reached. Moreover, finding promissory notes consisting of lactate, phosphorus compounds, and other components is a very difficult job because you have to "catch" them. Some of them disappear very rapidly and some may appear more slowly than the others. At best one can get only a rough estimate of the amount on the promissory notes. Moreover, is it not possible that a prolonged increase in resting metabolism may be considered also as an *indirect* cost of some cellular im-

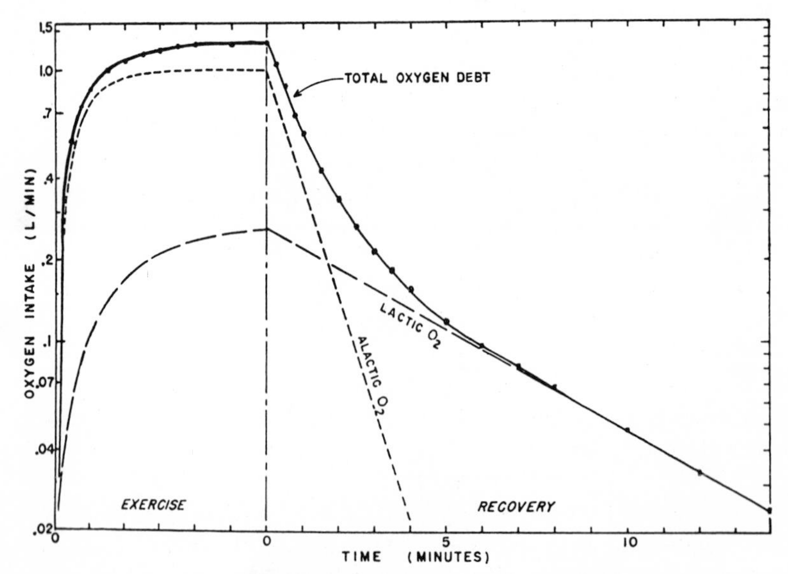

Figure 39. Relation between the total oxygen debt and its alactic and lactic components. (Henry and DeMoor: J. Applied Physiol. *8*:610, 1956.)

provement? A stimulating discussion of the oxygen debt may be found in reference 498.

Frequently, after severe exercise, the resting metabolism of the body remains higher than the pre-exercise rate for many hours. This does not seem to have anything to do with real oxygen debt; that is, it does not appear to provide energy for the resynthesis of substances which broke down during the exercise. For this reason, the determination of oxygen debt is to be completed within an hour and a half after exercise. For mild exercise, such as walking, less than twenty minutes is sufficient; and after some calisthenics, it may be less than ten. The greatest part of the debt is made up during the first few minutes of recovery.

OXYGEN INTAKE AND OXYGEN DEBT AS LIMITING FACTORS IN EXERTION

Energy requirements during all-out sprints are met almost exclusively through anaerobic metabolism with the accumulation of an oxygen debt. The alactate portion is realized within 20 seconds, and the lactate portion within 40 seconds, when energy is expended at a maximum rate.[363] The athlete will be exhausted at this point, and unable to continue. Consequently, when the distance becomes too far to complete within 40 seconds, the speed must be decreased, in order to provide more energy via aerobic routes. Success in longer races, therefore, depends upon the optimal utilization of both aerobic and anaerobic energy sources.

It is sometimes possible to make a reasonable prediction of how long strenuous work can be continued on the basis of ability to take in oxygen during work and to accumulate an oxygen debt.

Suppose a man's maximum oxygen intake is 4 L./min. and the maximum oxygen debt is 15 L. How long can he continue an exercise requiring 5 L./min.? The man needs 5 L./min. and can take in only 4 L./min.; therefore his deficit is 1 L./min. Since the limit of his credit (oxygen debt) is 15 L., he can keep on running for 15 minutes.[237] This technique has been used to make predictions for runners[196, 451] and for swimmers.[196]

The principle of this prediction is based on the assumption that both aerobic and anaerobic energy sources will be utilized maximally during a maximum effort. The maximum oxygen intake during exercise, the maximum oxygen debt, and the oxygen requirement per minute for various speeds are determinations which should be made on the person whose time is being predicted. From a narrow, practical point of view, this is an "impractical" test, because it is much simpler actually to time the man while he is running or swimming. From a pedagogical point of view, it is an impressive test because it shows that man's performance in breaking records conforms to mathematical equations.

As an example, let us predict the running time for one mile. Suppose an athlete has a maximum oxygen intake of 5.34 L./min. and a maximum oxygen debt of 19 L. If he runs at top speed for four minutes, his "available" oxygen will be $(5.34 \text{ L.} \times 4) + 19 = 40.36$ L. In the same manner, we can calculate available oxygen for 3:50, 3:40 and 3:30 as shown in Figure 40. Next, we have to determine the cost of running a mile in these time intervals. We find speeds corresponding to these times and make the athlete run once for 10 to 15 seconds at these speeds, determining the cost of running per second. Then the costs for the whole mile are also plotted on the graph. In this experiment, the two lines crossed at 3 minutes 56.6 seconds. This represents the best time for this particular runner.

Predictions made according to this procedure are quite accurate as long as the assumption is correct that the runner utilizes his maximum oxygen intake and his maximum oxygen debt. In practice, however, the maximum intake is not maintained indefinitely. On the basis of laboratory tests and speed during competitive running, Costill[104] estimated that marathon runners utilized 75 per cent of their maximum during a race requiring approximately 2400 kilocalories of energy. Very little lactic acid accumulates unless such a distance is finished by a sprint.[105]

One apparent advantage gained from training is the ability to utilize a greater portion of the aerobic capacity without producing lactic acid. In untrained subjects, Williams and co-workers[544] found an increase in the blood lactate when the exercise required as little as 40 to 50 per cent of the maximum oxygen intake, while this did not occur in

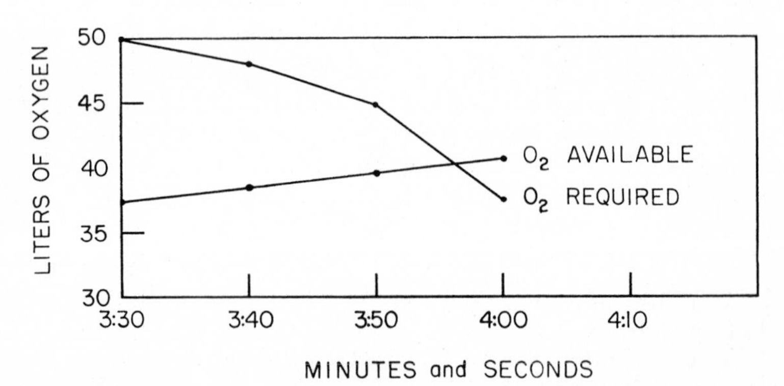

Figure 40. Prediction of running time of 1 mile, based on the amount of O_2 required and O_2 available. The upper curve shows the amount of O_2 required to run the mile in corresponding times shown on the abscissa. The lower line shows the total amount of O_2 available (oxygen intake plus maximum oxygen debt) during those times. The predicted time corresponds to the intersection of the two lines.

well trained men until 55 to 60 per cent of the maximum was reached. Costill[104] found that highly trained distance runners were capable of utilizing more than 90 per cent of their aerobic capacity during 25 to 30 minutes of exercise, with only a moderate amount of lactate appearing in the blood.

Another factor to consider is the way in which we accumulate the oxygen debt. In the example related above, it was assumed that the runner accumulated his oxygen debt at a constant rate over the course of the run. Laboratory tests have shown that the potential energy expenditure is greatest when the runner waits until the end of his run to go deeply into his oxygen debt, by finishing with a sprint.[433] This is substantiated in practice by examining the "lap" times of good milers. An example is Ryun's record of 3:51.1 run at Bakersfield, California on June 23, 1967. His 440-yard times were: 59.0, 59.9, 59.7, and 52.5 seconds.

Assuming that champion athletes have perfect techniques in performance, the availability of oxygen will be the decisive factor in the winning of a race. Plotted curves of swimming and running records have shown that these records follow definite mathematical formulas.[237, 352] On this basis, it was predicted that certain records could be broken. The records fell.

It is well known that world records continue to fall. There are several reasons for this: intensification of competition, better training, lower age of athletes in activities, and betterment of equipment (shoes, fiberglass poles). Craig[109A] compared 1920 and 1962 running records, and 1910, 1930, and 1962 free style swimming records. Invariably the 1962 records for both types of activity were the highest. He also predicted new records in both events.

INTERMITTENT EXERCISE

When exercise intensity is beyond the capacity of the body to maintain a steady state, exhaustion soon sets in, accompanied by high blood lactate levels and a large oxygen debt. I. Åstrand and co-workers[26] found, however, that when such an exercise was divided into short periods of work and rest, each lasting 30 to 60 seconds, it was well tolerated for over an hour. Under these conditions, the circulatory and respiratory responses, as well as the lactic acid accumulation, were of a magnitude typically expected during submaximal exercise. When the alternating work and rest periods were increased to two or three minutes, the physiological responses were more nearly like those expected during severe continuous exercise with early exhaustion.

The responsible physiological mechanism is not completely understood. Myoglobin, which is contained in the muscle, may act as a storage depot for oxygen, allowing exercises to be performed aerobically until the depots are depleted.[27] In this case, the brief recovery periods would allow the stores to be replenished from the blood. When the intermittent periods of exercise and rest are increased, the stores might not be adequate to maintain aerobic conditions, and anaerobic glycolysis would lead to the gradual accumulation of lactic acid and exhaustion.

More recently, Margaria and co-workers[366] investigated the roles of the alactacid and lactacid oxygen debt portions during severe intermittent exercise. Subjects performed series of runs lasting 10 seconds at an intensity that was known to exhaust them in 30 to 40 seconds, with intermittent rest periods lasting 10, 20, or 30 seconds. If the oxygen debt developed during exercise was entirely due to alactacid phosphate mechanisms, it could be repaid during the rest period due to rapid replacement rate. On the basis of known half-reaction times for alactacid repayment, it was calculated that the minimum rest interval necessary to prevent fatigue in all-out intermittent work is approximately 25 seconds.

This response, whatever its physiological basis, is important to consider when heavy work must be performed. By making very short, all-out efforts, followed by adequate recovery, we can tolerate such tasks for long periods of time. This phenomenon may also play a role in interval training, where brief periods of intense exercise are interspersed with brief periods of recovery.

FACTORS DETERMINING THE RATE OF OXYGEN INTAKE

Several factors determine the rate at which oxygen may be supplied to active tissue, and these must be properly coordinated and integrated with the work of the muscles if the body is to attain its highest efficiency. Four of these factors may be discussed as follows:

The first is the *ventilation of the lungs*. Lung ventilation ordinarily increases proportionately to the increase in the load of work. By deep breathing, the partial pressure of oxygen in the alveolar air may be slightly increased. Since the rate of the passage of oxygen into the blood is determined by the pressure of this gas rather than by its percentage, more oxygen will be picked up by the blood.

The second factor is the *oxygen-carrying capacity of the blood*, determined by the hemoglobin content of the blood.

The third factor is the *unloading of oxygen at the tissues*. The oxygen capacity of the blood of people at sea level ranges between 18.5 and 22.5 cc. per 100 cc. of blood. Usually about 5.5 cc. of oxygen per 100 cc. of blood are taken up by the tissues during a rest period. During activity this may be increased by two or two and a half times.

The fourth factor is the "*minute-volume*" *of the heart.* The rate of blood flow through the body as a whole depends upon the amount of blood the heart pumps per minute. As a rule, the blood output during exercise runs practically parallel with the consumption of oxygen.

QUESTIONS

1. Where and how much oxygen is present in the body for use in an emergency?
2. How much oxygen is used per minute by an average man at rest?
3. How much O_2/min. may be used up during exercise?
4. Define normal load, crest load, and over load.
5. Define a steady state.
6. Define oxygen debt.
7. What is the largest O_2 debt that has been reported?
8. What are alactacid and lactacid O_2 debts?
9. What must be known in order to predict proficiency in performance?
10. What is the relation between a maximum intake of O_2, oxygen debt, and success in running short and long distances?
11. Suppose there are two athletes of the same body build and weight. One has an O_2 intake of 4 L./min. and a maximum O_2 debt of 10 L. The other has a maximum O_2 intake of 3 L./min. but an O_2 debt of 15 L. Who can presumably run faster for 1 minute? For 20 minutes?
12. Describe the physiological method of prediction of time for running one mile.
13. What is the effect of training on the oxygen debt? On aerobic exercise as it relates to oxygen debt?
14. Describe how an athlete might use the concept of intermittent work.

Chapter Seven

WORK, ENERGY AND MECHANICAL EFFICIENCY

It is frequently said, and rightly, that the human body is a machine, and that its activities should be explainable by the known facts of physics and chemistry. The mechanical engineer long ago admitted that a living body is like a machine when he adopted the "horsepower" unit as a measurement of the power of the man-made machine.

The muscular power and the mechanical efficiency of the body, together with the conditions which modify and control these, are topics that appeal to individuals from different points of view. The athlete, the trainer, the physical educator, the physician, and the employer of labor are all interested in ascertaining the maximum power, the physiological cost of labor in calories, or the effectiveness of the human machine. The athlete and his trainer and teacher desire to improve the working capacity of the body by suitable diet and other means. In the medical world the restoration of an impaired function usually demands attention, but the physician is being called upon more and more to evaluate the fitness of a patient. The industrial manager, with the assistance of the industrial physician, must assign the laborer to some appropriate form of activity to protect him from strains that would be detrimental to health. The problem of finding the right job for a person who is below par is likely to become one of the measures of industry's service and worth in the community. For all these interests cited, an understanding of work units, of sources of energy and of measures of efficiency is desirable.

WORK

The term work may often be either vague or unsatisfactory. Sawing wood, digging a ditch, or examining a patient is called work. Even playing a professional sport is work. At other times, the same activities, used for fun, are not considered work.

Physics defines work rigidly, as a product of force and the distance through which this force acts. Thus, lifting 10 pounds to a height of 5 feet will constitute 50 foot-pounds of work. Pushing an object horizontally for a distance of 5 feet and applying 10 pounds of pushing force throughout this distance will also result in 50 foot-pounds of work.

This definition is often unsatisfactory and unfair from the standpoint of physiology. For example, if a man holds 10 pounds in his hand while his arm remains motionless in the horizontal position, he is not doing any work (according to physics), yet he quickly gets tired.

Further confusion results when the work done is expressed only in terms of external work. For instance, work done in running may be calculated by measuring the sum total of elevations of the center of gravity and multiplying it by the weight of the body. One can easily detect the inaccuracy of this method, because, besides the obvious work of body elevation, which for all purposes may be called external work, there is additional work involved in swinging the limbs forward and backward, starting and stopping, and overcoming the friction of various tissues against each other. According to Fenn,[165] in running, the work done in lifting the body is only 3.4 per cent of the total mechanical work.

These illustrations should make it clear that measuring the total mechanical work done by man is difficult, and that the usual references to work represent, at best, incomplete estimates. For this reason, physiologists prefer to measure human physical activities in terms of the amount of energy used.

METHODS OF MEASURING THE AMOUNT OF ENERGY USED IN PHYSICAL ACTIVITIES

Positive work and *negative work* are terms commonly used in physiology. Mathematically, both represent a product of force and distance. In positive work, muscles contract *concentrically* (shorten); in negative work, they "contract" *eccentrically* (elongate). Going upstairs or lifting weight represents positive work; going downstairs or lowering a load involves negative work. The amount of energy used for an activity can be found directly by measuring the amount of heat produced, or indirectly by calculating it from the amounts of oxygen absorbed and carbon dioxide eliminated.

The direct method consists of placing a man in a specially built chamber called a calorimeter. The heat liberated by the subject is absorbed by water circulating around the chamber. If the amount of water and its temperature on reaching and leaving the chamber are known, the total amount of heat absorbed by the water, that is, produced by the subject, can be calculated. The direct method, although it is the basic method for research in energetics, is rarely used because it requires elaborate equipment and a large staff of workers. It is also not applicable to activities of short duration. For these reasons, most investigations on human beings have been made by the indirect method.

Indirect Calorimetry. There are two subdivisions of this method: the closed circuit and the open circuit.

Closed Circuit. (See Fig. 41.) The subject inhales oxygen from a

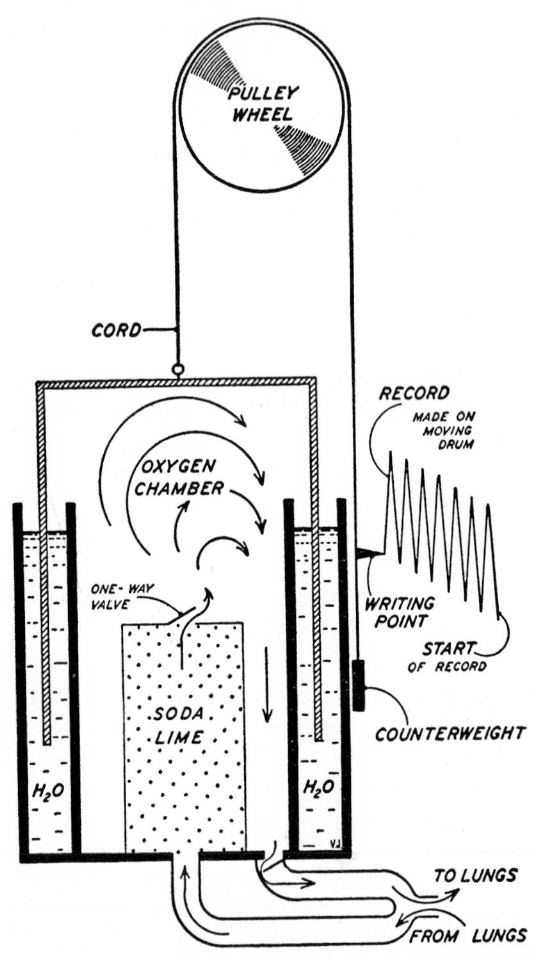

Figure 41. Diagram of a closed circuit type of apparatus for measuring metabolism. The subject breathes oxygen from the spirometer. Expired air returns to the spirometer and passes through soda lime where the carbon dioxide is absorbed. Movements of the upper cylinder of the spirometer are recorded on paper placed on the revolving drum. These recordings indicate the amount of oxygen used. (Carlson and Johnson: The Machinery of the Body. University of Chicago Press, 1941.)

special spirometer. The exhaled gases pass through a carbon dioxide absorbent and go back into the spirometer. Subtraction of the amount of oxygen after the experiment from that before gives the amount used up during the experiment. The subject either wears a special mask, or has a special mouthpiece. In either case, there are two valves which allow the movement of the gases in one direction only. Graphic recording of the oxygen consumption is indispensable to guarantee accuracy with this method (Fig. 42).

Open Circuit. (See Fig. 43.) In the open circuit method, the subject breathes atmospheric air. The exhaled air is collected in a very large spirometer or in a special airtight bag, often called a Douglas bag, where it is accurately measured. Samples of expired air are analyzed for their oxygen and carbon dioxide content. Since the composition of atmospheric air is constant, being 20.93 per cent of oxygen, 0.03 per cent of carbon dioxide and 79.04 per cent of nitrogen, and since the amount of expired air is known, it is possible to calculate the total amount of oxygen consumed and carbon dioxide given off. Special instruments have been developed for automatic analysis of expired air. When a subject is directly connected with such an analyzer, a small fraction of expired air flows continuously through this apparatus, and is analyzed at frequent intervals.

In measuring the energy expended in activities such as swimming and track running, the use of a large stationary spirometer for the collection of the expired air is impractical. Therefore, the bag method is preferred. The expired air is collected in bags (of 100- to 200-liter capacity) and later is measured either by a spirometer or by passing it through a gas meter.

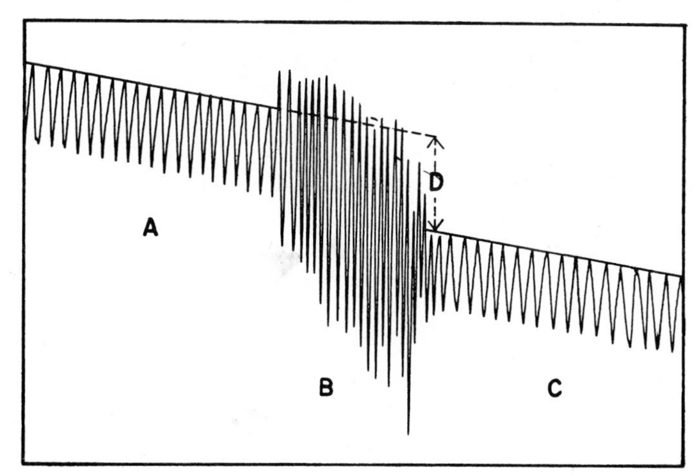

Figure 42. Graphic record obtained with a closed circuit method of measuring metabolism. *A*, Rest; *B*, exercise; *C*, recovery; *D*, amount of oxygen used for exercise.

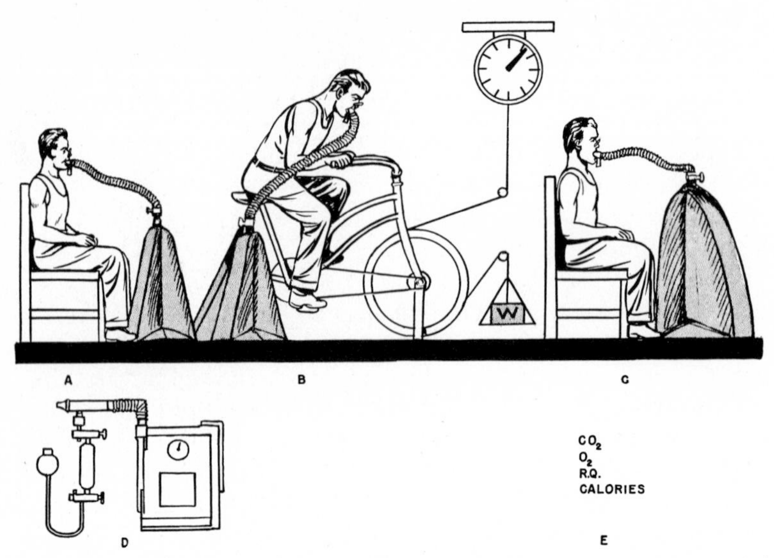

Figure 43. Open circuit method of measuring metabolism. The subject inhales atmospheric air and the expired air is collected in the Douglas bag. *A*, Rest; *B*, exercise; *C*, recovery; *D*, measuring the amount of collected air and taking samples for analysis; *E*, finding the amount of oxygen used, carbon dioxide eliminated, R.Q., and calories used.

Since the quantity of a gas (by weight) in a unit of volume depends on its pressure and temperature, it is customary to reduce the volumes of gases to standard conditions, which follow: temperature, 0° C.; pressure, 760 mm. of mercury; and water vapor, absent (dry). The usual notation for this is STPD (Standard Temperature and Pressure, Dry). In special studies, where the body temperature has to be used in final calculations, one should clearly indicate this, using the notation BTPD, where B stands for Body. In instances where it is necessary to know the total volume of air moved, volumes are corrected to BTPS (Body Temperature, Atmospheric Pressure Saturated).

CALCULATION OF ENERGY

This calculation would be simple if the number of calories produced whenever 1 liter of oxygen is used remained the same. The caloric equivalent of a liter of oxygen, however, varies from 4.686 calories to 5.04 calories, depending on the type of foodstuff being oxidized: carbohydrate, fat, protein, or any combination of them. Since, in exercise of short duration, oxidation of proteins may be disregarded, the caloric value of 1 liter of oxygen will depend on the relative amounts of carbohydrate and fat used. When the respiratory quotient (R.Q.) is known, the caloric value of 1 liter of oxygen may be found from special tables (see Chapter 5, Fig. 34).

By now it must be clear to the reader that, for an accurate appraisal of the amount of energy used, it is necessary to have data on both oxygen and carbon dioxide. In the closed circuit, where all calculations are based only on the amount of oxygen used, the respiratory quotient is assumed to be 0.82 in the post-absorptive state (empty stomach), and 0.85 on an ordinary mixed diet. Therefore, the values of 1 liter of oxygen will be 4.83 and 4.86 calories, respectively.

GENERAL PROCEDURE FOR MEASURING THE AMOUNT OF ENERGY USED

The subject should be tested preferably in a basal state or, at least, two or three hours after a meal (if the test is of short duration). He should not have indulged in physical exertion or smoking before the test, because exercise does, and smoking may, increase metabolism.

The procedure is as follows:

1. The subject rests for thirty to forty-five minutes, during which time he is given a chance to get used to breathing through the mask or mouthpiece. After this, the following determinations are made:
2. Oxygen used per minute at rest in a sitting position.* This test is usually continued for ten minutes, and the total is then divided by 10 to obtain the *resting O_2 consumption* per minute.
3. Oxygen used during the activity under investigation: *work O_2*.
4. Oxygen used during the period of recovery after exercise: *recovery O_2*.

Suppose one desires to find how much oxygen is used in doing full squats; in other words, one desires to find the energy cost of this exercise.

After a rest of thirty to forty-five minutes in the sitting position, the oxygen consumption in that position is measured for ten minutes. Suppose that the total amount consumed is 3000 cc. Dividing this by 10, we obtain 300 cc., which represents the *resting O_2 consumption per minute*.

The subject now does full squats for two minutes, controlling the cadence by a metronome. Suppose that 4000 cc. of oxygen are consumed during this period. This constitutes the *activity O_2 consumption.*

The subject sits down again to recover from the effects of the exercise, and the amount of oxygen consumed during fifteen minutes is measured. Suppose that the total amount consumed is 5500 cc. This is the *recovery O_2*.

Now for the calculation of the cost of the squatting exercise:

---

*Sometimes lying, standing, or other positions are required, depending on the type of activity investigated.

If we add the amounts of oxygen consumed during exercise and recovery, we obtain 4000 cc. + 5500 cc. = 9500 cc., which is called the *gross cost of exercise*. To obtain the actual or *net cost*, it is necessary to subtract from the gross cost the amount of oxygen which the subject would have consumed if he had remained resting during an equal period of time. (2 min. of exercise + 15 min. of recovery = 17 min.) Thus the net cost is:

17×300 cc. (resting O_2/min.) = 5100 cc.
$9500 - 5100$ = net cost of exercise
$4400 \div 2$ = net cost per min.

and the general formula is: net cost of exercise per minute =

$$\frac{(\text{work } O_2 + \text{recovery } O_2) - \text{resting } O_2/\text{min. (min. of work + min. of recovery)}}{\text{min. of work}}$$

During some exercises, such as fast running on a track or swimming in a pool, it is impossible to collect expired air during the activity without affecting the speed of motion. For this reason, in determining the net cost of such exercise, item 3 of the procedure is omitted, and the subject performs the exercise while holding his breath. Expired air is collected only during the recovery. The calculation of the next oxygen requirement will then be as follows:

$$\frac{\text{recovery } O_2 - \text{resting } O_2/\text{min. (min. of work + min. of recovery)}}{\text{min. of work}}$$

Therefore, in this case the determination of the total net cost of an exercise is based entirely on determination of the oxygen debt contracted during an activity. One should remember, however, that the energy cost of an activity determined exclusively by the oxygen debt is greater than the cost of the same activity performed aerobically, since no steady state can be reached.

When exercise is not strenuous and a steady state is attained, the net cost of exercise may be found without measuring the recovery oxygen. If, however, exercise is strenuous and short, the oxygen consumption during an apparent steady state is not a true measure of oxygen requirement (Fig. 44) and oxygen debt should be measured.[96]

In the preceding three problems, we determined the net oxygen consumption, which is found by subtracting the resting oxygen consumption from the gross work oxygen. One may see that the gross oxygen consumption is definite only when it is determined during a steady state. After prolonged, strenuous exercise the end of recovery may be indefinite, because, as a sequel to exercise, the resting metabolism may rise and remain elevated for many hours. This rise cannot be considered a part of the direct cost of the exercise. For this reason, a determination of the oxygen debt should be concluded within one hour and thirty minutes after exercise.

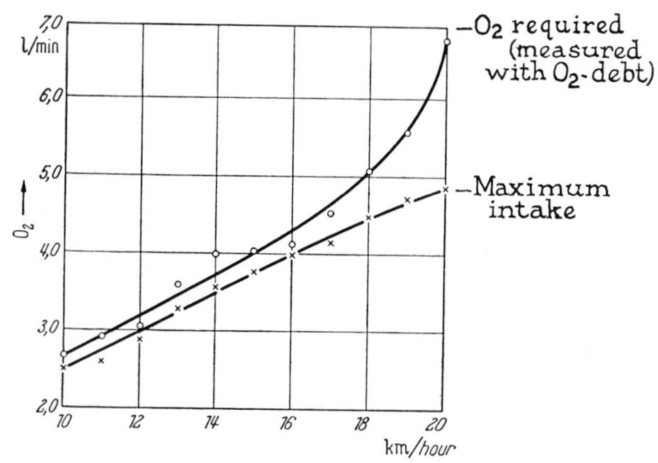

Figure 44. Relation between oxygen intake during a steady state and oxygen requirement (intake plus debt). (Christensen and Högberg: Arbeitsphysiol. *14*, 1950.)

In some activities, which require considerable amounts of fast running and jumping in unpredictable directions, it is impossible to estimate the amount of energy used by collecting expired air in a Douglas bag. One cannot play tennis, basketball, or soccer with a Douglas bag strapped to his back. For this purpose, another indirect method has been suggested. It is based on the observation that there is a linear relationship between the oxygen consumed and the pulse rate.

In order to use this method, a special graph should be prepared for each subject,[259] showing the relation between oxygen consumption and pulse rate during some easily controlled activity. The subject may be asked to run on a treadmill or to ride a bicycle ergometer at various rates of intensity; and then the oxygen consumed is plotted against the pulse rate (Fig. 45). This graph then becomes an instrument for estimation of the energy cost of other physical activities performed by the same subject. Suppose, for example, that it is desired to estimate the energy cost of rope skipping. For this, the subject is asked to skip the rope for a minute or more; then his pulse is taken for 10 seconds immediately after skipping (this period is reliable for finding the final exercise pulse rate[106]), and converted into a minute rate. Then the rate of oxygen consumption per minute, corresponding to this pulse rate, is found on the graph. This is the energy cost per minute of the rope skipping.

This linearity between oxygen consumption and pulse rate, however, usually holds only for the submaximal pulse rate, because on approaching the maximum, the pulse rate begins to level off while the oxygen intake continues to rise. Therefore, if maximum oxygen consumption is calculated on an assumption of linearity, it will be smaller than the actual value, as can be observed from Figure 45.

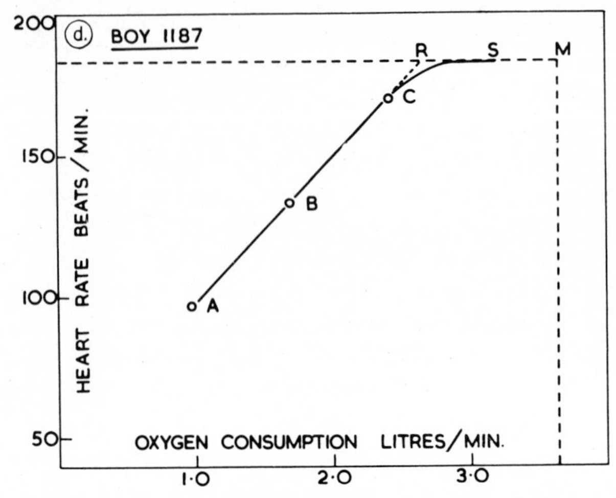

Figure 45. Relation between pulse rate and oxygen intake during work on a bicycle ergometer. This curve shows a bias toward oxygen intake values higher than would be indicated by extrapolation of straight line to maximum heart rate values. (Wyndham et al.: J. Applied Physiol. *14,* 1959.)

Work Classification. The energy cost of an activity may be expressed in various ways: in calories, in liters of oxygen used, or as a ratio of the work metabolic rate to either the basal metabolic rate or the resting metabolic rate.

The first ratio,

$$\frac{\text{Work metabolic rate}}{\text{Basal metabolic rate}}$$

was first used by Dill for classification of the intensity of exercise. Work may be considered moderate when its cost is three times that of the basal rate. When the metabolic rate increases eight times, it is considered hard work, but can be maintained for eight hours.

Karpovich and Weiss[305, 527] used the second ratio,

$$\frac{\text{Work metabolic rate}}{\text{Resting metabolic rate}}$$

in prescribing exercises to convalescent patients. They found that it was easier to make an average patient understand that a certain exercise was so many times "harder" than just resting while sitting in a chair, than to go into an explanation of the basal state.

Wells and co-workers[530] proposed a classification based on the pulse rate per minute and the lactic acid content of the blood: for light work, pulse rate under 120, lactic acid normal; for heavy work, pulse rate 120 to 160, lactic acid 20 to 40 mg. per cent; for severe work, pulse rate over 160, lactic acid up to 100 mg. per cent and more.

Maximum Oxygen Intake. The maximum oxygen intake is the capacity of the body to transfer oxygen from the atmosphere to the tissues. The student may find the terms maximum oxygen intake, aerobic power, or physical working capacity used interchangeably.

A high maximum intake is dependent on the optimal functioning of the cardiovascular and respiratory systems and the transfer of gases between the blood and tissues as well as between the blood and lungs. For this reason, it is frequently measured in exercise physiology laboratories to determine the functional capacity of these systems. It also enables us to determine the amount we are challenging a subject relative to his capacity, when studying a given response to submaximal exercise.

There are several approaches to measuring the maximum oxygen intake. Most of these utilize some kind of graded work load. An example is the test described by Balke,[36] in which the subject, who is attached to air collection equipment, walks on a treadmill at 3.5 miles per hour, or 90 meters per minute. During the first minute the treadmill is run on the level. Starting with the second minute, it is raised to 2 per cent grade, and every minute thereafter, raised 1 per cent. (*Per cent grade = 100 × vertical ascent/distance of belt travel.*)

The oxygen consumption is measured during the last 40 seconds of each period according to the open circuit procedures previously described. The pulse rate and blood pressure are taken each minute. The test is terminated when the subject shows signs of *impending exhaustion.* It is not necessary to push the subject to collapse, but the challenge must be taxing. Exact judgment of the point where the test is to be terminated depends somewhat on the experience of the tester.

Several objective criteria can be used to determine whether or not the maximal oxygen intake has been reached. Issekutz, Birkhead, and Rodahl[267] have suggested several. A work increase should be accompanied by a plateauing of the line showing the relationship between oxygen intake and work load. Acceptable values would be an increase of no more than 100 milliliters per minute, and at least 150 milliliters per minute below the values expected on the basis of previous work increments. The pulse rate should reach the maximum value expected for the subject's age group (see Chapter 12). The R.Q. is expected to be over 1.15 and the blood lactate at least 60 mg. per cent above the pre-exercise level, both of which levels indicate anaerobic metabolic reactions are being used to meet increased work demands. Motivation of the subject is essential to attain maximum intake. Such criteria do not provide the motivation, but they do provide a check on how well he did.

Dill[123] has described an adaptation of this procedure for a bicycle ergometer. Subjects pedal at a rate of 50 revolutions per minute according to a metronome. The first minute no load is applied, but thereafter it is increased by 150 kilogram meters per minute every minute to exhaustion. The same techniques for measuring oxygen intake and criteria for attaining the aerobic capacity are used as with the treadmill test. It has been our experience, when working with athletes, that the capacity of the subjects often exceeds the potential of the ergometer to provide work. Increasing the rate to 60 revolutions per minute and the work increments to 180 kilogram meters per minute solves this problem.

Other methods involve giving a series of tests over a period of several days,[500] the use of longer exercise periods with intermittent rest periods,[378] all-out runs on a level treadmill, or changing speed rather than grade. Newton[392] compared several approaches and found that grade walking and grade running gave higher maximums than either running or bicycling to exhaustion.

More recently, Hermanson and Saltin[232] compared maximum oxygen intakes measured using the treadmill and the bicycle, and found the former to give, on the average, 7 per cent higher values. There were no differences in work time, pulmonary ventilation, blood lactate, and heart rate. They also found that maximal running uphill gave a maximum value 0.20 liter per minute higher than maximal running on the level, while pedal frequencies of 60 or 70 revolutions per minute during maximal exercise gave values 0.10 liters per minute higher than frequencies of 50 or 80 revolutions per minute. It was postulated that, since step frequency is less on an incline, there is probably an optimal movement frequency for measuring the maximum intake.

Other applications involve the use of step-tests[306] and special adaptations to measure the aerobic capacity during performance of specific activities. Magel and Faulkner[354] described two approaches to measuring aerobic capacity during swimming, while Ferguson, Marcotte, and Montpetit[168] have applied the technique to skating.

The maximum intake is always corrected to STPD. It may be expressed either in liters per minute or milliliters per kilogram per minute. The latter gives a better index of cardiorespiratory function, since the large person will tend to have a higher value by virtue of his larger muscle mass.

There is always a possibility that danger may be involved in driving a subject to the limit. Consequently, some arbitrary termination point based on heart rate is frequently used. Balke,[36] in the test described above, stopped the test when the heart rate reached 180 beats per minute. Such procedures must be adjusted for age differences in maximum rate (see Chapter 12). Use of such an arbitrary end point can lead to error in the measure of the maximum intake. A number of submaximal tests of the physical working capacity have been developed. These are discussed in Chapter 18.

MECHANICAL EFFICIENCY OF THE BODY

A college student asked to define efficiency almost invariably defines it as "getting the most with the least effort." That might correctly describe the extent of ambition of a student in college, but it is incorrect as a definition of efficiency in general, because it is applicable only to *maximum efficiency.*

Generally speaking, however, efficiency is just a ratio of work done to amount of energy used. Even an "inefficient" worker may have some degree of efficiency; but, since his efficiency is low, his work becomes too expensive, and he is fired. Usually, efficiency is expressed as a per cent of total energy used.

$$\text{Gross efficiency} = \frac{\text{Work done} \times 100}{\text{Gross energy used}}$$

As in the case of gross cost, gross efficiency is definite only when the steady state can be reached. Obviously, work and energy should be expressed in the same units of measure, such as calories, foot-pounds or horsepower. Examples:

$$E = \frac{5 \text{ cal.} \times 100}{20 \text{ cal.}} = 25 \text{ per cent}$$

$$E = \frac{15{,}000 \text{ ft.-lbs.} \times 100}{20{,}000 \text{ ft.-lbs.}} = 75 \text{ per cent}$$

$$E = \frac{0.5 \text{ HP} \times 100}{5 \text{ HP}} = 10 \text{ per cent}$$

These equations, when used for human work, give a measure of so-called *gross efficiency.* But, since at any time of day a considerable amount of the energy expended by the body is being used merely to maintain life, the measured gross efficiency does not fairly represent the mechanical efficiency of the muscles of the body that are used for the physical work being appraised. To determine the *net efficiency* of the working body, the energy required for the maintenance of the body when it is at rest should be deducted from the total expenditure of energy. Thus the net efficiency is obtained with the equation

$$\text{Net efficiency} = \frac{\text{Work done} \times 100}{\text{Net energy used}}$$

Efficiencies may also be compared without the exact knowledge of work done. If, for example, in walking a given distance, a man uses a calories in one test and $2a$ calories in a second test, it is evident that walking during the first test was twice as efficient as during the second test.

Repeated lifting of objects weighing from 2 to 10 kg. to a height ranging from 5 to 44 cm. is less economical when it is done in a continuous rather than in an intermittent fashion.[446]

Suppose an object is lifted and lowered in a continuous manner 20 times per minute. To do this, the subject should follow a metronome set at 40 beats per minute.

In an intermittent technique, the subject makes a lift at a faster cadence of his own choice, and then rests the balance of the half-minute before making another lift and resting again. In an actual test, a subject spent 17 seconds on a lift and 33 seconds on rest. The net expenditure of energy during the intermittent lifting was about one half of that using the continuous method. This difference is also reflected in the heart rate, which is slower during intermittent work. However, when the number of lifts per minute becomes high, 35 or over, the difference in energy cost of the two techniques tends to disappear.

EFFICIENCY DURING AEROBIC AND ANAEROBIC WORK

If intensity of exercise is sufficiently low that oxygen intake is adequate for the aerobic phase, the efficiency of this work is higher than when work depends on the anaerobic contractions and a large oxygen debt is contracted. An athlete can draw a practical conclusion from this statement: He should avoid unnecessary spurts of speed except the final spurt to the finish line. Whenever speed is increased, an additional oxygen debt is incurred.

Robinson and co-workers[434] have shown experimentally that in middle-distance running, the athlete will do better if he runs the first part of the race a little slower than his average speed and then runs the last minute as fast as he can. If the order of speed is reversed, he will probably lose the race.

On the other hand, in winning a contest, an athlete may allow himself the luxury of sudden spurts if he is sure that his opponent is excitable and will try to pass him and waste his energy. Coaches use this device, and sometimes purposely sacrifice a runner to upset the pace of the opposing team.

APPARATUS EMPLOYED IN MEASURING WORK OUTPUT

Various machines are used to measure work output. When used to measure work done they are called ergometers. If they graphically record work done they are called ergographs. Ordinarily, an ergograph is also an ergometer. These machines usually provide a resistance against which the muscles have to work. The resistance is supplied by spring, weight, friction, or magnetic pull. The classic illustration of an ergometer-graph is the finger ergograph of Mosso. Differences in the construction of various apparatus depend on the group of muscles to be tested. It may be worth-while to warn a beginner that the testing of work done by an isolated group of muscles is extremely difficult, because of the reflex synergetic action of other groups of muscles.

In constructing an ergograph, one may avoid unnecessarily complicated calculations by keeping the direction of the resisting force constant. This may be achieved by attaching the resistance to a pulley (Fig. 46) placed on the ergograph shaft. In this manner, the direction of the resistance always remains at 90 degrees to the radius of the pulley; and, therefore, the moment of resisting force is always equal to $r \times R$, where r is the radius and R is the resisting force.

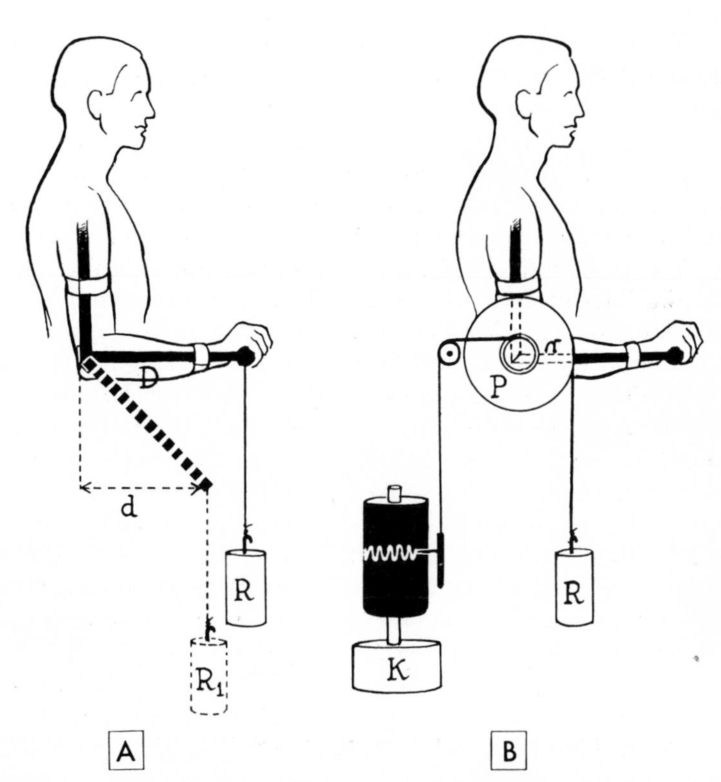

Figure 46. Practical considerations in constructing an ergograph. In *A* the *moment of resistance* varies from DR to dR_1, depending on the position of the lever to which resistance is attached. In *B* the moment of resistance remains equal to rR, regardless of the position of the lever (r is the radius of the pulley P). The black upright is the upper arm stabilizer and is immovable. It may be attached to the base of the ergograph or even to the wall of the laboratory.

If a number of smaller pulleys are placed on the same shaft, they can be utilized for graphic recordings of work, and for operating various other instruments such as counters, work-adders, and signal markers. Before constructing an ergometer one should consult an article by Zoth,[553] from which valuable practical ideas may be borrowed. These ideas may be utilized even in constructing more flexible electronic instruments.

In order to obtain a measure of the work of most muscles of the body while the subject remains in one place, investigators use either a treadmill or a bicycle ergometer, or, rarely, a rowing machine. The treadmill is usually first choice for several reasons, the main being (1) it allows the use of natural motions, such as walking and running, and therefore no time is wasted on developing the special skills required on the other two machines; (2) the subject works against a natural load—his own weight; and (3) no attention of the subject is required to keep the pace.

The treadmill consists of an "endless" conveyor belt, which is operated by an electric motor and provides sufficient space even for fast running. One end of the treadmill may be elevated so that an upgrade locomotion is possible. On some treadmills the movement of the conveyor can be reversed, making a downgrade locomotion possible.

Bicycle ergometers are of three types. They all require preliminary training of the subject to make the test reliable.

In the *friction type* (Fig. 47), one of the wheels of a bicycle is converted into a 30- to 40-pound flywheel by welding around it a heavy metal rim. A brake belt around the rim is attached at one end to a weight and at the other to spring scales. When the flywheel is not moving, and the chain from the sprocket wheel is removed, the reading on the scale is equal to the suspended weight. When the flywheel is in motion, the brake belt causes friction. As the result of the friction the belt will move up in the direction of the rotation, reducing the reading on the scale by an amount equal to the force of the friction. Thus the actual resistance (or the lifting force due to friction) will be equal to the weight attached minus the reading of the scale. Resistance does not depend upon the speed of the flywheel motion. This type of ergometer is easy to construct, and does not need any calibration. The only correction to be made (which is ordinarily disregarded) is 3 per cent of the resistance caused by the friction of the various parts of the bicycle, mainly that of the chain.

The other two types—where resistance is provided by magnetic brakes or by an electric generator—are more complicated, and require careful calibration.

The amount of work done on a friction-type bicycle ergometer is

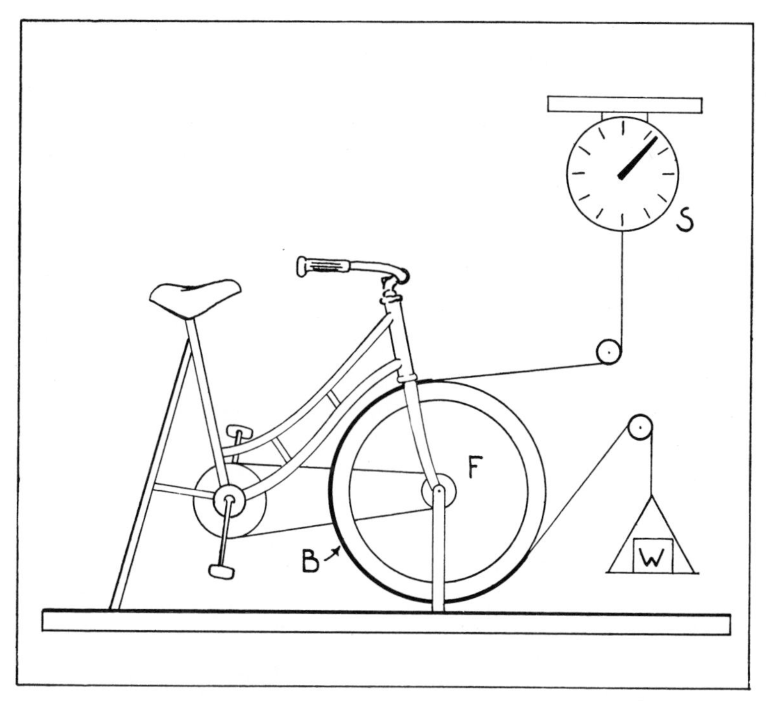

Figure 47. Bicycle ergometer, friction type. *F*, flywheel; *B*, brake, providing friction; *W*, weight; *S*, spring scale. The amount of friction is equal to the weight minus the reading on the scale. The distance traveled by the flywheel and the number of pedal revolutions are automatically recorded by special devices (not shown here).

calculated in a conventional manner:

$$w = f \times d$$

where w is work, f is equal to frictional resistance in pounds and d is the distance traveled by the rim of the flywheel. This distance may be found by multiplying the circumference of the flywheel by the number of its revolutions:

$$d = 2\pi r \times n$$

Details concerning construction and operation of this ergometer may be found in reference 290.

MAXIMAL MUSCLE FORCE DURING ISOTONIC CONTRACTION

Hitherto, maximal muscle force has usually been measured isometrically. The method of testing "breaking power," initiated by Martin, became popular. Recently, one of us (P.V.K.) has developed a method which permits the measurement of maximal muscular force not only isometrically, but also continuously *during* isotonic movements, both concentric and eccentric.[137]

Figure 48 illustrates a dynamometer for testing flexor and extensor muscles of the forearm.[479] The parts are as follows: *a*, control switch; *b*, electric motor; *c*, gear box; *d*, lever arm; *e*, strain gauges; *f*, elbow rest; *g*, wrist yoke; *h*, shoulder yoke; *i*, back supports; *j*, adjustable seat. This dynamometer has evolved from an earlier model reported by Doss and Karpovich.[137] In that machine, the tester provided resistance through a windlass arrangement, while the subject either supplied force during a concentric or isometric contraction or resisted force during an eccentric contraction.

In the present machine, movement of the lever arm is provided by the electric motor through the gear reduction box. The subject tries either to accelerate the action of the motor during the concentric phase or to brake it during the eccentric phase, by applying force through the wrist yoke. The amount of force applied by the subject is "sensed" by the strain gauge and recorded on photosensitive paper. An electrogo-

Figure 48. Dynamometer for measuring forearm flexor and extensor strength during isometric and isotonic contractions. See text for description. (Singh, M., and Karpovich, P. V.: J. Appl. Physiol. *21*:1435, 1966.)

niometer mounted on the end of the lever-arm axle provides a simultaneous recording of the joint angle throughout the movement. Isometric strength is recorded by placing the lever arm at a predetermined angle and applying a maximal contraction. The flexors are tested by placing the wrist yoke over the wrist, the extensors by placing it under the wrist. An improved lever arm has recently been described.[299]

Graphs showing strength curves derived from data obtained using the dynamometer are shown in Figure 49. The data on flexors com-

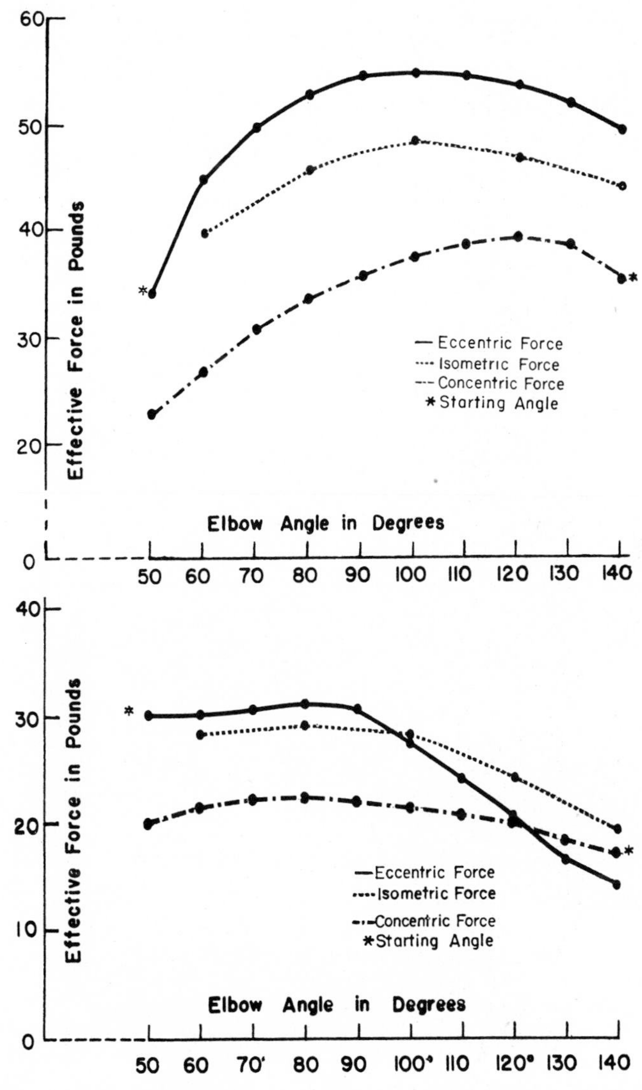

Figure 49. Curves of the maximum concentric, isometric, and eccentric force of the flexors and extensors of the forearm. Upper curve, flexors; lower, extensors. (Singh, M., and Karpovich, P. V.: J. Appl. Physiol. *21*:1435, 1966.)

pared favorably with those reported by Doss and Karpovich[137] on the earlier machine. The dynamometer has also been used to make similar measurements of the strength of preferred and non-preferred arms in men and women,[480] as well as the effect of eccentric training of agonists on antagonistic muscles.[478] Because of the sensitivity of the strain gauge recording system, it was also found useful in studying the duration of a maximal isometric contraction.[387]

One may wonder why force exerted by the same muscles is different during isometric, eccentric, and concentric conditions. The answer may be found in the relation between the length of a muscle and the tension it can develop. Hill and Howarth[240] found that a contracted muscle resists an applied stretch with a force which may be twice that of isometric maximal force. On the other hand, Asmussen[11] found that, if a muscle in a state of isometric contraction is allowed to contract, the isotonic force is much less than that of the isometric contraction. The student is reminded that the electromyogram showed fewer motor units were recruited during an eccentric contraction than were recruited during a concentric contraction of the same amount of force (see Chapter 1).

QUESTIONS

1. Define work and give the simplest formula for work.
2. What is negative work?
3. How much work is done when an object is *supported* by an arm extended horizontally?
4. How much work is done in the following cases: A 200-pound man walks upstairs a vertical distance of 10 feet; he descends this same distance; someone pulls him on a sled for 10 feet on ice?
5. Describe the indirect method of calorimetry.
6. Describe a closed and an open circuit for measuring metabolism.
7. How is the amount of energy measured by means of calorimetry?
8. Describe, step by step, the procedure used in measuring metabolism.
9. Give a formula for the calculation of the gross and net energy used in exercise.
10. How can the oxygen debt method be used for measuring energy cost of exercise?
11. If you wish to explain the energy cost of an activity to an uneducated man, how will you go about it?
12. Define mechanical efficiency, gross and net.
13. When is efficiency greater, during anaerobic or aerobic work?
14. Discuss the principles underlying the construction of a friction type bicycle ergometer and recording ergograph.
15. How is the work done on a bicycle ergometer measured?
16. How can maximum muscular force be measured during isotonic contraction?
17. What are the relative values of concentric, eccentric, and isometric maximum forces of the same muscle?

Chapter Eight

ENERGY COST OF VARIOUS ACTIVITIES

We observe an ever-increasing concern over the fact that many people lead a life which requires little physical activity. The sedentary life is blamed for various ailments, and physical activities are recommended as a prophylactic measure.

In prescribing, one should know why and how much. Usually, the basis is nothing but a guess. It is similar to fine cooking: a fine art, but a crude science. An artist-chef, by adding a pinch of this and a pinch of that, can prepare an exquisite dish to be relished by a gourmet regardless of whether it is good for him. But can everybody using this "pinch" recipe prepare the same dish? Of course not. Art obeys only an artist; an average man has to depend on infallible science.

Someday physical education will become a science, and on that day physical activities will be prescribed in measured amounts according to the needs of the individual. On that day, doctors and physical educators will use a book which will be called "Pharmacopeia of Physical Activities." Such a book is already in the making and this chapter has some of its elements. More of the same may be found in a review prepared by Passmore and Durnin.[405]

BASAL ENERGY REQUIREMENTS

Just to exist requires energy. During a period of fasting, while a person remains at rest, the metabolic rate gradually falls, until, ten or twelve hours after the last meal, a basal value characteristic of the individual is reached. This rate varies with body surface; thus each individual has his own basal metabolic rate which is required to maintain the "fire of life."

From 210 to 295 cc. of oxygen per minute are required to provide for the "basal" metabolism of adults. If it be assumed that the basal usage of oxygen is 250 cc., then we may conclude that the body requires, on an average, 1.20 calories of energy per minute for the maintenance of life.

POSTURE

As more muscles are used in standing than in sitting, the demand for energy increases as one moves from the reclining to the sitting, and from the sitting to the standing position. It is *possible*, however, to sit in a comfortable chair so economically that there will be no perceptible difference in metabolic cost between lying and sitting. It takes, on the average, about 9 per cent more energy to keep the body standing than reclining. More energy is required to stand at attention than at ease, because of the greater muscular effort involved in the former. On the other hand, a relaxed, slouching stance, in which the center of gravity shifts continuously, is more expensive than a rigid stance, in which body oscillations have been reduced.

The comparative costs of various body positions and the effect of a light and of a heavy meal are given in Table 5.

WALKING ON A HORIZONTAL PLANE

The energy cost of walking depends on the speed of walking and the weight of the walker. For this reason, when the cost of walking is discussed, the speed and the weight should be stated. If they are omitted, as is often the case, one should presume that the speed and the weight were *average*, whatever that means.

Passmore and Durnin[405] combined the data obtained in five different countries and found the results from each to be in excellent agreement. Since the subjects did not vary much in weight (60 to 75

TABLE 5. *Cost of Various Body Positions*

	Cal. per Min.		*Cal. per Min.*
Lying	1.14	Standing relaxed	
Sitting	1.19	after a light meal	1.45
Standing		after a heavy meal	1.56
Relaxed	1.25		
At attention	1.30		

(Benedict and Murschhauser: Carnegie Inst. of Washington, Publ. No. 231, 1915.)

kg.), they were able to derive a formula showing a relation between energy cost and speed of walking, regardless of weight: $C = 0.8\ V + 0.5$, where C = Cal./min. and V = km./hr.

From another source,[355] they took a formula showing a relation between the number of calories used and the body weight in kilograms: $C = 0.047\ W + 1.02$.

Using these two equations, Passmore and Durnin prepared a table showing the energy cost for different weights and for speeds from 2 to 4 m.p.h.

The authors have utilized their table and prepared a graph (Fig. 50) showing this cost.

There is still uncertainty regarding the effect of sex upon the energy cost of walking. Booyens and Keatings[64] have reported that, in walking, women expend less energy per unit of body weight than men. Passmore and Durnin found no such difference. Ralston also found no superiority in women, and prepared a formula applicable to both sexes:

$$E_W = 29 + 0.0053\ V^2$$

E_W is energy cost in small calories per kilogram of body weight per minute and V is velocity in meters per minute.

Erickson and co-workers[158] showed that the reliability of the determination of the energy cost of walking on a treadmill is high. For

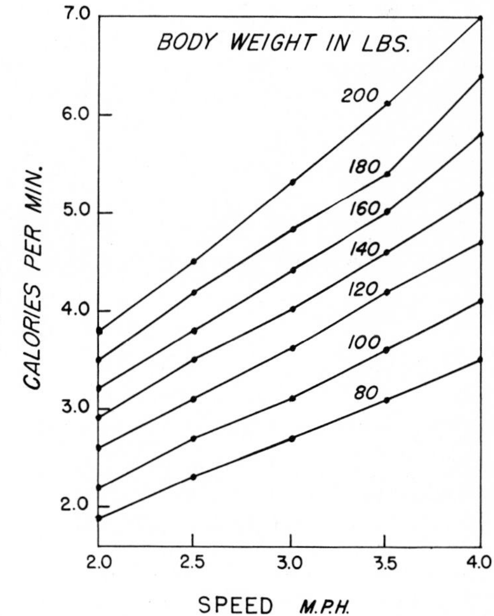

Figure 50. Energy cost of walking for men and women weighing between 80 and 200 pounds. (Made from data of Passmore and Durnin: Human Energy Expenditure. Physiol. Rev. *35*:801, 1955.)

speeds between 2.5 and 4.0 miles per hour and grades of 0 to 10 per cent, the variability of their measurements was 2.95 per cent of the grand mean. (It is only slightly higher than the experimental error, which was 1 per cent.) The interindividual variability is less at higher speeds than at lower ones. At 3.5 miles per hour and a 10 per cent grade, the interindividual standard deviation was 9.37 per cent* of the mean, whereas at 2.5 miles per hour it was 15.1 per cent.

The same investigators also observed the effect of daily practice in walking for periods of one to one and a half hours on the cost of this exercise. Training periods, which ranged from two to 240 days and involved a group of fifty-five men, caused a "trifling" decrease in the energy cost. This observation is at variance with the previous findings of Knehr, Dill, and Neufeld,[319] who reported a significant decrease in the cost due to training. A possible explanation of this variance may be found in the difference of walking surfaces on the treadmills used in these two investigations.

The treadmill used by Erickson had a walking surface made of a rubber belt sliding on a rigid slipway, whereas Knehr and co-workers used a treadmill with a smooth leather belt supported by a series of small rollers which required more skill on the part of walkers. Skill improves with training: extraneous movements are eliminated, and walking, therefore, becomes more economical.

One cannot expect much decrease in the cost of walking, because it is a natural type of activity, and little can be done about improving walking habits. A possible exception is competitive, fast walking. On the other hand, training in swimming leads to a great improvement in efficiency, because of the extensive learning process.

In rapid walking there is commonly great arm movement, the arms being swung violently back and forth with each step. Walking will be made more economical if the extraneous movements are reduced to a minimum.

Walking on a Treadmill versus Floor Walking

The question is frequently asked whether the energy cost of treadmill walking is the same as that of floor walking, because walking on the treadmill seems to be unnatural. Daniels and co-workers,[118] reported that walking in combat boots on a treadmill required 10 per cent less energy than walking on asphalt or cinder road. A similar observation was made by Ralston[420] when his subjects walked on a treadmill and on a smooth linoleum floor. However, when Ralston's

*When calculated per kilogram of body weight, this variability became 3.99 per cent.

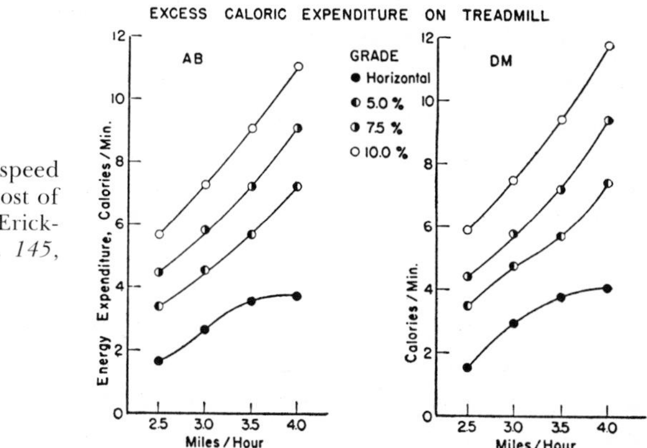

Figure 51. Effect of speed and grade on the energy cost of walking on a treadmill. (Erickson et al.: Am. J. Physiol. *145*, 1946.)

subjects wore rubber-soled shoes, the energy costs of walking on the treadmill and on the floor were the same.

PACK CARRYING

Most research regarding pack carrying has stemmed from military reasons—from a desire to determine the maximum load the foot soldier can carry without losing too much of his fighting efficiency, and from concern about the best way of carrying this load.

The classic investigations by Zunts and Schumburg[554] in 1901 demonstrated that the energy cost of carrying a load on the back is proportionate to the weight of the load, and that, under certain conditions, such a load may be carried even more economically than a corresponding amount of body weight. Additional experiments,[68] which have been sufficiently substantiated, reveal that a load up to 30 per cent of body weight can be carried at least as economically as the body weight.

Although the latter statement is correct, it is applicable only to laboratory conditions. If a man weighing 120 pounds straps to his back a load of 36 pounds and attempts to compete in some sport with a man weighing 156 pounds, he will quickly discover the difference between the extra load and live weight.

The difference in energy cost between high and low packs is very small, yet most soldiers prefer a high pack.

Cathcart and his co-workers[89] recommended 45 pounds as the optimal load for a soldier. War experience, on the other hand, has shown that, only too frequently, the soldier has to carry a much greater load. It costs 2.3 to 4 times more energy to carry a load on the feet than on the back.[139]

CLIMBING AND GOING DOWN STAIRS

Going down stairs is less costly than climbing and involves only 33 per cent of the energy used in going up stairs.

Housewives living in comfortable two-story cottages often complain of the numerous steps they have to make in going up and down. If a woman, during the course of a day, goes upstairs twenty times, it still will amount to less than a mile of horizontal walking—hardly a valid reason for complaint.

A woman weighing 150 pounds who is determined to use exercise for losing one pound will have to climb a 10-foot flight of stairs about 1000 times. Here is the proof: Work done per climb is 1500 ft. lb. Assuming that her climbing efficiency is 20 per cent, the energy used for one climb will be:

$$1500 \times \frac{100}{20} = 7500 \text{ ft. lb.}$$

The energy needed for one descent will be $7500 \div 3 = 2500$ ft. lb., or one third of the energy needed for a climb. The total amount of energy for going up and down once will be 10,000 ft. lb., or 3.24 calories. Since one pound of body fat is equivalent to 3500 calories,* the number of climbs needed will be $3500 \div 3.24 = 1080$. A rather discouraging proposition! The energy cost of climbing, as one may see from Figure 51, depends on the speed of walking and the grade. By walking faster, the number of climbs needed to lose a pound of fat will be reduced, of course.

Some well-meaning but poorly informed persons advocate "spot reduction"—exercising those parts of the body which have undesirable accumulations of fat. Tests conducted by Schade and co-workers[452] have shown that, no matter what exercise one performs, "spot" or general, the fat will be affected equally. The only advantage that "spot" exercises have is that subjects work more willingly because they know what they want to remove and attack it directly.

RUNNING

Because of the wide range of speeds obtainable, running affords a striking illustration of the close relationship between the speed of motion and the energy cost. Sargent[451] tested an experienced athlete, weighing 63 kg. (139 pounds), who could run 100 yards in 10.2 seconds. The energy expenditure in running for a given time varied as

*Although pure fat has a value of 4220 calories per pound, the value, if allowance is made for connective tissue and water content, becomes 3500 calories.

the 3.8 power of the speed. The energy required to run a given distance varied as the 2.8 power of the speed.

It is obvious that a high rate of speed in running would be impossible without oxygen debt: The limit of the maximum oxygen intake in exceptional cases may be as high as 5.88 liters per minute. For most people it is much lower. The runner studied by Sargent, at his highest speed, needed about 29 liters per minute, which was eight to nine times greater than his maximum oxygen intake, if he were to breathe while running. As a matter of fact, during this experiment the runner held his breath while running, and resumed breathing only during recovery. Expired air was then collected into a battery of Douglas bags, immediately after the run. Thus all calculations of the energy cost were based on oxygen debt. Incidentally, at his top speed, this runner developed about 14 horsepower of energy (29 × 0.48* = 13.92 H.P.).

This last figure is close to that determined later by Fenn,[165] who ascertained that the rate of expenditure of energy during running at maximum speed is about 13 horsepower for an average man. He also analyzed the rate of work done and the mechanical efficiency of running, as follows:

	Horsepower
Production of kinetic energy in arms and legs	1.67
Deceleration of limbs	0.67
Work against gravity	0.10
Work against wind resistance	0.13
Friction contact of foot with the ground	0.37
Total	2.94

Thus the mechanical efficiency in this case was:

$$\frac{2.94 \times 100}{13} = 22.6 \text{ per cent}$$

Margaria and co-workers[362] measured the efficiency of running at different grades in two athletes, while keeping the speed within the aerobic capacity of the subjects so that no lactacid oxygen debt developed. (Note that, in the studies of Sargent[451] and Fenn,[165] the oxygen debt technique was used to make the measurements.) The net energy expenditure per kilogram per kilometer was independent of speed at each incline, that is, efficiency was constant. For level running, the net cost was approximately 1 kcal./kg. for every kilometer run. Athletes were only 5 to 7 per cent more efficient than non-athletes, which indicated that the vast difference in performance capacity is a product of the athlete's ability to deliver oxygen to the cells rather than a greater efficiency.

The nomogram presented in Figure 52 was constructed on the

*1 liter of oxygen per minute in this experiment was taken to be equivalent to 0.48 H.P.

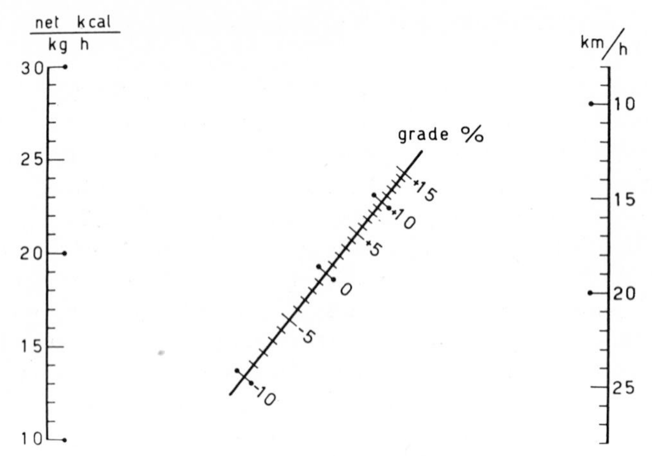

Figure 52. Nomogram for calculating the energy expenditure in running at different speeds and inclines. Place a straight edge between the running speed on the right ordinate and grade on the middle, oblique line. Read the net kcal./kg./hr. where the straight edge crosses the left ordinate. Add three per cent if untrained, subtract three per cent if trained. Multiply by the weight to get total cost per hour. (Margaria et al.: J. Appl. Physiol. *18*:367, 1963.)

basis of data obtained from athletes.[362] It is useful in obtaining estimated energy expenditures while running, a procedure of concern to many people as a result of current interest in jogging or low-speed running as a physical fitness activity. The nomogram also has value in the laboratory for estimating energy expenditure at different treadmill grades and speeds. The metric units can easily be converted to pounds and miles per hour by use of the conversion tables on the inside of the front cover of this book.

Knuttgen[321] showed that the cost of running is much higher when the step length at various speeds is kept constant (77 cm.) instead of allowing the runner to adjust it to the speed (77 to 145 cm.). In the first case, 3.99 L. O_2/min. is reached at 11.66 km./hr., whereas in the second case, 3.93 L. O_2/min. is reached at 15.00 km./hr.

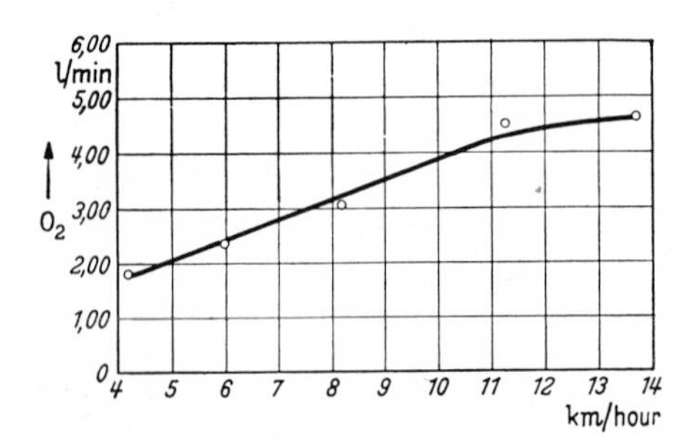

Figure 53. Relation between the speed of running on skis and oxygen consumption. (Christensen and Högberg: Arbeitsphysiol. *14*, 1950.)

SKIING AND SNOWSHOEING

Skiing represents the most efficient type of locomotion over snow. The amount of energy used on the level depends on the condition of the snow, increasing when snow is loose, and decreasing when snow is well packed. Using the Douglas-Haldane method, Christensen and Högberg[95] obtained data on the relative cost of skiing at various speeds. These data are presented in Table 6. As may be seen from this table, oxygen consumption reached its maximum of 5.24 liters per minute at 9.1 miles per hour. This large consumption indicates that the subject was an exceptional athlete. The corresponding pulmonary ventilation was 139.4 liters per minute, close to the maximum obtainable.

Frequently a skier, especially in the Army, has to carry a pack. This naturally requires extra energy. As one may see from Figure 54, for loads up to 35 kg. (77 pounds), the expenditure of energy rises in a straight linear relationship with the load carried.

The question is often raised whether it may not be more economical to carry the pack on a special light sled rather than on the back.

Experiments have shown that carrying a pack on the back is more economical than pulling it on a sled. The amount of energy thus saved may vary from 4 to 33 per cent. Yet it may be advisable sometimes to use a sled, because one can transport bulky objects more easily on a sled than on the back. Stopping and resting is also easier with a sled.

Since hunters in Alaska and Canada use snowshoes instead of skis, it is of interest to find the comparative energy cost of both types of locomotion. As one would expect, on level ground skis are more economical than snowshoes. A man, either without a load or carrying a 45-pound pack, will spend 8 to 24 per cent less energy if he uses skis instead of snowshoes. To the surprise of the investigators, even hill

TABLE 6. *Comparative Energy Costs of Skiing at Various Speeds*

Velocity			*Oxygen Used*				*Pulmonary Vent.*	*R.Q.*
mile p.h.	*km. p.h.*	*meter p. min.*	*L./min.*	*cc./min.-/kg.*	*Per one horizontal kg.-meter* (cc.)	(cal.)	(L./min.)	
2.6	4.16	69.3	1.69	20.34	0.293	0.00137	30.1	0.77
3.9	6.27	104.5	2.34	28.16			40.9	0.76
5.2	8.40	140.0	3.02	36.34			56.6	0.82
7.2	10.67	178.0	3.14	37.79	0.213	0.0010	57.9	0.83
8.1	13.05	211.5	4.14	53.07			115.0	0.95
9.1	14.73	245.5	5.24	63.06	0.2563	0.00128	139.4	1.01

The subject was a well-trained man: weight, 83 kg.; vital capacity, 6.95 liters (at 37° C.). Snow was loose. (Christensen and Högberg: Arbeitsphysiol. *14*, 1950.)

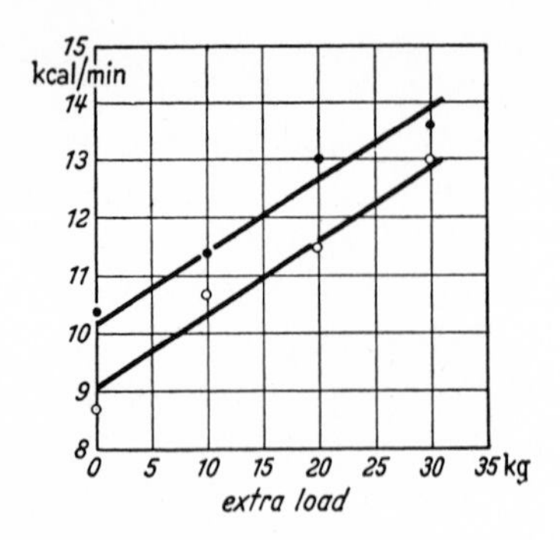

Figure 54. Energy cost of skiing with a pack on a level track for two men. Upper 5.8 km./hr.; lower, 5.4 km./hr. (Christensen and Högberg: Arbeitsphysiol. *14*, 1950.)

climbing is more economical on skis than on snowshoes, the difference being from 8 to 18 per cent.[95]

SWIMMING

Unlike walking, running, and bicycle riding, the energetics of swimming have been studied very little, mainly for three reasons: (1) it has had no practical military value as, for instance, has marching; (2) wide interest in competitive strokes is only of recent origin; and (3) technical difficulties are involved. Swimming has to be done outside the laboratory, and requires the development of special equipment.

Because of this neglect of research on swimming, this activity will be discussed here in greater detail than the others.

The energy used in swimming any stroke depends on the speed used, as can be seen in Figure 55. In the crawl stroke, the amount of energy spent in a given time is roughly proportionate to the square of the speed. For the other strokes the exponent is more than two, and has not yet been calculated.[302]

All strokes may be arranged, in the order of increasing energy cost, as follows: crawl, back, breast, and side. This relationship holds true for any speed. The butterfly stroke, however, has certain peculiarities. It is the least economical of the five strokes when the speeds are under 2.5 feet per second. Above this speed, it is more economical than the side stroke, and, at 3 feet per second, it is more economical than the breast stroke. The greater fatiguing effect of the butterfly stroke as compared with the breast stroke is caused by the local fatigue of the shoulder girdle muscles.

The economy of a stroke also greatly depends on the skill of the subject. One can observe in Figure 55*A* that, for a speed of 2.5 feet per second, a poor swimmer used five times more energy than an experienced one. Thus one can expect a marked improvement in the efficiency of swimming in the course of training. It is of interest to note

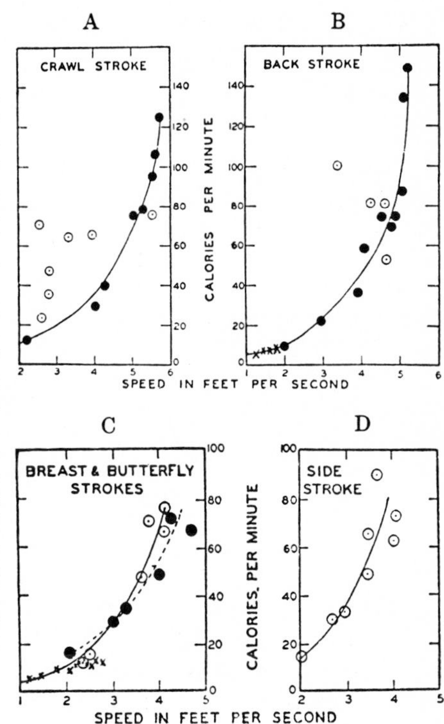

Figure 55. *A*, Energy expenditure in the crawl stroke ● = data, obtained on one good swimmer, from which the curve was plotted; ⊙ = other swimmers. *B*, Energy expenditure in the back stroke. ● = data, obtained on one good swimmer, from which the curve was plotted; ⊙ = other swimmers. × = data taken from Liljestrand and Stenström (1919). *C*, Energy expenditure and the speed of the breast and butterfly strokes. ⊙ = breast stroke; ● = butterfly stroke. One good swimmer was used for each stroke. × = data taken from Liljestrand and Stenström (1919). (Data obtained on four breast stroke and four butterfly swimmers are not shown here, to avoid confusion. They all were mediocre swimmers, and therefore the energy expenditure was a great deal higher than that indicated by the respective curves.) *D*, Energy expenditure and the speed of the side stroke. One good swimmer was tested. (Karpovich and Millman: Am. J. Physiol. *142*, 1944.)

in Figure 55*B* (back stroke) that one swimmer was able to reach a rate of energy expenditure of almost 150 calories per minute, or 14.07 horsepower, which is close to the figures obtained on runners by Sargent and Fenn (13.92 and 13.00, respectively). Thus one can see that swimming is not exactly a mild exercise, especially for beginners.

No complete analysis of the mechanical work involved in swimming, comparable to Fenn's for running, has yet been made. Only partial work, or the work of locomotion alone, has been determined for just two strokes—the crawl and the back crawl.[304]

In these two strokes, work done in locomotion is found from a formula, $W = R \times D = FD$, where W is work in ft. lb., R is the water resistance in lb., D is the distance in feet, and F is the propelling force in pounds. Obviously, when the swimming speed is constant, $F = R$; if speed accelerates, $F > R$, if speed decelerates, $F < R$.

Water resistance has been found experimentally by towing subjects in water by a rope attached to a windlass operated by an electric motor. The speed of towing and the necessary towing force were recorded automatically on a kymograph.[280]

Table 7 gives the relationship between body size, water resistance, and speed of swimming. The water resistance found for the body in gliding positions may be applied to the whole crawl and back strokes in

TABLE 7. *Relation Between Water Resistance and the Skin Surface Area*

	Skin Surface Area in Square Feet	*Water Resistance*	
		Prone Glide	*Back Glide*
Men......................	24–19	0.65† V^2*	0.75† V^2
Men and Women......	19–16.5	0.55† V^2	0.6† V^2

*V is the speed of the glide in ft./sec.; † are constants k.

the calculation of work done. Water resistance in the other strokes requires further investigation, because of the complexity of body movement. One can easily see the mechanical inefficiency, for example, in the breast stroke, when arms and legs are brought close to the body and act as water brakes. When the arms, after a pull, are raised out of the water, a new and faster stroke is produced—the butterfly stroke.

The propelling force in pounds in the crawl stroke may be found from a formula: $F = KV^2$, where V is the velocity of swimming in feet per second and K is a constant indicated in Table 7. The relationship between the maximum velocities obtained by the arms and legs in the crawl stroke is interesting. Let V_a and V_l be the maximum velocities developed when arms and legs are used separately; and F_a and F_l the corresponding forces. Thus, when arms alone are used, $F_a = KV_a^2$; and, when legs alone are used, $F_l = KV_l^2$. When arms and legs are used in a whole stroke, we have a summation effect: $F_a + F_l = F_w$, where F_w is the propelling force of the whole stroke.

Substituting velocities for the forces, we obtain $KV_a^2 + KV_l^2 = KV_w^2$, where V_w is the maximum velocity of the whole stroke. Thus $V_a^2 + V_l^2 = V_w^2$, which means that *the square of the maximum speed of the whole crawl stroke is equal to the sum of the squares of the maximum speeds developed with arms and legs separately*. This formula was found to be true also when swimming was done with fins.[484]

In the preceding discussion it was assumed that $F_a + F_l = F_w$. This is not a correct assumption, because, in the summation of maximal muscular forces, there is always some loss. Since, in practice, when the speeds developed with arms and legs are determined separately, the swimmer has to support the idle end of his body with a water polo ball (Fig. 56), this slightly reduces the speed and makes the equation true. The law of squares can be used for diagnosis of weakness of the arms or legs in swimming. Details may be found in the literature.[282] MacDonald and Stearns[353] have applied this formula to other strokes and found that for the breast stroke, the equation becomes $0.75\ (V_a^2 + V_l^2) = V_w^2$, and for the dolphin-butterfly it is $0.90\ (V_a^2 + V_l^2) = V_w^2$.

So far we have discussed the *average* speed only. If one wishes to find the actual speed, within a single stroke, he may determine it by means of a natograph.[6]

The human body is very inefficient in water. A large number of

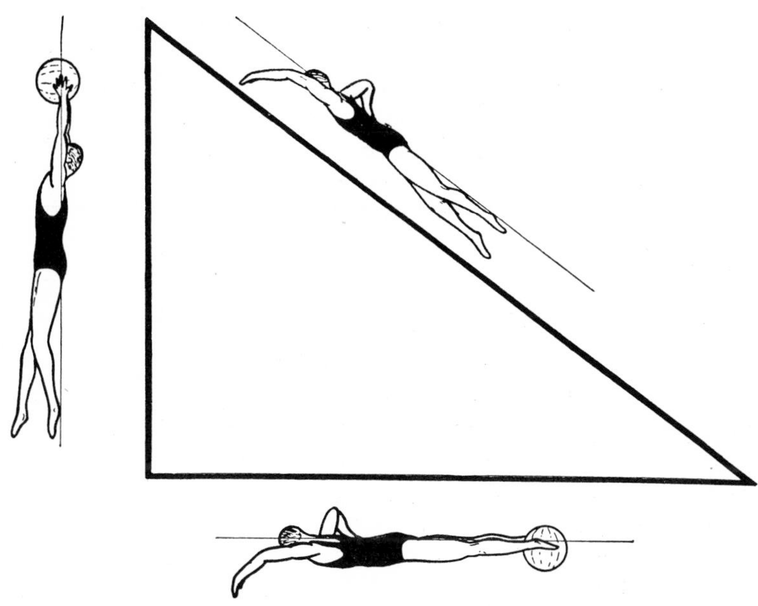

Figure 56. Pythagorean theorem applied to the crawl stroke. The square of the maximum speed of the crawl stroke as a whole is equal to the sum of the squares of the maximum speeds developed with the arms and legs separately. Swimmer begins from a dead start. Water polo ball supports the idle end during the test. (Karpovich: Scholastic Coach, Dec. 1937.)

varsity swimmers were tested with a spring dynamometer attached firmly to the edge of swimming pool. Their pulls varied from 26 pounds for the crawl stroke to 37 pounds in the breast stroke. Compare this figure with the propelling force developed by a dolphin. An animal weighing about 200 pounds can easily clear the surface by some 16 feet from a running start.*

The small propelling force that a swimmer can develop in water and the large expenditure of energy required make swimming a costly type of locomotion. Efficiencies of the crawl, trudgeon, and back crawl vary from 0.5 to 2.2 per cent.[304]

The use of the legs in the crawl stroke is apparently quite inefficient. Adrian, Singh, and Karpovich[3] found that the energy cost of the arm stroke alone was less than that of the whole stroke up to a velocity of 3.35 feet per second. The ranges of the efficiencies at the range of velocities tested were as follows: legs alone, 0.05 to 1.23 per cent; arms alone, 0.56 to 6.92 per cent; arms and legs, 1.71 to 3.99 per cent. Obviously, during long distance swimming, the legs should not be used more than the amount necessary to maintain normal balance in the water.

---

*This information has been kindly supplied by Mr. D. H. Brown, Curator of Mammals, Marineland of the Pacific.

Mechanical efficiency of fin swimming has been estimated to be 3 per cent.[184] During underwater swimming, in which fins are used, average efficiency varies from 2 to 5.6 per cent, depending on speed. The optimum speed is 1.2 feet per second.[490] Incidentally, it was found that underwater swimming, with a speed of 2 ft./sec. (1.2 knots), was an ordeal; and, in one experiment, only one man out of six could maintain it for 20 minutes[184] (swimmers carried oxygen tanks).

CALISTHENICS

The mean energy cost of various exercises used by the British Army varies from five to 6.5 times that of the basal metabolic rate of the men. Marching and running raise metabolism from seven to 10.5 times.[311]

Missiuro and Perlberg[378] found the cost of an average class in Swedish gymnastics to be 3.5 times the basal metabolic rate.

Weiss and Karpovich[527] prepared a cost list for forty-two different exercises which can be used for the prescription of any desired dosage of calisthenics. The cost of an exercise is expressed as a ratio of the work metabolism rate to the resting metabolism rate. Figure 57 shows some of these exercises and their relative costs.

Cadence is an important factor, since a change in the rate of movement will cause a change in the cost. One should be warned, however, that this cost list is merely a guide, and that the cost will vary somewhat from individual to individual.

ROWING

One can easily see that the amount of energy spent in rowing depends on the type of boat and the speed of rowing. Henderson and Haggard estimated the upper limits of energy cost in rowing from a study of the racing crew that represented the United States at the Olympic games in Paris in 1924,[223] and established a world's record for the 2000-meter race. The maximal rate of work by five of the participants ranged from 0.45 to 0.55 horsepower per man, or from 4.8 to 5.9 calories per minute. The lower power was maintained by these men for twenty-two minutes during a 4-mile race.

FOOTBALL

A direct study of the energy used in football is impossible. One cannot take part in a scrimmage with a bag attached to his back and a

Figure 57. Energy cost of exercise. CPM = counts per minute. Note: The first figure in each exercise shows the starting position for that exercise. All exercises are performed in four counts, except exercise 8, which is performed in eight counts. For example, in exercise 1 the count is: 1, trunk bend right; 2, return; 3, trunk bend left; 4, return. In exercise 5, two figures represent each count: 1, elbows back and return; 2, 3 and 4, are repetitions of 1. In exercise 8 the alternate figures show the side view. (Weiss and Karpovich: Energy Cost of Exercises for Convalescents. Arch. Phys. Med. *28*, 1947.)

tube in his mouth, or while holding his breath. For this reason, an estimate of the cost of football training and actual playing has been made through a study of diet.

Edwards and co-workers[151] found that the Harvard football players maintained a weight balance on a diet of 5600 calories a day. Subtracting from this figure 2800 calories, which are spent on basal metabolism and light activities, and 300 calories eliminated with the urine and feces, we obtain a figure of 2500 calories, which represents the daily cost of football practice and training. After taking into consideration the fact that any severe exertion causes an increase in resting metabolism that may last for hours, Edwards and his co-workers assumed that, during actual football play, metabolism is increased to 13.3 calories per minute.

BICYCLING

Most information regarding the energy cost and efficiency of bicycling has been obtained from experiments on stationary bicycle ergometers. Riding a bicycle requires not only skill but also specific training of the muscles of the legs. A beginner, for example, riding on an ergometer with work output of 6000 foot-pounds per minute, may last only five minutes. After several weeks of training, he will be able to continue this work output for hours.

The net mechanical efficiency of riding a bicycle may be calculated in two different ways, depending on the method of estimating net energy. In one method, the net amount is found by subtracting resting metabolism from gross work metabolism. This method gives efficiencies varying from 14.5 to 24.9 per cent, the average being 21 per cent. In the other method, net energy is found by subtracting the amount of energy used in riding the bicycle without the load (free wheeling) from gross work metabolism. The second method gives efficiencies ranging from 22.4 to 41.0 per cent, the average being about 30 per cent.

The question of the optimal rate of pedaling seems not yet settled. Grosse-Lordemann and Müller[198] found experimentally that the optimum lies between 40 and 50 pedal revolutions per minute at a rate of work output of 4338 foot-pounds per minute. This is in close agreement with previous findings by Hansen. Karpovich and Pestrecov,[303] however, observed that subjects as a rule disliked rates of revolutions less than 60 per minute, although the rate of work was 5000 foot-pounds per minute or more. Professional bicyclists also prefer higher rates, 70 being the choice of some professional road riders.

The amount of force necessary to propel a bicycle on a good road is equal to the frictional resistance of the bicycle, the air resistance, plus or minus the force of the wind. For a man weighing, together with his bicycle, 90 kg., frictional resistance has been found to be 0.016 kg. per kg. of weight. It may vary from 0.005 to 0.03, depending upon the surface of the road. Thus it takes about 1.44 kg. of force (3.17 pounds) to overcome this resistance. Air resistance varies in proportion to the square of velocity – $0.03V^2$ kg. At a speed of 18 kilometers per hour, air resistance is equal to one-third of the total force. At a speed of 24 kilometers per hour, air resistance is equal to one-half of the total resistance. One can easily see why a bicycle racer tries to reduce resistance as much as possible by streamlining his dress and body position.

Of interest in this connection is an estimate made by Carpenter[86] of the amount of energy used by a professional rider who rode an average of 20.77 hours a day for five days during a six-day race. He computed that 3366 calories were used for work per day, or 2.70 calories per minute.

Karpovich and Pestrecov[303] had a subject ride without stopping for over six hours, with a work output of 7150 foot-pounds per minute. The total amount of work done was 2,659,800 foot-pounds, and the amount of energy expended was 7.07 calories per minute, of which 5.72 calories were converted into work. The subject could have ridden longer, but, for reasons beyond the experimenters' control, the test had to be terminated. Racing bicycles have narrow tires because the common wide tires require more energy to overcome a greater friction against the ground.[124]

WEIGHT LIFTING

Although a weight lifting training session may last from seventy-seven minutes to two hours, the time actually spent on lifting may be only two to six minutes. During this period, an average weight lifter will lift from 2640 to 3300 pounds, and a man in a champion class may lift from 10,000 to 20,000 pounds.

Work done in lifting may be calculated by multiplying the weight by the height to which it has been lifted. Since this height varies from 6.25 to 7.17 feet, the total amount of work will vary proportionately. If we assume that the average lifting height is 6.7 feet, the amount of work done during one session may vary from 17,688 to 134,000 foot-pounds. For an efficient lifting, it is important to have rest periods of three to five minutes between each two trials. If a rest period is too short, muscles do not recover sufficiently. If a rest period is too long (up to eleven minutes), weight lifters seem to lose their trigger timing, and their proficiency decreases.

The three classic lifts employing barbells are the *press*, the *clean and jerk*, and the *snatch*. All these lifts are executed rapidly: press – 7.21

seconds; clean and jerk – 6.62 seconds; and snatch – 4.98 seconds. Lowering the barbell in each of these maneuvers takes two seconds.[331]

The muscular force of a weight lifter depends on his body weight, because muscles make up at least 40 per cent of this weight. For this reason, weight lifters are divided into seven classes.

It is a rather common belief that weight lifters are "muscle bound." This belief originated probably because some excessively fat strong men have been greatly impeded in their movements. Well trained men may have large muscles, but, if they are lean, their flexibility remains normal.

Another common belief is that weight lifters' muscular contractions become slower. Experiments conducted by Zorbas and Karpovich[552] showed just the opposite. They designed a machine which automatically recorded the time required to execute twenty-four complete revolutions with the arm. The arm was chosen for this test because weight lifting especially affects the muscles of the arm and shoulder girdle. Three hundred non-lifters and three hundred weight lifters were studied. Among the latter were men in the champion class, not only from the U.S.A., including Hawaii and Puerto Rico, but also from Australia and Canada. The lifters had the fastest movements, and students from a liberal arts college the slowest.

Chui[97] found that three months of training in weight lifting (15-pound dumbbells and 125-pound barbells) made participants more proficient in the Sargent jump, standing broad jump, shot put, and 60-yard dash.

WRESTLING

It is obvious that the amount of energy used in wrestling will depend on the intensity of the wrestling. If, however, the time spent in this activity is sufficiently long, the intensity has to be relatively low; otherwise, wrestlers will quickly become exhausted.

In order to measure the amount of energy used in wrestling, the subjects must wrestle inside a special respiratory chamber that allows complete freedom of action and from which samples of air may be taken for analysis and calculation of consumption of oxygen and production of carbon dioxide. Tests conducted on two men showed that metabolism rate increased approximately twelve times.[199]

SNOW SHOVELING

After every snow storm one either reads in the papers or hears over the radio about men who drop dead, either while shoveling snow

Figure 58. The energy used in snow shoveling. If snow is dry, the energy spent is equivalent to that needed for climbing to the third or fourth floor in one minute. If the snow is wet, energy expended equals that of climbing to the seventh floor in one minute.

or soon after. The victims usually are cardiac patients who foolishly decided that they had to remove the snow to take their car from the garage, or thought that snow shoveling was good exercise for them. These individuals ordinarily had been cautious in undertaking physical activities, and frequently even had avoided stair climbing.

Work done in snow shoveling depends on two factors: the condition of the snow and the eagerness of the worker. The threat of death lurks behind the eagerness. A snow shovel weighs 5 pounds, and the

TABLE 8. *Energy Used in Snow Shoveling*

*Weight of Snow and Shovel**	*O_2 Used on Lifting, Making Two Steps and Throwing Snow*	*Rate of Shoveling per Minute*	*O_2 Used per Minute*	*Equivalent to Climbing to*
lbs.	*cc.*		*L.*	
8.75	153	10	1.53	3rd floor
13.75	220	10	2.20	4th floor
22.50†	370	10	3.70	7th floor

*Shovel alone weighed 5 pounds.
†Shovelful of wet snow.

weight of the snow varies from 3.75 to 17 pounds, depending on whether it is dry or wet, and whether only half a shovel or a full load is lifted. If we assume that only ten shovels of snow are removed per minute, this activity requires energy output equal to that in climbing to either the third or the seventh floor, respectively, in one minute (Fig. 58). Many a man will think twice before doing the latter. Table 8 shows a comparison between snow shoveling and stair climbing, calculated for a subject weighing 160 pounds.

ENERGY SPENT ON HOUSEWORK

Men ordinarily underestimate the amount of energy spent by women while housekeeping. Housework often goes unappreciated and unnoticed, unless it goes undone. The amount of energy spent by a woman on housework obviously depends on the size of her family, her house, the amount of help given by other members of her family, the number of labor saving devices, and her aptitude for housekeeping. While a perfectionist seems never to end her work, the opposite type seems never to start anything. However, since most women take good care of their homes, the amount of energy spent by an average woman is considerable.

In order to measure objectively this energy, a study was conducted on three wives, from two-, five-, and six-member families.[141] The Douglas-Haldane method was used. It was found that the average number of calories spent by each woman on daily work was about 1600, varying from 881 calories on Sundays to 4378 calories on a wash day. The number of hours spent on work varied from nine to eleven and one-half hours.

Table 9 gives the range of duration and the amount of energy

TABLE 9. *Amount of Energy Spent in Housework*

Type of Work	*Duration in Min.*	*Calories*
Making beds	8.8– 19.5	38.8– 72.3
Preparing beds	1.1– 6.8	4.9– 43.0
Shining shoes	*2.5– 4.9	*3.8– 13.2
Ironing laundry	1.1– 58.2	2.3–122.2
Light picking up	107.6–172.4	226.0–372.5
Heavy picking up	2.7– 29.5	10.4–103.0
Dusting	11.8– 47.2	31.8–127.4
Mopping	1.8– 11.8	5.7– 30.6
Scrubbing on knees	8.2– 22.1	36.9–108.2
Window washing	0.3– 6.2	.06– 17.5
Cooking	22.2– 87.1†	25.2–182.9†

*This lady shined only her own shoes.
†This lady had the biggest kitchen.

spent on various types of housework. The ranges, rather than averages, were used because of a considerable fluctuation in data obtained.

EFFECT OF TRAINING ON BASAL METABOLIC RATE (B.M.R.)

There is much uncertainty regarding changes in basal metabolism during training. Observations made by Schneider and Foster[460] indicated that, of ten athletes, eight had a drop in their basal metabolic rate, one had an increase, and one had no change. Of seven non-athletes in their control group, three had a fall in their basal metabolic rate, two had a rise, and two had no change.

Steinhaus,[492A] in his five-year-long observations on dogs, the diet and training conditions of which were well controlled, noticed a slight decrease in the basal metabolic rate. Most of the confusion regarding the effect of training upon the basal metabolic rate has resulted from an insufficient number of tests of basal metabolic rate on each subject. It is well known that even two consecutive tests on a person in one morning may show a large difference. Moreover, after strenuous exertion the basal metabolic rate may be raised for longer than twenty-four hours. Variations in the diet affect the basal metabolic rate.

Morehouse[383] carried out investigations on twenty athletes in and out of training, taking care to meet the criticism leveled against other researchers. He concluded that the B.M.R. did not change.

Clarence De Mar, the great marathon runner, at the peak of training, had a basal oxygen consumption of 211 cc. per minute; and the number of calories he expended, per square meter of body surface per hour, was 37.

EFFECT OF TRAINING UPON WORK OUTPUT AND EFFICIENCY

Training, especially if it demands the learning of new skills, such as those in swimming or bicycle riding, considerably increases work output. Figure 59 illustrates a steady rise in the endurance of stationary bicycle riders. The range in improvement varies, depending upon several factors, including:

1. Individual differences, which often are hard to explain.
2. The degree of physical condition at the beginning of training. Men in better condition have a smaller margin for improvement.
3. The intensity of exertion. For example, with rates of work ranging from 0.159 to 0.261 horsepower, bicycle riders who trained for seventeen to twenty-two weeks, five times a week, improved their rid-

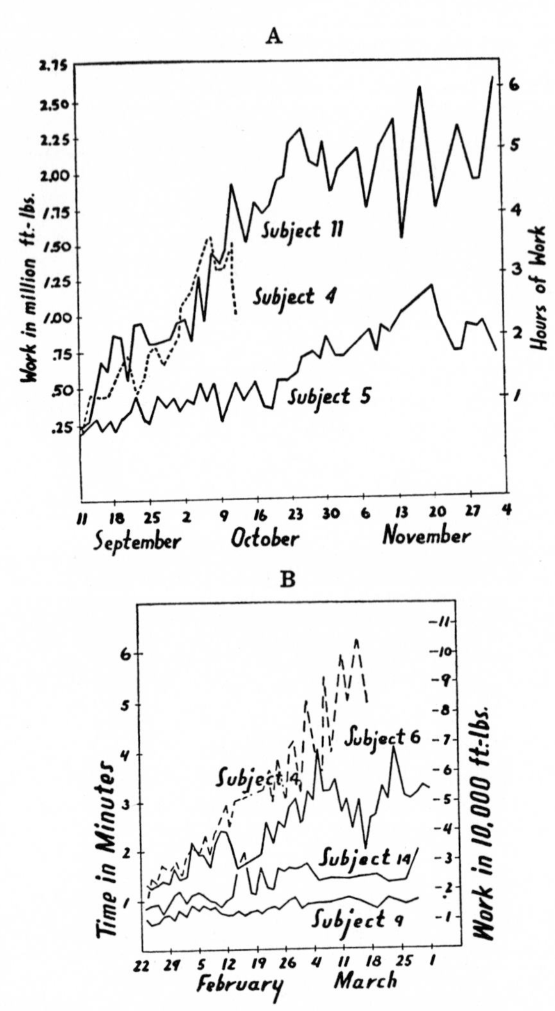

Figure 59. *A*, Typical curves of prolonged work on bicycle ergometers. Subjects, jail inmates. Subjects 4 and 11, 0.217 horsepower; subject 5, 0.182 horsepower and subject 7, 0.170 horsepower. Subject 4 had a knee injury which forced him to discontinue participation in the experiment.

B, Typical curves of short intensive work on bicycle ergometers. Subjects, college students. Rate of work is 0.506 horsepower; pedal revolutions, 117 per minute. A drop in performance in curves 6, 14 and 9 was the effect of college term examinations. Of the sixteen subjects, only three were not affected by the examinations. Subject 4 is one of them. (Karpovich and Pestrecov: Am. J. Physiol. *134*, 1941.)

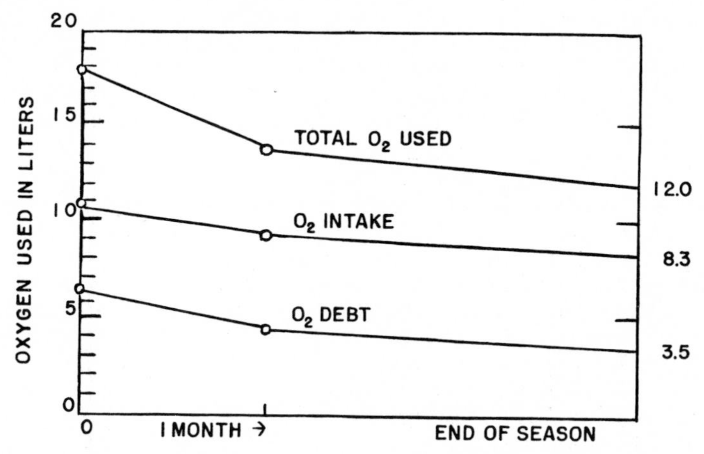

Figure 60. Effect of training in skiing upon oxygen consumption. The test consisted of skiing 690 meters in five minutes. (From data quoted by Krestovnikoff: Fiziologia Sporta, 1939.)

ing times from 75 to 4420 per cent of the original. On the other hand, with a rate of work equal to 0.506 horsepower, the improvement in ten weeks was from 49 to 334 per cent. The difference in the length of training could not account for the difference in the degree of improvement, because, when the best man in the second group continued training for nine more weeks, the total improvement was only 463 per cent.[286]

These observations coincide with those made by practical coaches—that training produces more marked improvement in activities requiring endurance rather than speed. Efficiency is especially improved in activities requiring the mastering of skills, as, for instance, in swimming and skiing. Examination of Figure 55*A* shows that some poor swimmers used five times more energy than skilled, trained men.

In skiing, learning also may reduce the amount of energy used for the same distance and speed—as much as three and one-half times.[331] Even skilled skiers show improvement in efficiency after training. Figure 60 shows this effect on three men. It may be seen that, for the same activity, the amount of oxygen used at the beginning of training was 17.2 liters, and, at the end of the season, 11.9 liters. The oxygen intake during the run increased by 2.4 liters, thus reducing the size of the oxygen debt.

QUESTIONS

1. What is the difference between basal and resting metabolism?
2. What are the most important factors which affect the cost of walking?

3. How much energy is used in climbing? How can you calculate this outside the laboratory?
4. Assuming that efficiency of climbing is 20 per cent, how high must a 150-pound person climb to lose one pound of fat (3500 calories)?
5. How does the cost of negative work compare with that of positive work?
6. Which swimming stroke is the most economical? Why?
7. What is the Law of the Squares, as applied to swimming, and how was it derived?
8. What is the mechanical efficiency of swimming without and with fins?
9. What factors affect the cost of an exercise?
10. Why are we interested in knowing the cost of various physical activities?
11. How much energy is spent during football practice?
12. Why do racing bicycles have narrow tires?
13. How much horsepower may be developed in weight lifting? How does this compare with that developed in other sports?
14. What is the mechanical efficiency of weight lifting?
15. Are weight lifters muscle bound?
16. How can we calculate the amount of work done and energy used in snow shoveling?
17. What effect has altitude on weight lifting?
18. How much energy is used in housework?
19. What effect has training upon (a) basal metabolism, (b) work output, (c) mechanical efficiency, (d) oxygen consumption?

Chapter Nine

RESPIRATION

GENERAL CONSIDERATIONS

The purpose of respiration is to provide oxygen for metabolism of the body cells and to eliminate the carbon dioxide resulting from oxidation. Therefore the respiratory mechanism is so adjusted that its function corresponds directly to changes in metabolism.

The volume of air breathed varies with every change in bodily activity: sleeping, sitting, walking, and running. The amount of air breathed cannot be estimated from mere inspection. A considerable increase, 200 and sometimes even 300 per cent, may not be noticed by the breather himself or by a casual observer. The total volume of the air inhaled and exhaled during a certain period rarely can be exactly controlled by the will, but is automatically adjusted to maintain the interior atmosphere of the body as nearly constant as possible.

Respiration essentially plays a twofold part in the body during physical exertion. On the one hand, it supplies the oxygen required by the muscles; on the other hand, it serves to keep the acid-base balance of the blood constant within certain narrow limits.

Gaseous exchange takes place between the blood in the lung capillaries and the air in the alveoli. The process of diffusion of gases is so rapid here that a virtual equilibrium of the partial pressure of every gas is established in less than a second between the blood in the capillaries and the air in the alveoli. The efficiency of this exchange may be better understood if one realizes that, even during strenuous exercise, only about 1 pint of blood per second passes through the lungs. In the lungs, this blood is spread out in capillaries over a surface of approximately 100 square meters, an area equal to about one-half of a tennis court for singles (Fig. 61). A pint of blood sprinkled on the tennis court would certainly be rapidly aerated.

It has been shown that champion swimmers have a greater capacity for transferring oxygen across the alveoli than non-athletes, average swimmers, and even long-distance runners.[386]

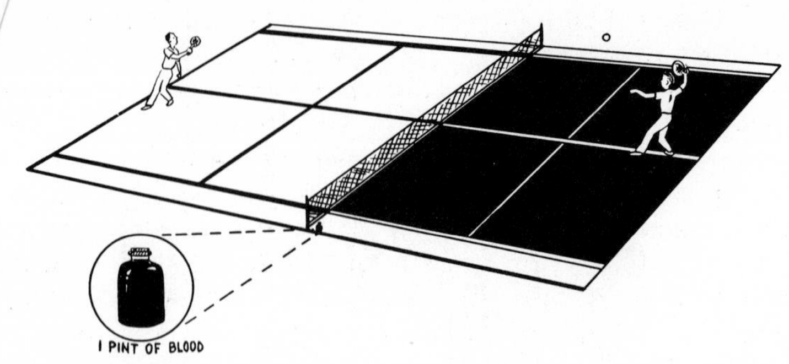

Figure 61. Efficiency of blood aeration. The surface area of the capillaries in the lungs of an athlete is about equal to the area of one-half of a tennis court for singles. The amount of blood that passes through the lungs during strenuous exercise is about 1 pint per second.

PULMONARY VENTILATION IN THE SEDENTARY INDIVIDUAL

That people differ in the manner of breathing has been shown by a number of investigations. In shallow breathing, only a relatively small amount of fresh air gets past the dead space to mingle with the air in the alveoli of the lungs. The deeper the breathing, the greater will be the amount of fresh air that reaches the alveoli, and hence the greater will be the supply of oxygen for exchange with the blood.

The range of the respiratory rate in adults is rather wide: from four to twenty-four breaths per minute. The average, however, has been accepted as sixteen.

The amount of air taken in with each breath is called the *tidal* air, and is also referred to as the *depth* of respiration. This may vary from 300 to 1500 cc. in healthy adults at rest, the average being 500 cc.

MINUTE-VOLUME OF LUNG VENTILATION DURING PHYSICAL WORK

The total amount of air taken in during one minute is called the minute-volume of respiration, or the minute-volume of lung ventilation. Obviously, the average minute-volume may be found by measuring all the air inhaled during a certain time and dividing it by the number of minutes, or by multiplying the average rate per minute by the average depth of inspiration.

$$\text{Minute-volume} = \text{rate} \times \text{depth}.$$

At rest it varies from 3000 to 10,000 cc., the average being 16 ×

500 = 8000 cc. = 8 liters. During strenuous work, it may be as high as 25 × 4600 = 115,000 cc. = 115 liters or more.

During physical work, metabolism is increased; therefore more oxygen is required. This leads to intensified respiration. A linear relationship between the minute-volume of lung ventilation and the rate of oxygen consumption up to over load has been well established (Fig. 62). This relationship is true not only during work, but also during recovery after work. Thus the intensity of an activity with a normal load may be roughly appraised by measuring the corresponding lung ventilation rate (Table 10). With an over load, this proportionality is disrupted, and ventilation increases in excess of the oxygen consumption increase.

A sustained ventilation of the lungs of 100 to 110 liters a minute is unusual. It is a rare occasion when one is called upon to maintain the maximum lung ventilation of 200 liters for as much as a minute or two.

The time required for breathing to return to its pre-exercise condition is determined by the relative severity of exercise. This means that individual differences and state of training markedly affect the post-exercise minute-volume curve. What may be strenuous for one man is moderate for another. After exercise, the volume falls off more rapidly than the respiratory rate. The drop is especially rapid during the first two minutes.

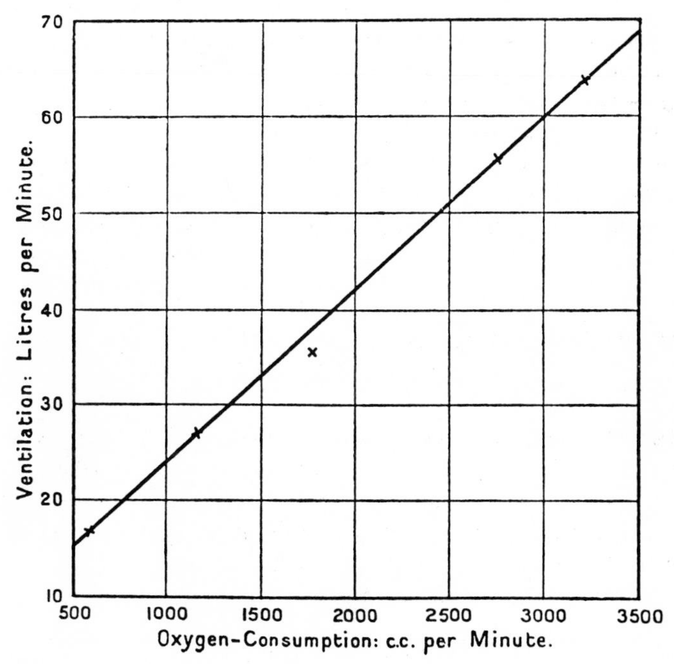

Figure 62. Relation between the pulmonary ventilation and oxygen consumption. The ratio of ventilation to O_2 is called oxygen ventilation equivalent. It is also known as ventilatory equivalent. (Constructed from Lindhard's data in Bainbridge, Bock and Dill: Physiology of Muscular Exercise, 3rd ed. Longman's Green and Co.)

TABLE 10. *Oxygen Consumption and Lung Ventilation during Various Activities*

	Oxygen Consumption: Liters per Minute at 0° C. and 760 mm.	*Volume of Air Breathed: Liters per Minute at 20° C.*
Rest in bed, fasting	0.240	6
Sitting	0.300	7
Standing	0.360	8
Walking 2 miles per hour	0.650	14
Walking 4 miles per hour	1.200	6
Slow run	2.000	43
Maximum exertion*	3.000 to 4.000*	65–100*

*Note: 5.88 of O_2 and 200 liters of air have been recorded (personal communication from Dr. L. B. Rowell).

FREQUENCY AND DEPTH OF RESPIRATION DURING WORK

As soon as work begins, the rate and the depth of respiration increase. This may be observed with the first inspiration (Fig. 63). Recently, Craig and co-workers[112] showed experimentally that this sudden increase in respiration at the beginning of work occurs only in people accustomed to a particular activity. For instance, people not accustomed to running on a treadmill or riding an ergocycle may not show any increase during the first 10 seconds because of the inhibiting effect of the "complexities of the work situation."

When the excitement of competition or emotion is involved, an anticipatory increase in breathing may occur even before the work starts. However, at the start of a sprint race, participants usually suspend respiratory movements after the command "Get set." The beneficial effects of this practice can easily be explained: (1) alertness to sound is thus facilitated; and (2) the propelling force of the legs and arms can be greater when the chest and abdomen are immobilized.

Roughly, the number of breaths per minute increases proportionately to the load of work up to the crest load. When the exertion is moderate and steady, the minute-volume and frequency of breathing continue to increase for several minutes, and then more or less level off. Ordinarily, the frequency of breathing first reaches a steady state within two to four minutes. The depth and, consequently, the minute-volume need three to five minutes to become steady.

When exertion is severe (over load), both the minute-volume and the frequency of breathing continue to be augmented throughout the entire period of work, although the depth may decrease.

The commonly observed upper limit of the respiratory frequency in activities on land is about 30 per minute. In swimming, the respiratory rate frequently rises as high as 60. This is probably because

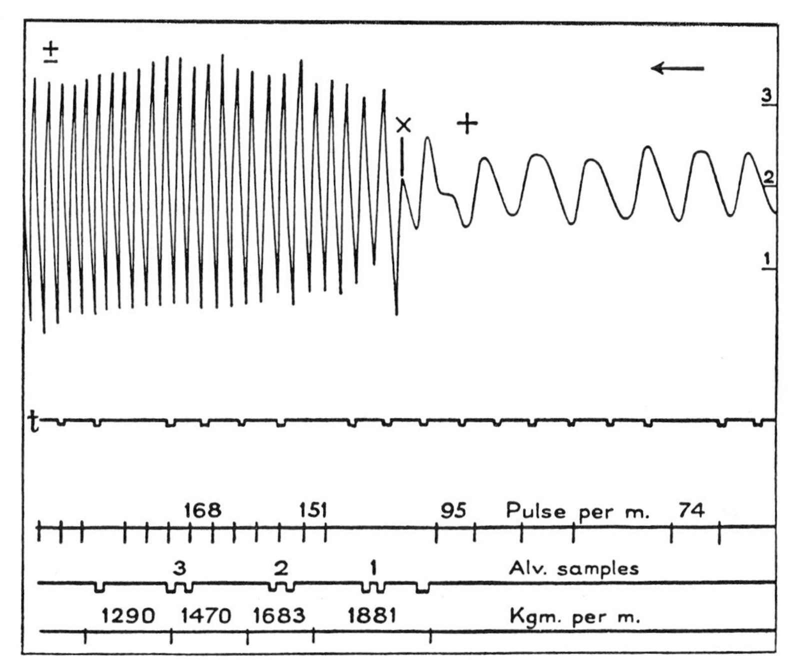

Figure 63. Depth and rate of respiration in transition from rest to work. Scale, upper right hand, in liters; t, time in six-second intervals. +, ready; ×, begin; ±, stop. The first breath, less than a second after beginning of exercise, is already modified. (Krogh and Lindhard, from Bainbridge, Bock and Dill: Physiology of Muscular Exercise, 3rd ed. Longman's Green and Co.)

swimmers try to breathe with each stroke. A rate as high as 75 per minute has been reported for the crawl stroke.

Herxheimer recorded a respiratory depth of 4.6 liters at a rate of twenty-five breaths per minute in a man riding a bicycle. This was, however, an exceptionally strong man, with a vital capacity of 5700 cc. In swimming, the respiratory depth at the same rate is considerably smaller. Probably this may be explained by the pressure of water upon the chest and abdomen. A man lying in a swimming position in water may have his chest girth reduced by 0.75 inch, and his vital capacity by 350 cc.

A strong man, exerting an inspiratory suction equal to −70 mm. of mercury, would be able to inhale 2334 cc. of air in 0.4 second (respiratory rate of 70 per minute).[285] Such a feat can be performed only a few times in succession, however, after which the depth is reduced considerably. According to Agostino and Fenn,[4] the air flow to the lungs is limited by the ability of the muscles to mobilize chemical potential energy for work. When contractions become too fast, respiratory muscles, like any other skeletal muscles, can develop little force.

An excessive respiratory rate is fatiguing and inefficient. It imposes a great strain upon the respiratory muscles and reduces the depth of respiration. A coach may draw the conclusion that an excessive rate of respiration cannot increase the efficiency of his athletes.

There is one more reason why fast, shallow breathing is inefficient. Not all the air inhaled during an inspiration reaches the alveoli. That air which fills the dead space does not take part in the gaseous exchange with the blood, and therefore represents a respiratory waste. The ratio of this waste is greater in shallow than in deep breathing. Thus, to bring the same amount of air to the alveoli, the total amount of air inhaled with shallow breathing must be greater than with deep breathing.

Since it costs about 0.8 to 2 cc. of oxygen to move a liter of air, some oxygen will be expended on the act of respiration rather than on the physical work. This is particularly important when the emphasis is on winning an athletic competition. In taking advantage of this source of efficiency, some coaches, especially in swimming, have trained their men either to hold the breath too long or to make as few inhalations "as possible." This technique has certain disadvantages. Although a man can swim a short distance faster without breathing, because respiratory movements decrease the maximum force of the muscles attached to the chest, prolonged breath holding may cause such an oxygen deficit that the man, at the end of a long race, will have to swim more slowly to "catch his breath."

In running 60 to 100 yards, respiration may be entirely suspended. Most runners prefer, however, to make at least one inhalation, thus preventing the development of great discomfort. In general, however, it must be admitted that "talks" about respiratory control in athletics are mostly fanciful and futile attempts to interfere with a wonderfully adjusted breathing mechanism.

When, during physical exertion, the respiratory depth begins to decrease and the rate to rise excessively, one may experience a sensation which has aptly been called intolerable. Incidentally, the sensation of distress in intolerable breathing apparently comes from the respiratory muscles themselves. One may observe old people climbing stairs and stopping to place their hands on the rails, or athletes at the end of a race bending forward to place their hands on their knees. In both cases, by immobilization of the arms, the muscles which are attached to the chest and the humerus can help in inspiration and relieve the strain upon the regular respiratory muscles—intercostals and diaphragm.

It should be noted that when a runner quits, "out of breath," there is still more oxygen in the lungs than he can use. A popular notion that lungs should be trained usually means something else—that the respiratory capacity of the circulating blood should be increased.

VITAL CAPACITY OF THE LUNGS

The vital capacity of the lungs is measured by the largest quantity of air which a person can expel from his lungs by a forcible expiration after the deepest possible inspiration. It consists, therefore, of reserve inspiratory, tidal, and reserve expiratory airs.

Vital capacity in normal people varies from 1400 to 6500 cc. For the adult male the average may be accepted as 4000 cc., and, for the woman college student, 3400 cc. The only correct and easy way of measuring the vital capacity is by a spirometer. When air is collected in the spirometer, its temperature and, therefore, volume decrease. For this reason, in precision investigations, the volume data have to be corrected for the temperature drop and recalculated at body temperature.

A maximum lung inflation is rarely used; yet there are reasons for believing that its magnitude bears an important relation to the physical fitness of a person. It should be borne in mind, however, that it is not safe to pass judgment on the respiratory function of different individuals by merely comparing the absolute figures of their vital capacities, because vital capacity is related to body weight and skin surface area. According to Dreyer,[140]

$$\text{Vital capacity in cc.} = \frac{(\text{Body weight in gm.})^{0.72}}{0.69}$$

Nevertheless, there is some relationship between participation in physical work and vital capacity. Thus in 1920 West[532] showed that the ratio of the vital capacity to the skin surface area is greatest in athletes and least in sedentary women:

	Men	*Women*	*Athletes*
Vital capacity, cc. per cm. of height...	25.0	20.0	29.0
Vital capacity, cc. per square meter of body surface............................	2500	2000	2800

A number of studies conducted during the past 20 years have substantiated this observation.[494]

There have been exceptional athletes with low vital capacities and poor ones with large capacities. Probably this means only that an athlete should have at least a normal relation between vital capacity and body size. A large vital capacity does not make a champion. One should not be greatly surprised that a notion still exists that a greater vital capacity is an unmistakable index of greater physical fitness. Notions are given up slowly.

In the study of marathon runners, it was observed that vital capacity fell immediately after the race, but returned to normal within twenty-four hours. The fall in vital capacity has been explained on the basis of a greater amount of blood present in the lungs at the end of the race.

Schneider[458] found the coefficient of correlation between breath holding and vital capacity for 127 men to be 0.24 ± 0.06. This indicates that there is some relation between the two, but it is too low a correlation to make vital capacity a highly important factor in determining the length of time breath can be held. Among these men, vital capacity averaged 4330 cc.; the smallest was 3090 cc., and the largest 6620 cc.

PAIN IN THE SIDE

The cause of the stabbing pain in the side, called by some a "stitch," is not definitely known and is not easy to investigate, for the physiologist cannot be sure of its appearance when he is prepared to study it.

One of the authors observed cross country runners during practice runs. They ran along the road so that he could follow them in his car. During every practice period, one or several members of the team had severe cramp-like seizures, which sometimes occurred on either side of the chest, but most often were felt in the subcostal arch, extending deep in and down. Sometimes this pain lasted for more than twenty-four hours.

Previously it was believed that the stitch was caused, at least in part, by cramps of the diaphragm or intercostal muscles. No objective evidence has been presented, however, regarding the diaphragm. As to the intercostal muscles, there is some evidence. One of the authors lately was so fortunate or unfortunate as to have muscle cramps in the intercostal muscles. Whether they were in the internal or external intercostals, the only treatment possible was to hold the breath until they disappeared. The cramps never lasted too long, and several painful breaths were possible. These cramps may occur in athletes, but they are acute and of very short duration. The common stitch lasts too long to be considered a cramp. The explanation of the stitch should be looked for elsewhere.

Recently, Evdokimova[159] studied 162 athletes, most of whom were distance runners, skiers, or bicyclists. She found that, after a race, 95 of them complained of a pain in the area of the liver and 67 did not. Those who complained had enlargement of the liver more frequently than did the non-complainers, the ratio being 49:14. She also reported that athletes often bend forward and press upon the liver area with a hand in order to "squeeze" the excess of blood from the liver. Pain usually diminishes. Rowell and co-workers[444] have shown that during intensive exercise the hepatic blood flow may be reduced as much as 80 per cent. Evidently this reduction does not preclude the engorgement of the liver. The question remains: what happens when the pain is felt on the left side? Possibly the spleen is involved. This explanation was given to the author some 50 years ago, and Evdokimova mentions it again in her review of the literature.

ALVEOLAR AIR

This term is used to describe the air, in the depths of the lungs, which is in contact with the respiratory epithelium and in position to carry out gaseous interchange with the blood. We live, so to speak, in our alveolar air atmosphere. The alveoli afford a steadying influence. They contain, not atmospheric air, but an atmosphere of different composition: one that is quite constant.

The expired air is a mixture of air from the alveoli and the air which remains in the respiratory tubes (the dead space) at the end of inspiration. Not all the alveolar air can be exhaled. Even after the deepest expiration, about 1.5 liters of air, the so-called "residual volume," remain in the lungs. An ordinary expiration leaves in the alveoli about as much more, known as the reserve expiratory air. The reserve expiratory plus the residual air constitute the so-called "stationary air." The tidal air of quiet breathing, as we have seen, ranges from 300 to 1000 cc., with an average of about 500 cc. Of this tidal air, only two-thirds reaches the alveoli, the other third (about 150 cc.) remaining in the dead space, where it does not participate in respiratory exchange. In the deepest possible inspiration, an addition of approximately 1.5 liters, called the reserve inspiratory air, can be drawn in.

If the volume of the dead space and the composition of the air expired are known, we can calculate the average composition of alveolar air (see Table 11).

As may be seen from this table, the sum total of the partial pressures is 713 mm., or 47 mm. less than the average atmospheric pressure. The difference results from subtraction of the partial pressure of water vapor. Alveolar air is saturated with water vapor, which at body temperature has a tension of 47 mm. of mercury.

Under normal conditions, there is always a sufficient reserve of

TABLE 11. *The Composition of Dry Alveolar Air in Man at Rest*

	Volume per Cent	*Tension in mm. Hg (Partial Pressure)*
Oxygen	14	100
Carbon dioxide	5.6	40
Nitrogen	80.4	573
	Total	713

oxygen in the alveolar air almost to saturate the hemoglobin of the blood. As a rule, the oxygen tension in the alveolar air during rest remains constant. If we compare the oxygen tension of the arterial blood with that of the alveolar air, we find the tension of oxygen in the blood to be approximately 5 mm. less than that in the alveoli. The arterial blood is approximately 95 per cent saturated with oxygen. The function of respiration is to maintain an atmosphere in the alveoli in which the tension of carbon dioxide and oxygen in the arterial blood as it leaves the lungs shall be nearly constant.

During physical work the tidal air increases. The gain, however, is made chiefly by the use of the reserve inspiratory air space and, in a smaller degree, by more complete expirations. Thus the ventilation continues to be the mixing of fresh air with the 3000 cc. of stationary air, with the result that the composition of alveolar air is kept remarkably close to its mean.

ALVEOLAR AIR CHANGES DURING WORK

The variations that may occur in alveolar air composition were well shown in a series of experiments by Dill and collaborators, in which ten men ran for twenty minutes on a motor-driven treadmill. In three of the men, alveolar carbon dioxide tension fell as much as 1.7 to 6.5 mm.; in the other subjects it rose 0.3 to 5.2 mm. The oxygen likewise showed variations up and down; in five cases alveolar oxygen tension fell 2 to 13 mm., and in the other five it rose 1 to 8 mm. In another study made by Dill and collaborators while at an altitude of 10,000 feet, at Leadville, Colorado, alveolar oxygen tension during work always rose 4.2 to 13.8 mm. Without the presence of stationary air, the drop in tension of the alveolar gases could have been considerably higher. In the case of carbon dioxide, the tension might have been reduced to approximately 2 mm., causing a subsequent drop in blood carbon dioxide content and producing an inhibitory effect on respiration.

The rise in oxygen tension of the alveolar air after cessation of intensive exertion is convincing proof that breathlessness or "air hunger" is not caused by failure of the lungs to supply oxygen, but by failure of the blood to transport it. Training improves transportation.

NASAL VERSUS MOUTH BREATHING

At rest, the healthy man breathes through his nose. The inspired air is warmed, moistened, and cleansed of foreign particles as it passes over the moist surface of the nose and nasopharynx. During vigorous exercise mouth breathing tends to replace nasal breathing, for thus there is less resistance to the entry and exit of air, and exposure of the moist vascular surface of the mouth to the air assists in cooling the body.

A common misconception, especially among women teachers, is that one must breathe through the nose no matter what the exercise. This must provide an interesting challenge to the student when performing a test such as the 600-yard run for time. During intense exercise the mouth is the air conduit of choice, and the nasal passages have little if any significance at such times.

Although this relationship usually holds true for athletes in the United States, war veterans who spent some time in the subtropics know that their inspired air did not have to be warmed during a hot summer day; it had to get cool during inhalation.

REGULATION OF RESPIRATION

The movements of breathing are regulated by a center in the brain located in the medulla oblongata. To the respiratory center run afferent nerve pathways from various parts of the body and also from the higher nerve centers. Some of the afferent nerves bring impulses to the center only occasionally; others more or less continuously. From the respiratory center efferent nerves pass to the motor neurons of the muscles concerned in breathing. These include the phrenic nerves to the diaphragm and the intercostal nerves to the intercostal muscles of the chest.

The regulation of the respiratory center is both reflex and chemical. The initial increase in the minute-volume of breathing, which occurs with the onset of exertion, and the anticipatory increase, which sometimes appears just before work, are responses to stimuli from higher brain centers and to reflexes from working muscles. Some time must elapse after the beginning of exercise before chemical changes in the blood, brought about by the activity of the muscles, can act upon the respiratory center.

The part played by the vagus nerve during bodily activity has not as yet been completely determined. However, it is known that the vagus fibers in the lungs are stimulated by the distention of the alveoli in the act of inspiration. As soon as the lungs collapse, inhibition of the center ceases, and it again responds to chemical stimulation by initiating another inspiration.

THE CONTROL OF BREATHING THROUGH THE BLOOD SUPPLY OF THE RESPIRATORY CENTER

For a long time, it was believed that the normal stimulus to the respiratory center was the partial pressure of carbon dioxide in the arterial blood. Experimentally it can be shown that, when blood is shaken up with carbon dioxide and injected into a carotid artery of an animal, an immediate increase in the minute-volume of breathing occurs. If the injection is made into the jugular vein, there may be no effect upon breathing; or, if there is an effect, it will be delayed and slight, owing to the loss of carbon dioxide as blood flows through the lungs. It has been shown also that, if a little acid, such as lactic or butyric, is injected into the blood as it flows toward the brain, respiration will be increased in the same manner as when carbon dioxide is added to the arterial blood. This leads to the conclusion that the respiratory center is sensitive to changes in hydrogen ion concentration in the arterial blood, and that carbon dioxide affects the center only in so far as it influences hydrogen ion concentration.

The question arises whether carbon dioxide has any specific action on the respiratory center, or acts solely by virtue of its effect on the hydrogen ion concentration of the blood. An addition of carbon dioxide to blood apparently causes greater effect upon respiration than do acids. This has been interpreted as meaning that carbon dioxide acts as a respiratory stimulus, not solely because it dissolves in the blood to form carbonic acid and thus gives rise to hydrogen ions, but because it has some specific action apart from this property.

Considering that one of the purposes of breathing is to satisfy the oxygen needs of the body, it is to be expected that oxygen deficiency will act as a potent respiratory stimulus. While the respiratory center will respond to a lack of oxygen, it is not so sensitive to this lack as it is to an increase of carbon dioxide. Oxygen content must drop to 13 per cent before a noticeable effect is obtained.

The response mechanism of the respiratory system to lack of oxygen in the blood became clear after the discovery by Heymans of the function of the carotid and aortic bodies. Near the carotid arteries and the aorta, there are small, round bodies having a rich network of capillaries. They are stimulated by a lack of oxygen and, to a lesser degree, by an increase of carbon dioxide. They are called chemoreceptors. When the tension of oxygen in the blood is lowered, the carotid and aortic bodies are stimulated and send impulses to the respiratory center through the glossopharyngeal and vagus nerves, respectively. Under normal conditions, chemoreceptors are not of great importance.

The temperature of the blood also influences respiration: an increase of body temperature in a normal person, caused by exposure to an unduly high environmental temperature, results in an augmentation of breathing in excess of that which would be expected from the increased metabolism caused by the high body temperature. In bodily activity body temperature frequently rises and must, therefore, be an additional cause of increased ventilation of the lungs.

That carbon dioxide is not the sole stimulus acting on the respira-

tory center was demonstrated by Barcroft and Margaria,[41] who found that a moderate amount of physical exercise produced a greater frequency of respiration and a greater total ventilation than could be induced by breathing the highest percentage of carbon dioxide that could be tolerated. The maximal minute ventilation produced by the highest concentration of carbon dioxide which could be breathed for a quarter of an hour was about 60 liters, whereas a volume of 200 liters has been recorded during exercise.

On the other hand, it has been found that during muscular exercise the carbon dioxide content of the alveolar air may decrease, whereas lung ventilation will increase.[138] The following figures clearly illustrate this fact.

	Rate of Walking in Miles per Hour	
	4	*5*
CO_2 tension in the alveolar air, mm. of Hg	45.7	43.5
Lung ventilation, liters per minute	37.3	60.9

This experiment indicates that carbon dioxide was not responsible for the increase in lung ventilation when the intensity of muscular exercise was increased.

One may raise the question whether the increase in lung ventilation during muscular work may be caused by a rise in hydrogen ion concentration of the blood, brought about by the production of lactic acid. Dill and collaborators have shown experimentally that some men are able to maintain a metabolic rate during work ten times higher than during rest, without any appreciable increase in the lactic acid content of the blood. Thus, a rise in hydrogen ion concentration cannot be responsible for the rise in lung ventilation during muscular work which is not strenuous for the individual.

CONTROL OF RESPIRATION THROUGH REFLEXES FROM WORKING MUSCLES

Harrison and his associates[211] have shown that, even if a tourniquet is placed around the arm, hand movements will cause an increase in lung ventilation. Since the blood circulation has been stopped, stimulating agents cannot reach the respiratory center by way of the blood. Therefore stimuli must reach the respiratory center through the nerves. If a leg of an anesthetized animal is practically severed from the rest of the body, leaving only the sciatic nerve intact, passive movements of the severed limb will cause an increase in lung ventilation.

Experiments on man have shown that passive movements of one leg in the knee joint at a rate of 100 times per minute will increase respiratory minute-volume 40 per cent, although the blood circulation in the leg is stopped by a tourniquet.[103] Dixon and co-workers[136] have shown that some passive movements affect respiration more than others. Torso movements, for example, produce hyperventilation in excess of metabolic demand. This probably explains hyperventilation in pilots who fly high-velocity, low-level aircraft and are subjected to much jolting.

Asmussen, Christensen, and Nielsen[14] have demonstrated the effect of reflexes from actively contracting muscles upon the respiratory center. Their experimental subjects rode on a bicycle ergometer, and the amounts of oxygen consumed, lung ventilation, and rate and depth of respiration were recorded. After those data were collected, the subjects continued to ride with pneumatic cuffs placed around each thigh to stop the blood circulation. The comparison of data thus obtained showed that, although shutting the blood from the legs lowered the oxygen consumption by 20 to 50 per cent, lung ventilation remained the same. Thus it is obvious that, during physical work, muscles exert reflex control upon the respiratory center.

These findings were substantiated again in 1964.[17] That this stimulation is of a reflex nature and does not originate in the motor area of the cortex was also demonstrated by Asmussen and his associates.[18] Their subjects performed a certain amount of work on a special leg ergograph (Fig. 64) and the lung ventilation and the tension of carbon dioxide in the alveolar air were determined. After these calculations had been made, work was continued at the same rate as before, but contractions of the leg muscles were induced artificially by stimulating them rhythmically with electric current. The indifferent electrode was placed on the subject's back, and the active electrodes were placed on

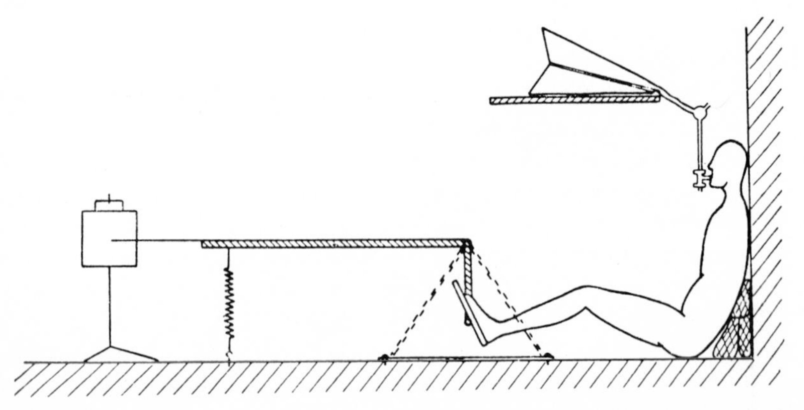

Figure 64. Leg ergograph. Subject's feet are strapped to two pedals. Work is done when legs are alternately flexed and extended. The writing point records movements on the kymograph. Expired air is collected in a Douglas bag. (Asmussen et al.: Acta Physiol. Scandinav. *6*, 1943.)

the calves and on the anterior surfaces of the thighs. Closing the circuit caused stretching of the legs. The experiment demonstrated that in both kinds of exercise—active and passive—the rate of lung ventilation was the same. Since, in the electrically induced work, motor stimuli did not originate in the cortex, it must be concluded that maintenance of lung ventilation during muscular work does not depend on cortical control of the respiratory center.

Experiments on decerebrate dogs showed that, during electrically induced work of the muscles, respiration was not any lower than in a dog with an intact cortex.[279]

As has been indicated in this book, it is impossible to explain all the changes in respiration by just one factor: carbon dioxide, or the hydrogen ion concentration of the blood. Therefore a number of physiologists have expressed the belief that not a single factor but the interaction of many factors determines the behavior of respiration.

A suitable quantitative explanation of the interdependence of various stimulants for the respiratory center was lacking until Gray[195] proposed his multiple factor theory of the control of respiratory ventilation: *Although a number of factors exert independent effects upon respiratory ventilation, the total effect is determined by the algebraic sum of the partial effects of the separate agents.* This theory helps in understanding the mechanism of respiratory changes during physical activity.

It has been shown that, during severe exercise, lactic acid rapidly accumulates in the body. In intense muscular effort it may be set free at the rate of about 3 gm. per second. While lactic acid is accumulating in the blood, a certain amount of carbon dioxide will be displaced from its combination with bicarbonate and driven off from the blood. The increase in H ion concentration, however, more than offsets the inhibitory effect of the decreased carbon dioxide tension upon the respiratory ventilation, and the sum total of the partial effects increases. The total lung ventilation increases. This explanation clarifies the observation made by Douglas and Haldane[138] in the course of their experiments with walking. When there was an increase in speed of walking, from 4 to 5 miles per hour, lung ventilation rose, in spite of a drop in carbon dioxide tension in the alveolar air, because of a possible rise in the lactic acid content of the blood and an increased intensity of reflexes from the muscles.

"SECOND WIND"

During violent exercise, such as running or rowing, a feeling of distress frequently develops which is associated with considerable breathlessness. If, however, the exercise is continued, this distress disappears and may be replaced by a sense of great relief. When this change occurs, we say we have our "second wind."

The symptoms that precede "second wind" are varied. There may be a look of distress on the face, often thought of as an anxious expression. Breathing is rapid and comparatively shallow; the pulse is rapid and fluttering or irregular. The person may feel a sense of constriction around the chest; his head may throb and "swim"; but outstanding among the symptoms is a feeling of breathlessness. Muscle pains sometimes occur. The minute-volume of breathing and the percentage of carbon dioxide in the alveolar air are higher before than after "second wind" has occurred. With the onset of "second wind," breathlessness and discomfort sometimes disappear suddenly. The look of distress vanishes; the head becomes clear; and the muscles seem to act with renewed vigor. Breathing becomes easier; the minute-volume is usually reduced, the frequency decreased, and depth increased. Even the heart action may change, its beat becoming slower and more regular. Some observers find that sweating also accompanies "second wind." The man can now continue his exertion with comparative comfort. There are individual differences in the way adjustments are made: in some persons the sensation of relief coming with "second wind" is definite, while in others it may be so indefinite as to pass unrecognized.

It has been observed that the sensation of "second wind" comes earlier during exercise of great intensity than during exercise of low intensity. A higher outside temperature also hastens the appearance of this phenomenon. When the subject is lightly clothed and works in a cold room, no "second wind" may be recognized. Use of fans blowing on the subject during work may postpone the onset of "second wind."

Widimsky and co-workers[540] have found that if two sessions of identical exercises are given, the pulmonary vascular resistance during the second session is markedly lower than during the first, because of a possible opening or dilation of lung vessels. This phenomenon may be considered a contributing factor in "second wind."

The presence of "second wind," whether it comes on dramatically or not, indicates that the organism has been wholly mobilized to meet the needs of greater physical activity. This mobilization includes better circulation through working muscles, higher efficiency of chemical processes involved in muscular contraction, better buffering of acids in the blood, and finally an improved peripheral circulation, which aids in heat dissipation. These changes tend chiefly to restore the acid-base balance of the body. One may just as well add that in this mobilization there are adjustments in other organs, including the endocrine glands, because such undoubtedly do take place; but to use more words would only serve to cover up our ignorance.

Why, in some athletes, the onset is dramatic and, in others, is gradual and unnoticeable, we do not know. On the other hand, in some people no adequate adjustment ever occurs, and they have to slow down, quit, or collapse.

EFFECT OF TRAINING ON RESPIRATION

Training brings about well-defined changes in the respiratory mechanism and its functioning. The expansion of the chest is increased; the rate of breathing is slowed, and its depth is augmented. In sedentary individuals, large portions of the lungs may be physiologically closed off from the air inhaled; while, with training, the entire lung volume easily becomes accessible, exposing the blood to oxygen over as much as 100 square meters of surface, instead of a fraction thereof.

In comparing the types of breathing of 200 men, Hörnicke[252] found that, in inefficient, untrained subjects, the diaphragm moved little, and respiratory frequency was eighteen to twenty breaths per minute. Those trained in sports had, on the other hand, a deep diaphragmatic breathing, with six to eight breaths per minute. Lung capacity cannot always be judged by external measurements, since they take no account of the movement of the diaphragm. With limited chest movement and good breathing, the diaphragm must move freely.

Youth is the time for the development of the chest. Exercise will result in enlargement of the chest during the period of growth, but will not have much influence on size thereafter.

A group of sixty-eight adolescent boys who took regular exercise gained 130 cc. in vital capacity in four months, while a group of fifty boys who did not exercise had a gain of only 20 cc.[467] College students who took part in physical activities during their college course gained 625 cc., while their sedentary colleagues gained only 295 cc. Occasionally, however, even special exercises fail to influence vital capacity. In a group of twenty-two freshman girls in college, eleven showed no improvement, while the others improved by 210 to 600 cc.

It should be noted that Gordon, Levine, and Wilmaers[190] found that marathon runners had only an average vital capacity, which indicates that prolonged vigorous training does not necessarily increase the breathing space of the lungs. They found no important relationship between the vital capacity of the lungs and the order in which the runners finished a 25-mile race.

The trained man breathes more economically than the untrained. For the same task, he needs less air because he can utilize a greater portion of its oxygen than can the untrained one. This difference becomes pronounced when heavy loads of work are carried. The effect of training shows itself so gradually that only after weeks may a slight evidence be observed. The maximum, however, may be reached after seven weeks of training.

In one experiment on two subjects, the minute-volume of pulmonary ventilation decreased by 15 to 23.5 per cent, while absorption of oxygen increased by 12.0 to 18.5 per cent.[462] No wonder that the same work was performed more easily after training. When subjects discontinued their training, they were practically back to a pretraining condition in four weeks.

The greater the learning factor involved in exercise, the greater the reduction in minute-volume after training. Therefore the smallest change will be observed in walking, but there will be considerable change in ice skating and, especially, in swimming. For example, in the crawl stroke at 2.5 feet per second, a trained swimmer may have a minute-volume five times smaller than that of a beginner.

It seems clear, therefore, that the trained man ventilates his lungs, both during rest (although this difference may disappear under basal conditions) and in exercise, more economically than does the untrained. This is particularly advantageous during exercise, for exertion then causes an increased utilization of oxygen without an exorbitant increase in the minute-volume of breathing.

After exertion, the untrained man will often breathe in the "Cheyne-Stokes" manner. Individual breaths wane and then increase in rhythmic fashion. This condition, however, is absent in physically well-trained men.

Besides causing more efficient respiration, training increases the endurance of the respiratory muscles. A feeling of discomfort and tightness around the chest, experienced by the untrained, is greatly reduced and may be absent in the trained person.

It has been found that champion swimmers have a higher pulmonary diffusion capacity during the steady state than normal subjects of the same age or long-distance runners.[386] Whether this is due to training or to the innate characteristics of the men is impossible to tell.

RESPIRATORY GYMNASTICS

No impressive objective evidence has ever been presented that a normal person needs special respiratory gymnastics. Whenever a group of muscles is engaged in activity, the respiratory organs commence their normal and useful gymnastics. An indiscriminate use of artificial respiratory gymnastics causes a disturbance in circulation and respiration, as evidenced by a sense of giddiness after several consecutive deep inhalations. The only indication for this type of exercise is when it is needed for corrective or therapeutic purposes. If, for example, an increase in the vital capacity of the lungs is desired, it may be achieved through respiratory gymnastics. A bedridden patient possibly may profit by respiratory gymnastics. No medical attendant should, however, attempt to prevent blood stasis in the lungs by respiratory gymnastics instead of by changing the position of the patient.

QUESTIONS

1. Name the parts into which the process of respiration may be subdivided.
2. Define lung ventilation.
3. What is "dead space"?
4. How large is the surface area of the capillaries in the lungs?
5. How is minute-volume of respiration determined?
6. How large is the minute-volume of pulmonary ventilation at rest and during work?
7. What is the relation between pulmonary ventilation and oxygen consumption? This relation is used as an index. Name this index.
8. How much oxygen is used at rest and during a maximum exertion?
9. Describe the changes in respiration at the start of a sprint race.
10. How large may tidal air volume become during a great exertion?
11. What is the average respiratory rate at rest?
12. What is the cost of respiration in terms of O_2 consumed?
13. Is there any relation between vital capacity and athletic performance?
14. Is there any relation between vital capacity and body size?
15. Is there any relation between vital capacity and ability to hold the breath?
16. Explain what causes a "stitch" in the side.
17. What per cent of O_2 is present in the alveolar air at rest? Does percentage increase or decrease when a runner feels respiratory distress?
18. Which is more efficient, nasal or mouth breathing? Why?
19. Discuss the mechanism of control of respiration in general.
20. What is the most important factor controlling respiration during work?
21. Does the cerebral cortex control respiration during work?
22. Discuss "second wind."
23. What is the effect of training on respiration?
24. Is there any value in respiratory gymnastics?

Chapter Ten

BLOOD COMPOSITION AND TRANSPORTATION OF GASES

The blood may be looked upon as a fluid tissue. It has both a definite structure and particular duties to perform, among which are those of a carrier and a buffer. It carries nutriments to and wastes away from all parts of the body. Of these substances, oxygen, carbon dioxide, lactic acid, and glucose are of particular interest in this discussion.

As a buffer, the blood serves to prevent marked changes in the chemical reactions of the tissues. In this connection it is worth while to note that a sudden increase in the acidity of the blood, no greater than the difference between distilled water and ordinary rain water, would be sufficient to cause death. The blood can take up large amounts of acid or alkali without itself becoming more acid or alkaline. Its ability to maintain a fairly uniform hydrogen ion (acid) content is brought about by the presence of so-called "chemical buffers" found in both the blood plasma and the corpuscles.

BUFFER SUBSTANCES IN THE BLOOD

There are a number of substances in the blood which are all salts of weak acids. They constitute the so-called "buffer" substances, and are so named because they prevent a strong acid from at once neutralizing a weakly alkaline solution. The "buffers" form compounds which ionize or dissociate to a slight extent and, therefore, set free few H ions or OH ions.

It has been found that weak acids, like carbonic (H_2CO_3) and phosphoric (H_3PO_4), have the remarkable property of maintaining the reaction tolerably constant when they are present in a solution which also contains an excess of their salts. The H ion concentration of such a buffer solution is proportional to the ratio: $\frac{\text{Free acid}}{\text{Free salt}}$. Since large amounts of sodium and potassium bicarbonate are actually present in blood (enough to yield from 50 to 65 cc. of carbon dioxide per 100 cc. of blood), it is clear that we have in the ratio of carbonic acid to sodium bicarbonate a means which serves to damp down or buffer the effect on the H ion concentration when acids are added. The blood corpuscles contain a considerable amount of potassium acid phosphate (KH_2PO_4), which enables them to aid in maintaining the neutrality of the blood.

It has been known for a long time that proteins act as both acid and alkaline buffers, so that it is difficult to ascertain sharply, by means of ordinary indicators, the neutral point in a solution containing proteins. Proteins occur in both the plasma and corpuscles of the blood. In the red corpuscles the hemoglobin is outstanding in this regard.

Since carbonic acid is produced in large amounts in the normal process of metabolism, it stands at the head of the list of the acids that must be buffered. Variation in the excretion of carbon dioxide by the lungs is the most important mechanism for controlling temporary changes in the H ion content of the blood. Ordinarily the ratio of $\frac{H_2CO_3}{NaHCO_3} = \frac{1}{20}$. The breathing of a healthy man usually maintains this ratio in the blood. Hence, if lactic acid is poured into the blood stream by the working muscles, it will be neutralized by the sodium bicarbonate. This will free some carbon dioxide and the ratio $\frac{H_2CO_3}{NaHCO_3}$ will become greater and the chemical equilibrium will be upset. To bring it back to normal the excess carbon dioxide must be "blown off." Thus respiration becomes increased and brings the ratio of carbonic acid to sodium bicarbonate back to $\frac{1}{20}$.

In normal resting individuals the chemical reactions of the arterial and venous blood are almost identical. The blood has a slightly alkaline reaction which may be a little greater in the arteries. This reaction is kept nearly constant even under the most variable conditions of health. The normal *p*H of the blood is 7.36. During exercise it becomes more acid, 7.05. On the other hand, by deep and rapid breathing, it is possible to "blow off" some of the carbon dioxide from the blood and cause a shift of action to alkalinity (*p*H 7.85).

The preponderance of experimental evidence indicates that training does not change the alkaline reserve in men and animals.[431, 432, 492]

ORGANS RESPONSIBLE FOR THE REGULATION OF THE ACID-BASE BALANCE

During health the regulation of the reaction of the blood is carried out with great delicacy by at least three organs of the body: the lungs, the kidneys and the intestines. In this regulation the lungs deal rapidly with changes in the H ion content which are caused by varying production of carbon dioxide and of lactic acid during muscular exertion.

The kidneys respond to the minutest variation in blood alkalinity by secreting a more acid or a more alkaline urine. They also serve to keep normal the proportion of sodium, potassium and other crystalloid substances in the blood.

The intestines eliminate some of the phosphoric acid and thereby regulate it when necessary.

THE TRANSPORT OF OXYGEN

Only 0.22 to 0.7 per cent of oxygen is carried in solution in the blood plasma. The remainder is in chemical combination with the hemoglobin of the red blood corpuscles. The amount of hemoglobin in healthy men and women is 14.7 and 13.7 gm. per 100 cc. of blood, respectively. Since it takes 1.34 cc. of oxygen to saturate 1 gm. of hemoglobin, it follows that the arterial blood of a man may carry 19.7 cc. of oxygen per 100 cc. of blood; and that of a woman, 18.4 cc.

Besides the amount of hemoglobin, the oxygen content of the blood depends on several factors: mainly, partial pressure* of oxygen, acidity of the blood, and temperature. The relation between the partial pressure and the oxygen content of the blood may be seen in Figure 65. The series of curves in this figure shows how the amount of oxygen in the blood decreases with a decrease in its partial pressure. This is why one suffers from lack of oxygen at a high altitude. The partial pressure of oxygen in the alveolar air is the highest in the body. In man at rest it is equal to 100 mm. Hg and is sufficient to cause a 95 to 97 per cent saturation of the blood with oxygen. On the other hand, when the blood reaches the tissues where the partial pressure of oxygen is lowest in the body, being not more than 20 mm., oxygen diffuses from the blood and its percentage drops to 14 or less. The reader may recall that the curves in Figure 65 are called dissociation curves, if they are used for the purpose of determining the amount of oxygen dissociating from hemoglobin at different pressures. The same curves may be used

*Partial pressure of a gas in a mixture of gases is proportionate to its volume per cent, and therefore can be found by multiplying the total pressure by the figure denoting volume per cent of the gas in question. It is usually expressed in mm. Hg (mm. of mercury).

to determine how much oxygen can combine with hemoglobin at various partial pressures; for example, at pressures met in altitude flying. Then they may be called association curves.

It may also be observed from Figure 65 that an addition of carbon dioxide affects the shape of the curves. Pure hemoglobin, without any carbon dioxide, is an inefficient oxygen carrier. Although it readily absorbs oxygen from the lungs, it forms such a stable compound that little oxygen (3 to 4 per cent) is given off from the capillaries. The addition of carbon dioxide increases the ease of dissociation by raising the acidity of the blood so that more oxygen can be unloaded from the capillaries. In other words, better utilization of oxygen becomes possible.

Not all the oxygen picked up by the arterial blood in the lungs is unloaded in the tissues. A large part of it is brought back in the venous blood. Thus, at rest, there is about 14 volumes per cent of oxygen in the venous blood. Therefore, the efficiency of blood as a transportation system can be measured, not by its maximum oxygen-carrying capacity, but by its unloading ability. Thus, if the amount of oxygen in 100 cc. of arterial blood is 20 cc. and that in venous blood is 14 cc., the amount of oxygen consumed is 6 cc.

The time allowed for saturation of arterial blood in the lungs and the desaturation in the tissue capillaries is rapid, about 0.7 second at rest and only 0.35 during exercise. During exercise, more capillaries dilate, and the contact between the blood and the active tissue is increased. Rowell and co-workers,[443] however, found that, during exhaustive three-minute runs by excellent long-distance runners, oxygen blood saturation may drop to 85.2 per cent. One of the causes of this undersaturation may be that these athletes have a greatly increased

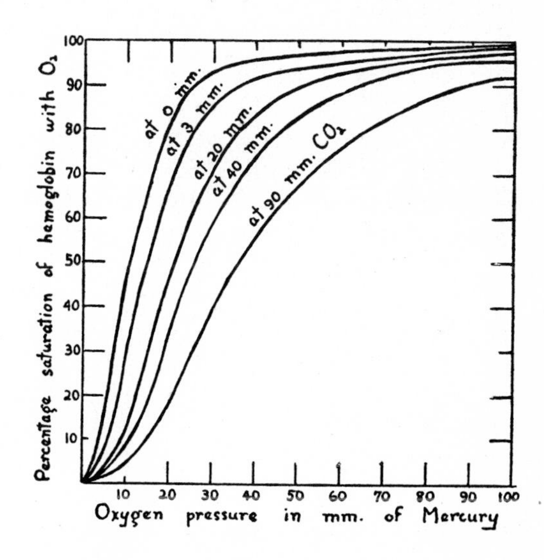

Figure 65. Dissociation or association curves of oxygen and hemoglobin at 37° C. The curves from above downward were obtained from blood exposed to oxygen mixed with different and increasing tensions of carbon dioxide. Note that the addition of small amounts of carbon dioxide moves the whole curve to the right and downward. Consequently, more oxygen dissociates from the hemoblogin; or, if viewed as association curves, less oxygen is taken up by the hemoglobin. (After Barcroft.)

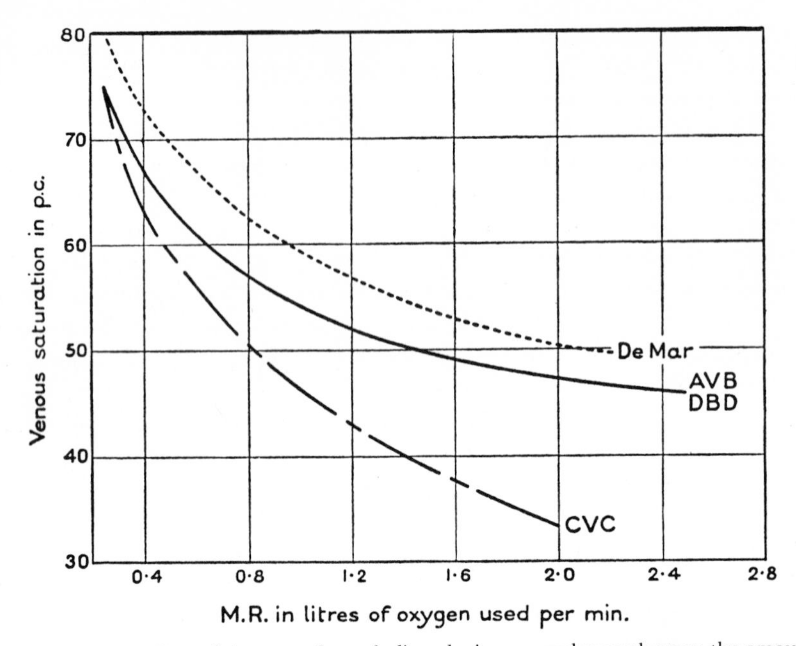

Figure 66. Effect of the rate of metabolism during muscular work upon the amount of oxygen in venous blood. The upper curve shows conditions in a champion runner. The lowest curve shows conditions in an untrained man. In the middle are two men in good physical condition. (Bock, et al., J. Physiol. *66*, 1928.)

cardiac output, which raises velocity of the blood flow through the lungs to such an extent that there is not time for complete saturation.

The same investigators observed a similar phenomenon for sedentary men who received three months of intensive training in running.

During exercise, the rate of oxygen use increases and therefore the oxygen tension in the tissues is lowered below 12 mm. Hg, thus increasing the pressure gradient between the oxygen in the blood and that in the tissues. The effect of this change is an increase in the oxygen utilization. If we suppose that the amount of blood flowing through working muscles increases six times, and that the utilization increases three times, the combined effect of these changes will cause an eighteenfold increase in the oxygen supply to the muscles.

A good illustration of the effect of the intensity of exercise upon the unloading of oxygen from the capillaries may be found in Figure 66. In each case the oxygen content of the venous blood falls off rapidly as the metabolic demand for oxygen rises up to 0.8 or 1 liter of oxygen per minute. In men in good physical condition, the venous saturation falls less rapidly as the load of work is increased. C.V.C., who had never had any experience in any form of physical exercise, showed a far greater reduction in the oxygen content of his venous blood than did DeMar, the marathon runner (Fig. 66). C.V.C.'s curve, however, does not tend to level off. An exercise which was comparatively easy for DeMar could not be carried on by C.V.C. without straining the unloading capacity of his blood.

After exercise, the amount of oxygen in the venous blood may remain low for some time, increase to almost the same level as in the arterial blood, and then return to normal.[351]

OXYGEN PULSE

During muscular work not only is the rate of oxygen utilization increased, but also the amount of blood discharged from the heart with each heart beat is augmented. The combined effect of these two factors results in an increased delivery of oxygen to the tissues. The amount of oxygen taken out of blood per pulse beat is called the *oxygen pulse,* and is obviously determined by dividing the amount of oxygen used during a certain period of time by the number of pulse beats during the same period.

During exercise, the oxygen pulse increases rapidly with acceleration of the heart, and in most cases reaches its maximal value of 11 to 17 cc. at heart rates of 130 to 140 beats per minute. With further acceleration of the heart, the oxygen pulse may even tend to decrease. However, an average oxygen pulse of 23 cc. has been reported during heavy work.

Table 12 gives a summary of the oxygen pulse of four men who worked with loads up to 10,999 foot-pounds on a bicycle ergometer. If the oxygen pulse is a reliable index of stroke volume, the conclusion follows that the stroke volume of W.C.'s heart increased with each step upward in the load of work. L.H.'s heart, on the other hand, reached its limit in output per beat with a load of 8000 foot-pounds; while H.M.'s practically reached its maximum output with a load of 6000 foot-pounds.

In a study of athletic young women, Radloff[419] found a close correspondence between the values for the oxygen pulse of young women during exercise and those of men at similar pulse rates when doing approximately 3600 foot-pounds of work per minute.

TABLE 12. *Effect of Load on Oxygen Pulse in Cubic Centimeters*

Subject	*Rest*	*2000 ft.-lb.*	*4000 ft.-lb.*	*6000 ft.-lb.*	*8000 ft.-lb.*	*10,000 ft.-lb.*
W.C.	3.5	8.5	11.1	12.3	13.4	15.3
L.H.	3.2	7.8	9.9	11.7	12.8	12.4
D.M.	3.0	6.9	8.8	12.4	13.8	14.4
H.M.	3.6	9.4	12.0	12.8	12.9	

THE TEMPERATURE OF THE BLOOD

In active tissue the temperature rises. The ordinary ways of determining body temperature show that it may rise to as high as 102° F. (38.9° C.) during muscular work and, in extraordinary cases, even to 105° F. (40.6° C.). The rise in temperature of muscles depends on the intensity of work, as may be proved by inserting needle thermocouples into the muscles. This change in muscle temperature is much faster than that recorded in the rectum.

The effects of a rise in temperature of the blood are threefold. First, there is a decrease in the affinity of hemoglobin for oxygen, which, as previously shown, results in an easier separation of oxygen from hemoglobin. Second, a rise in temperature, by reducing the absorption by the blood plasma, tends to expel the dissolved gases. Third, the rate of diffusion of gases is also speeded up.

THE TRANSPORT OF CARBON DIOXIDE

Normally, venous blood carries about 55 to 60 volume per cent of carbon dioxide. About 5 per cent is given off in the lungs, and therefore arterial blood still contains about 50 to 55 per cent. Only a small amount of carbon dioxide is present as carbonic acid—not more than 0.1 per cent of the total. The remainder is carried as follows: 5 per cent in solution in the plasma, as much as 20 per cent in combination with hemoglobin, and 35 per cent in the form of bicarbonates in the plasma.

As is true of any gas, diffusion of carbon dioxide is determined by its partial pressure. Thus, in active tissue, where carbon dioxide is produced, its pressure is greatest, varying from 43 to 50 mm. at rest and from 63 to 75 mm. during work. Therefore this gas enters into the capillaries, where the tension is less. Then, when brought into the lungs, where the tension of carbon dioxide is still less, it escapes into the alveolar air, where the pressure is least. Figure 67 illustrates the relationship between the percentage of carbon dioxide in the blood at various pressures. In the same figure one may observe that dissociation of carbon dioxide is facilitated by the presence of oxygen. Thus we see that oxygen and carbon dioxide exert upon each other helpful antagonistic effects. When oxygenated blood comes to the tissues, carbon dioxide, on entering the blood, forces an additional amount of oxygen out. On the other hand, when venous blood comes to the lungs, the entrance of oxygen into the blood forces some carbon dioxide out of the blood.

Under no circumstances does all the carbon dioxide leave the blood. Even after prolonged hyperventilation of the lungs, arterial blood contains a large amount of carbon dioxide, as is indicated by the

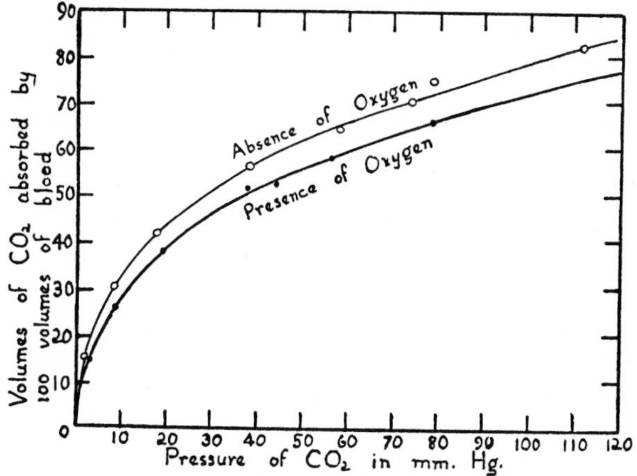

Figure 67. Dissociation or association curves of carbon dioxide and blood at 37° C. Note that the presence of oxygen moves the whole curve to the right and downward, and that the addition of oxygen causes an expulsion of carbon dioxide. (Christiansen, Douglas and Haldane: J. Physiol. *48*, 1914.)

partial pressure of the gas in the alveolar air. Even though the tension may be reduced from the normal 40 mm. to 15 mm. Hg, the blood still contains about 30 per cent carbon dioxide.

One may expect that there would be an increase in the carbon dioxide content of the blood during exercise because more carbon dioxide is produced by working muscles. As a matter of fact, the amount of carbon dioxide in the blood shows, if anything, a tendency to decrease during exercise. If the exercise is moderate and a steady state is attained, there will be no change in the carbon dioxide content. Whenever the intensity of the activity becomes so high that the supply of the oxygen is inadequate, excess lactic acid is produced in the muscles and escapes into the blood. Being stronger than carbonic acid, lactic acid combines with sodium bicarbonate, forcing out carbon dioxide, which is eventually blown off in the lungs, lowering the percentage of this gas the blood can hold. At this moment a temporary rise in the alveolar carbon dioxide may be observed, but since the carbon dioxide content of the arterial blood also decreases, and more than balances the drop in the venous blood, the total result is an increased ability of the blood to unload carbon dioxide into the lungs. Since the rate of blood flow is also greatly increased during intense activity, the excess of carbon dioxide in working muscles will be eliminated at a highly increased rate.

For the same head of pressure, carbon dioxide diffuses twenty or thirty times more rapidly than oxygen. For this reason the pressure of this gas in the alveolar air and in the arterial blood is almost always equal, the difference lying within 1 mm. Hg in most cases.

LACTIC ACID

As has been stated, lactic acid begins to appear in the blood whenever the supply of oxygen is inadequate. In many people engaged in physical work, this inadequacy occurs when the oxygen requirement is about 2000 cc. per minute, which corresponds to work of 6000 foot-pounds per minute.

Accumulation of lactic acid therefore depends on the relative intensity of exercise. A. V. Hill[238] has stated that as much as 3 gm. of lactic acid may be produced per second. The limit of tolerance is approximately 130 gm., which means that, theoretically, one can run for forty-three seconds at top speed without breathing. This calculation demonstrates why the 440-yard run is considered the most difficult sprint.

After vigorous exercise the amount of lactic acid in the blood may be considerably higher than before the exercise. A figure as high as 300 mg. per 100 cc. has been reported.[113] The normal content of lactic acid in the blood is about 10 mg. per 100 cc.

When strenuous exercise is discontinued, lactic acid continues to escape from the muscles into the blood for some time. Of special interest is the period of two to eight minutes immediately after strenuous exercise. During this period the high level of lactic acid in the

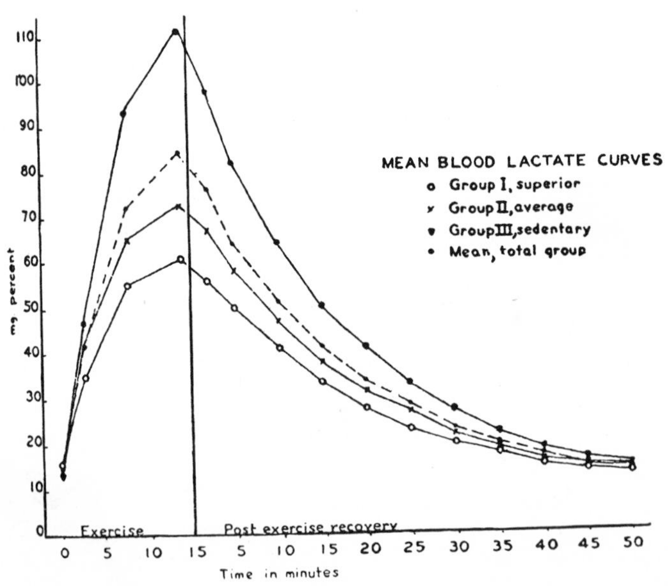

Figure 68. Relation between physical fitness and the postexercise lactate content of the blood. Men in best physical condition had the lowest amount of lactate in the blood. (Crescitelli and Taylor: Am. J. Physiol. *141*, 1944.)

blood remains unchanged, and then begins to decline, reaching a pre-exercise level in thirty to ninety minutes, depending on the intensity of the exercise. The presence of this initial "lag" in the reduction of lactic acid in the blood has been taken as evidence of an alactacid debt; yet it may also be interpreted as evidence that during that time the rates at which lactic acid enters and is removed from the blood are balanced.

Although the amount of lactate in the blood after identical types and amounts of work varies even in the same individual, this level is related to the degree of physical fitness. Figure 68 shows the relations between degrees of physical fitness and levels of blood lactate, and Figure 69 shows how training causes a decline in the level of blood lactate for a standard amount of work.

The exact fate of lactic acid in the blood (present in the form of sodium lactate) is a matter of speculation. While the main source of lactic acid is the muscle, the organ chiefly concerned with its removal is the liver, in which it is transformed into glycogen and then, as needed, is sent back as blood sugar to the muscle. Even during the period of exertion, whatever muscles in the body are inactive may be removing lactate from the blood stream. A considerable amount of lactate disappears from the blood during its passage through the inactive muscle and the heart. It is worthy of note that there is a higher carbon

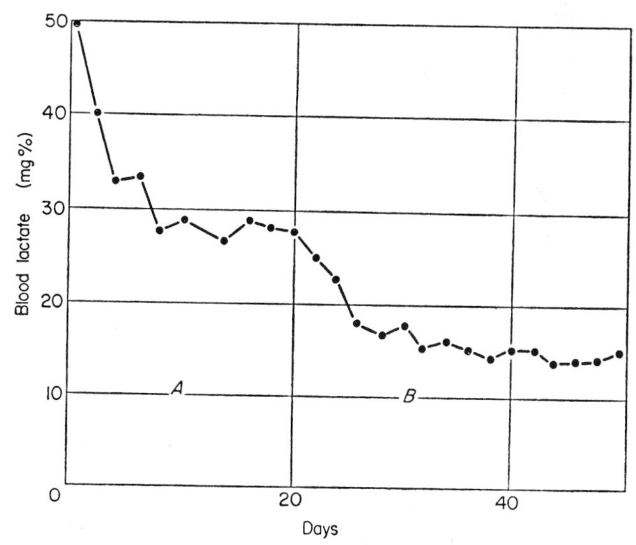

Figure 69. Progressive decrease of blood lactate for a standard amount of exercise: running on a treadmill at 7 m.p.h. for 10 minutes. During the first 20 days (Section A) training consisted of running daily on the treadmill for 20 minutes at 7 m.p.h. A steady level of blood lactic acid is reached around 28 mg. per cent. During the following 30 days, training is increased to running at 8.5 m.p.h. for 15 minutes daily (Section B). Blood lactic acid decreases further, and a new steady level is reached around 15 mg. per cent after the standard test. (Brouha, L.: Physiology in Industry. Pergamon Press, 1960.)

dioxide-combining capacity in the venous blood than in the arterial blood of the resting muscle. This increase is indicative of a return of alkali to the blood as it passes through the resting muscle. There is, nevertheless, a limit to the capacity of the resting muscle for reclaiming lactate from the blood stream, and, after a while, samples of blood obtained from working and idle muscles are practically identical in their alkaline reserve and oxygen capacity.[35]

As the reconversion process goes on, the base that has been used to neutralize the lactic acid is again combined with carbonic acid. Thus the bicarbonate or the alkaline reserve of the blood is restored to normal.

When exercise is sufficiently intensive, some of the lactic acid that enters the blood stream is eliminated by the kidneys. Ordinarily the amount that escapes through the urine is small. Sodium lactate is found in the urine for thirty to fifty minutes after vigorous exercise.

THE RED BLOOD CORPUSCLES

The number of red blood corpuscles is definitely affected by exercise. Even after a short bout of exercise, such as a 220-yard run, the number of red blood corpuscles increases. This increase depends on the load and duration of the exercise. The average number of red blood corpuscles in man at rest is about 5,000,000 per cu. mm. After exercise it may be 5,200,000 to 6,180,000, although a rise of more than 10 per cent is seldom observed. The increase in the number of red corpuscles for a given exertion is not constant, but seems to be modified by previous activity, stage of digestion and so on. The increase is of short duration. Within a few minutes after exercise, the number of corpuscles begins to diminish, and, within half an hour to two hours, will return to the pre-exercise level.

This increase in red corpuscles during exercise is regarded as a compensatory adjustment whereby the tissues are more adequately supplied with oxygen.

The Mechanism of Increase

The increase in the number of red corpuscles depends on two factors. There is a reserve or dormant supply of red corpuscles stored away, mainly in the spleen and accessory spleens. When there is a demand for a greater oxygen supply this reserve is put into circulation. The other factor consists of the loss of water from the blood during muscular work. This reduces the total volume of the blood and increases the concentration of the red blood corpuscles. Both of these factors have been experimentally substantiated.[42]

It has been estimated that in man the spleen can expel from 110 to

258 cc. of blood into the general circulation. The blood in the spleen is more concentrated and contains as much as 49 per cent more red blood corpuscles than normal blood. Therefore, in this respect, 258 cc. of blood from the spleen are equal to about 384 cc. of ordinary blood.[115]

Dill and his co-workers[132] tested two splenectomized subjects. The subjects had an average increase of 4 per cent in the number of red blood corpuscles induced by exercise. In these men the increase in concentration of red blood cells corresponded to an increase in concentration of blood serum protein and, therefore, could be explained on the basis of a loss of water from the blood.

Loss of water by the blood during exercise has been observed by a number of investigators. Keys and Taylor[313] considered the fluid exchanges through capillary walls in exercise and found that, while fluid left the vessels, reserve stores of red cells entered the active circulation at the same time. On the other hand, an hour of complete muscular inactivity in a recumbent position resulted in a significant decrease in the red cell count. These changes indicate the withdrawal of red cells into storage.

Red Corpuscles During Protracted Exercise

In more protracted exercise the blood picture may change. The increase in the number of red corpuscles per cubic millimeter of blood becomes gradually less pronounced as exercise is prolonged. This change depends on an increase in blood plasma and the destruction of the red blood corpuscles.[72] Clear evidence of injury to the red corpuscles was found by presence of dissolved hemoglobin and hematin in the blood plasma and urine of men after strenuous marching.[171]

Broun[72] found that, when dogs accustomed to an inactive life were vigorously exercised for a day or two, the destruction of corpuscles ranged from 12 to 30 per cent of the total mass. Recovery was slow. One animal completely restored its corpuscular volume during the first week after strenuous work, while another required as much as three weeks to make good the loss.

For some time after excessive destruction of corpuscles, an unusually large number of reticulated red cells appear in the circulation. These reticulated cells are youthful corpuscles and can, therefore, be used as a measure of the degree of activity of the red bone marrow.

Training and Red Corpuscles

There is no agreement about the effect of training upon the number of red blood corpuscles. Reports are conflicting. Increase, decrease, and no change have been reported.[461] Olympic athletes had an approximately normal corpuscle count, but a hemoglobin content below normal.[492]

Although additional investigation of the relationship between the number of the red cells and training seems to be in order, the student should not be greatly disturbed by the conflicting results. Even if training did cause an increase in the *absolute* number of red cells in the blood, it would not guarantee that an investigator would find an increase in the red count in a cu. mm. of blood while the subject is *at rest.* It is quite possible that the excess number of red cells may be diverted from the circulation and stored somewhere. It is also possible that the subjects have not recovered yet from the preceding vigorous activity, during which a large number of corpuscles had been destroyed.

There seems to be no disagreement among investigators that the bone marrow, under the influence of training, becomes redder, indicating an increased blood-forming activity. Therefore, we still may conclude that the rate of red cell production during training is increased.

Thus, the effect of training upon red bone marrow may lead to a higher degree of physical fitness of the individual. In a sedentary man, prolonged strenuous muscular effort results in an excessive destruction of red corpuscles, and the loss cannot immediately be made good. Consequently, some degree of anemia results for a period of several days to two or three weeks. Regular exercise, or a period of physical training, so develops the red marrow that any ordinary destruction of corpuscles is quickly made good during or soon after exercise, and the person is ready again for strenuous work.

It must be noted here that a blood transfusion given to a man resulted in lower pulmonary ventilation and pulse rate during work. This was especially noticed when the oxygen content of the inspired air was decreased to 14 per cent.[251]

Sedimentation Rate of the Red Cells

When a sample of blood is placed in a vertical glass tube and a chemical is added to prevent coagulation, one can observe that the red blood corpuscles will gradually begin to settle down, leaving at the top of the tube a clear layer of plasma not containing red blood corpuscles. By measuring the height of this clear plasma, one can measure the rate of sedimentation. This is referred to as the *erythrocyte sedimentation rate*, and is usually expressed in millimeters per hour, if one uses the Westergren method.

Hannisdahl[208] experimented on well subjects, who rode a bicycle ergometer and performed work of various intensities (the maximum was twenty times that of resting). Although the rectal temperature sometimes rose to 39.3° C., and the pulse rate to 180 beats per minute, there was no increase in the sedimentation rate during three hours of observation.

Black and Karpovich[57] used a five-minute Harvard step-up test (strenuous work) on thirty-nine convalescent patients, three of whom had an elevated sedimentation rate, and found that a statistically significant increase in the sedimentation rate developed five hours after the test. Twenty-four hours after the test the rate returned to the pre-exercise value. This observation may be interpreted to mean either that the exercise did not do any harm to the subjects, or that the sedimentation rate is not an index of possible harm caused by exercise. Since the clinical observation on patients showed no ill effects, it was concluded that the exercise was within the limits of the patients' physical fitness, and caused no harm.

CHANGES IN THE COUNT OF WHITE BLOOD CORPUSCLES AFTER MUSCULAR ACTIVITY

Muscular activity exerts a clear-cut influence upon the distribution of several kinds of white corpuscles. It should be recalled that the white corpuscles are differentiated according to their histological characteristics, as shown by stains. Polymorphonuclear leukocytes, so called because of their finely granular lobular nuclei, which vary in shape, are most abundant (70 per cent). These are active phagocytes. If they stain with neutral dyes, they are known as neutrophils; if they stain with basic dyes, such as methylene blue, they are known as basophils; and if they stain with acid dyes, such as eosin, they are known as eosinophils. Another group of white corpuscles, known as lymphocytes, is distinguished by a large round nucleus almost as large as the corpuscle itself. They are of two varieties, large and small, and normally form about 25 per cent of the total white corpuscular count.

Egoroff[153] and his co-workers, who probably studied the effect of exercise upon "the white count" more than any other group of investigators, subdivided blood changes into three phases on the basis of the relative increase in percentage of the various white blood cells: (1) the *lymphocytic phase*, characterized by an increase in lymphocytes, up to 55 per cent, (2) the *neutrophilic phase*, characterized by an increase in neutrophils, sometimes to 78 per cent, and (3) *intoxication phase*, during which the percentage of neutrophils may rise to 90, and lymphocytes may drop to 5. Egoroff suggested that this classification could be used to determine the degree of difficulty of an activity and, at the same time, the degree of physical fitness of the individual.

Numerous investigators have noted that, after short but strenuous exercise, one may observe the first lymphocytic phase. After prolonged exercise comes the neutrophilic phase. There is no agreement as to the intoxication phase, because, even after a marathon run, a man may be still in the neutrophilic phase.

The number of white blood corpuscles in 1 cu. mm. may rise to 27,000 from the normal 5000 to 7000. This rise requires some time. For instance, although after a 400-meter race a runner is considerably exhausted, the number of white corpuscles in his blood will be less than in a marathon runner who may not be exhausted.

The lymphocytic phase, observed after a sprint race, changes into the neutrophilic phase during recovery. The recovery period even after short but intensive exercise is rather long. It may take two hours for the blood count to come back to normal. The decrease in the number of white blood corpuscles may be accelerated if the subject lies down.

Although both the mechanism and the significance of exercise leukocytosis are not well understood, there seems to be general agreement that this increase is due to "washing out" of the white blood corpuscles from the storage places (bone marrow, spleen, liver, lungs) caused by the greatly increased circulation. Edwards and Woods[152] demonstrated that the lactic acid content of the blood, the blood sugar, and the blood pressure have no separate relations to exercise leukocytosis. Excitement alone has no effect upon leukocytosis. It has been demonstrated that there was no change in the white blood count of track athletes just a few minutes before an important race. Football players who did not play also had no change in white blood count.[150]

EFFECT OF TRAINING UPON WHITE CELL COUNT

Training seems to have little effect upon the white blood corpuscles. Hawkins[214] studied the effect of collegiate training in football, basketball and track. His findings may be summarized as follows:

1. The number of white blood corpuscles in 1 cu. mm. did not change.

2. The percentage of various white cells did not change, but there was an increase in the number of younger forms of neutrophils and in smaller lymphocytes.

3. After the cessation of training, the blood picture began to change to normal. The change was relatively rapid during the first eight to fifteen days and then slower. The time needed for a complete return to normal varied from ten to sixty-seven days.

BLOOD PLATELETS OR THROMBOCYTES

While the number of blood platelets in a unit volume of blood is changed as a result of muscular activity, nothing is known regarding the cause of the change. The normal number is variously stated as ranging between 200,000 and 700,000 per cu. mm. of blood. Immediately after muscular work there is either no change or a slight decrease in the number; then follows a period of rapid reduction, by 17 to 30 per cent, which may last thirty to sixty minutes. After this comes a period of rapid increase which results in an overproduction. Hence within an hour or two after exertion, the number exceeds the normal by 17 to 25 per cent. Eventually, in the course of several hours, there is a final return to the normal number.[461]

SPECIFIC GRAVITY OF THE BLOOD IN EXERCISE

The specific gravity of the blood in muscular activity has frequently been studied. It has usually been found to vary directly with the red corpuscles; therefore, it has been suggested that, if the specific gravity has been determined and if the normal number of red corpuscles is known, the increase in red corpuscles after a given exercise may be predicted fairly closely. Ordinarily, the specific gravity of the blood increases during exercise; then, after exercise, gradually falls until it is somewhat subnormal; and finally reaches normal. The degree of change is of the order of 0.001 to 0.0025.

EFFECT OF EXERCISE UPON BLOOD SUGAR

Various forms of mild exercise do not produce any significant changes in blood sugar; but, as the intensity of the exertion increases, the sugar content may show a marked rise. Even after violent exercise lasting thirty to forty minutes there is an excess in blood sugar.[76] If physical activity is performed under emotional stress, the increase is greater. The amount of sugar may rise to 244 mg. per 100 cc. of blood.[150] Obviously, samples of blood taken after practice exercises contain less sugar than those taken after contests.

If exertion is prolonged (three hours in one study[76]), a definite drop in blood sugar may be observed. Whether the sugar content of the blood falls below normal will depend upon the amount of carbohydrate stored within the body and the amount of fuel required. Hence various investigators report different findings.

Those who investigated the effect of a mild exertion reported no change. Those who investigated the effects of one hour of gymnastics or boxing in fasting men reported a decrease, no change, and an increase; and, of course, a steady drop was reported during prolonged fatiguing work.[468]

In this connection the classic study of the Boston marathon runners is of special interest. In the 1924 race, a close relation between the blood sugar content after the race and the physical condition of the runners at the finish was observed. While, before the race, their blood sugars ranged between 81 and 108 mg., at the finish, there was a marked difference. The winner had a comparatively normal blood sugar and was in excellent condition. Four of the runners had 50, 49, 47, and 45 mg., respectively. Three of them were completely exhausted, and the fourth was unconscious and had to be carried.

Before a similar race in 1925, the men were placed on a moderately high carbohydrate diet during training. They were advised to eat moderately large amounts of carbohydrates before the race, and were supplied with candy and tea containing sugar during the race. At the end of the race their blood sugar was practically normal and their physical condition far better than after the race of the previous year.

Observations on Olympic runners in 1930 also showed the importance of sugar in endurance activities.[53] An athlete who was exhausted had had only 55 mg. of sugar; he showed a loss of muscular coordination and low blood pressure, and was cyanotic. Runners who finished in fair condition had blood sugars around 100 mg.

PHOSPHATES IN EXERCISE

It has been shown[213] that there is a marked difference in the phosphate content of the blood after a given exercise between the trained and untrained man. After a given exertion the phosphates of the blood in the trained man do not fall so low as in the untrained man. It is suggested that during the progressively increasing activity of a muscle, brought about by training, a larger amount of phosphocreatine is laid down in the muscle.

ATHLETES AS BLOOD DONORS

Karpovich and Millman[300] investigated a number of cases illustrating the unfavorable effect of loss of blood upon athletes engaged, soon after, in strenuous contest. A gymnast and two wrestlers almost collapsed at the end of competitions. A cross country runner who usually finished a race in first or second place, running on the second day after giving a transfusion, came in last, and did not regain his previous endurance until ten days later.

On the other hand, the same investigators[300] have seen cases in which the men did not seem to be affected at all by the loss of one pint of blood. A sprinter and a short distance swimmer each equaled his respective record just a few hours after giving blood for transfusion. Moreover, it has often been observed that, although blood donors may feel somewhat faint, or even dizzy, after the withdrawal of a pint of blood, this condition is usually transient and soon may be succeeded by a state of euphoria.

Karpovich and Millman studied experimentally the effect of the loss of one pint of blood on sprint and endurance exercise. They used five subjects, three of whom had given blood twice. The exercise was a ride on a bicycle ergometer. In all cases but one there was a noticeable drop in endurance lasting ten to eighteen days. The exception was probably caused by an intensive degree of motivation. The subject tried to prove that loss of blood had no effect on him, and beat his record by fifteen seconds, but finished the ride in an utterly exhausted condition. On the following day his endurance was markedly decreased, and it took him about three weeks to regain his endurance (Fig. 70).

Other investigators observed that, 48 to 72 hours after 500 cc. of blood loss, there was no statistically significant lowering of performance on a treadmill.[38] The intensity of exertion, however, was less than in the author's experiments.[300]

In the sprint type of exercise, performance was not affected by loss of blood. The reason for an apparent immunity of sprinters to the ill effects of loss of blood can easily be explained: Sprinting depends mainly on the oxygen debt to be tolerated, and not on the current supply of oxygen.

It may therefore be concluded that an athlete in training should not act as a blood donor unless he is contributing at a time of humanitarian emergency, which is more important than winning a contest.

Schmid,[456A] in Czechoslovakia, observed 50 athletes who donated blood on the average of six times each. Although some of them apparently felt no deleterious effect and resumed training routine, his conclusions were similar to those of Karpovich and Millman.[300]

Howell and Coupe[255] conducted an investigation in which they demonstrated interesting psychological effects. Two groups of men

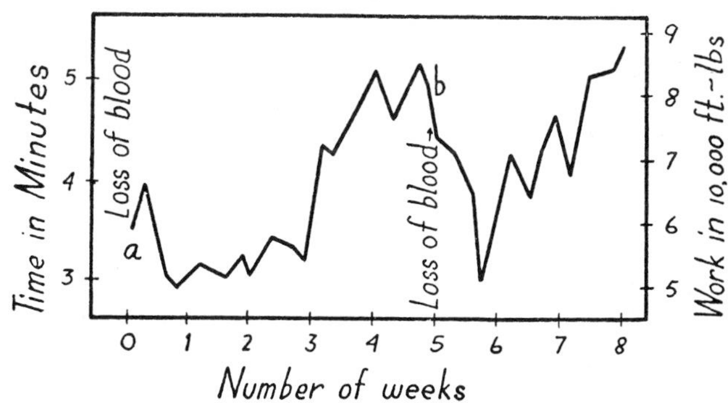

Figure 70. Subject worked on a bicycle ergometer at a rate of 0.507 h.p. At points *a* and *b* subject gave blood transfusions of 500 cc. It took three weeks to recover completely from the effects of the loss of blood. (Karpovich and Millman: Research Quart. *13*, 1942.)

were blindfolded, and 500 cc. of blood was withdrawn from those in the experimental group; no blood was taken from the controls, although a needle was inserted and they were told that blood was being drawn out.

A few days before the "blood donation," subjects were given the Balke treadmill test and divided into two groups equated on the basis of this performance. They were again tested 30 minutes, 24 hours, and 7 days after the "blood donation." Both groups showed a decrease in performance in the 30-minute test, but at the 24-hour test performance was better than before the "donation" and after seven days even better.

In the Balke test, performance is terminated when the pulse reaches a rate of 180 beats per minute; therefore, it is not an all-out test. Yet the test is a very strenuous one, approaching closely an all-out one. This experiment showed the great effect of a psychological factor not used in the experimental design of previous investigators.

QUESTIONS

1. What are "buffers" in the blood? What do they do?
2. Name the organs responsible for the acid-base balance.
3. What is alkaline reserve? How is it affected by training?
4. How many cc. of O_2 are found in 100 cc. of blood?
5. To what degree (per cent) is arterial blood saturated with oxygen at rest? During exercise?
6. Is there any relation between the oxygen-carrying ability of the blood and carbon dioxide content of the blood?
7. Give the percentages of oxygen and carbon dioxide in arterial and venous blood.
8. Define oxygen pulse.
9. How does blood temperature affect transportation of oxygen?
10. How is oxygen transported by the blood?
11. How is carbon dioxide transported by the blood?
12. What is the source of lactic acid in the blood?
13. How is lactic acid eliminated from the blood?
14. How much lactic acid may be produced per second during vigorous exercise?
15. Why is the 440-yard dash the most difficult sprint?
16. Is there any relation between physical fitness and the amount of lactic acid in the blood after a standard exercise?
17. What happens to the red blood corpuscle count during work? Explain the mechanism involved.
18. Why is it not advisable to engage in strenuous and prolonged exercise the day before an athletic contest?
19. What is the effect of training upon the red corpuscles?
20. What is the sedimentation rate of the red blood corpuscles?
21. What is the effect of exercise and training upon white blood corpuscles?
22. What is leukocytosis?
23. Explain the changes in specific gravity of the blood during exercise.
24. Discuss the effect of exercise upon blood sugar.
25. How may a loss of one pint of blood affect athletic performance?

Chapter Eleven

BLOOD CIRCULATION AND THE HEART

The circulatory system exists for the sake of its capillaries, through the walls of which the exchanges of oxygen, carbon dioxide, acids and other materials take place between the blood and the tissues. The heart, in large part, supplies the necessary force to propel the blood; the arteries, by their elastic and muscular tissue, maintain an adequate pressure for the period between the heart beats and thus provide a steady flow through the capillaries; and the veins conduct the blood away from the capillaries.

The cardiac output is the amount of blood pumped by the heart, usually expressed in liters per minute. During exercise it increases to meet the greater demands for oxygen and removal of metabolic end products. The cardiac output depends on the pumping ability of the heart and is, therefore, important relative to general cardiovascular dynamics and exercise capacity.

MEASUREMENT OF THE CARDIAC OUTPUT

The methods employed to determine the cardiac output in man have not been so simple, direct, and exact as those used for measuring respiration. All of them measure, at best, the amount of blood passing through the lungs in a certain period of time.

The Fick Method. We collect the expired air for a minute; it is analyzed later for the oxygen content. Simultaneously we obtain samples of arterial and venous blood, and the A-V difference in oxygen content is found by analysis. Since it is important to have thoroughly mixed venous blood, the sample is obtained from the right auricle by inserting a tube (catheter) through a vein of the right arm.

Suppose that the analysis has shown that 600 cc. of oxygen was consumed in one minute and the A-V oxygen difference was 6 per cent. This means that from each 100 cc. of blood, 6 cc. of oxygen was picked up from the lungs. The amount of blood which would pick up 600 cc. of oxygen can be calculated ($600/6 \times 100 = 10{,}000$). Suppose that the pulse rate during the test was 100 beats/min.; the stroke volume then was $\frac{10{,}000 \text{ cc.}}{100} = 100 \text{ cc.}$

Dilution Method. A certain amount of indicator substance, usually a dye or a radioactive substance, is injected into a vein, and a sample of blood is taken later from an artery. The concentration of the dye is measured, and one can calculate the amount of blood that caused the dilution while the dye was passing through the lungs. Dividing this amount by the number of heart beats, one obtains the stroke volume.

Bloodless Methods. Both of the techniques described above involve laboratory procedures which are very difficult to apply during exercise. Consequently, investigators have turned to "bloodless" methods, which utilize the exchange of gases between the blood and lungs as an index of pumping rate.

Two which are commonly used are the nitrous oxide inhalation[47] and carbon dioxide[167] rebreathing procedures. In the former, nitrous oxide is inhaled and the cardiac output computed on the basis of the rate of nitrous oxide uptake from the alveoli.

In the latter, the rate at which carbon dioxide is being produced is first measured. The arterial concentration is accepted as being equal to the concentration in the air expelled at the very end of a normal expiration. Then, the carbon dioxide in venous blood is determined by rebreathing from a bag containing a mixture of carbon dioxide and oxygen until equilibrium is reached between the incoming blood and the gas in the bag. The cardiac output is then calculated by dividing the carbon dioxide output in milliliters per minute by the difference between the venous and arterial carbon dioxide concentrations in milliliters per minute.

Neither of these procedures is simple. Both require accurate measurement with sensitive instruments and have certain shortcomings. However, their use broadens the application of the study of cardiac output.

INFLUENCE OF POSTURE ON HEART OUTPUT

There are surprising differences in the ease with which persons in good health make circulatory adjustments to changes in posture. Most of us frequently alter our position from one pose to another without realizing that changes in bodily posture are attended by important circulatory changes that are absolutely necessary if the new position is to be maintained with comfort. Apparently healthy persons not infrequently feel real discomfort when they change from a reclining position to an upright posture. The sensations, which are often difficult to define, range from a vague feeling to marked dizziness and sometimes to fainting. These latter reactions appear when the supply of blood to the brain is inadequate.

An immediate factor that causes adjustment by bodily mechanisms, when a change is made from a reclining to a standing position, is the effect of gravity on the circulation. There will be a natural tendency for the blood to stagnate in the lower parts of the body. To avoid this undesirable interference with circulation, the vasomotor mechanism must make such adjustments as will give sufficient tone to the vessels in the abdomen and legs so that blood and lymph will not pool in those regions. The adjustments should confine the blood to a small enough space so that the return of blood will fill the heart sufficiently to assure a normal output per beat. There is no doubt but that, when some unhealthy persons stand, the blood pools in the abdomen to such an extent that its return to the heart is much reduced.

The effect of varying postures on the heart output has been the subject of numerous studies. These studies have shown that the heart output is greatest in recumbency, less in sitting, and least in the standing position.

In man, the average heart output per minute is about 4.2 liters while lying down and 3.9 liters in the standing position. In woman, it is 3.4 and 3.3 liters respectively.[197] The stroke volume on a change from a recumbent to an erect position diminishes proportionately more than the minute-volume. It is only because of an increased heart rate that the minute-volume remains adequate, for the stroke volume may be greatly reduced.

During prolonged standing, however, the stroke volume may decrease so much that the minute-volume becomes inadequate in spite of an increase in the pulse rate, and the person may become dizzy and may faint.[515]

It may be concluded, therefore, that the circulatory adjustments to erect posture are better in physically fit than in physically unfit persons.

Individuals predisposed to circulatory embarrassment when standing indulge in deep sighs, which may be considered respiratory aids to the return of blood to the heart. They also appear to be restless. These restless movements are also an aid to circulation. During muscular contractions the blood in the veins is squeezed, and, because of the presence of valves in the veins, the blood is forced toward the heart. These same persons will feel very comfortable standing immersed in water, because water pressure prevents blood stasis in the legs and also reduces it in the abdominal cavity.

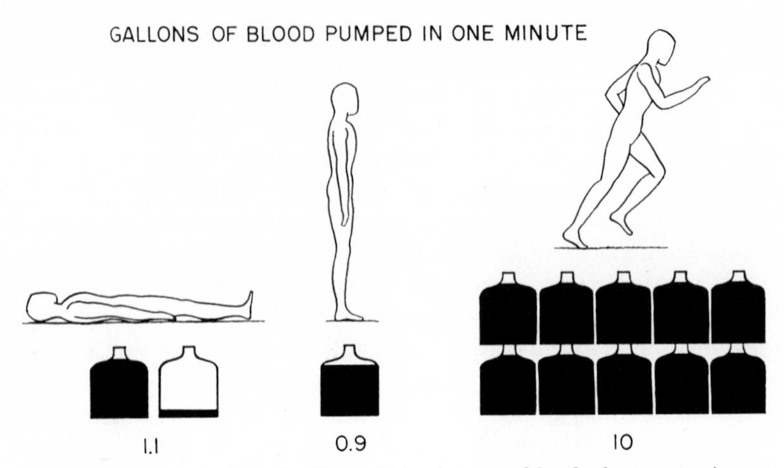

Figure 71. The number of gallons of blood pumped by the heart at various conditions: lying – 1.1, standing – 0.9 and running – 10.

EFFECT OF EXERCISE UPON THE MINUTE-VOLUME AND STROKE VOLUME

During exercise the cardiac minute-volume increases. In the average man it may rise from 4 liters to 20 liters, whereas in athletes values as high as 40 liters have been reported.[241] The volume pumped depends on the intensity of work and has a linear relationship with the amount of oxygen consumed, as shown in Figure 72. This relationship, which is expressed by the slope of the line, is affected also by the size of muscles and the rate of contraction. The measured cardiac minute-volume has two components: the heart rate and the stroke volume.

The amount of oxygen transported is determined not only by the cardiac output but also by the amount of oxygen extracted from the

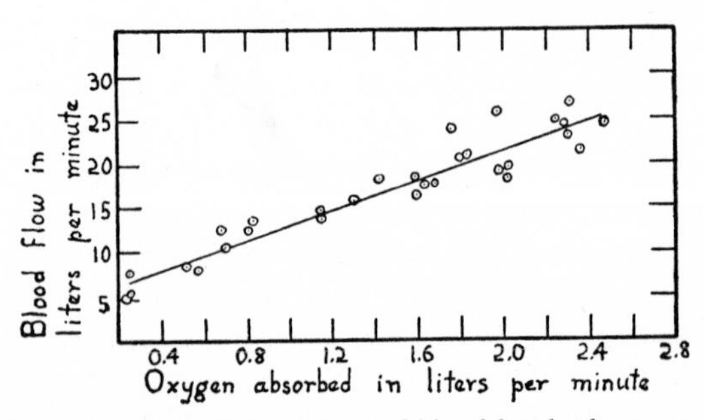

Figure 72. Determinations of the output of blood by the heart per minute during work on a bicycle ergometer. Note the linear relationship between minute-volume of the heart and the load of work (oxygen absorbed per minute). (Bock and co-workers: J. Physiol. *66,* 1928.)

blood. This is measured by finding the difference between the oxygen saturations of arterial and venous blood in milliliters (A-V difference). The interaction between these factors can be expressed as follows: *Oxygen intake = heart rate × stroke volume × A-V difference.* The way these change during exercise is of interest to us here.

The data presented by Åstrand, Cuddy, Saltin, and Sternberg[28] serve to illustrate typical response patterns. Figure 73 shows changes in stroke volume and heart rate compared to the percentage of the maximum oxygen intake utilized by the subjects. The maximum stroke volume was reached at 40 per cent of the maximum oxygen intake and a heart rate of 110 beats per minute. Any increases in cardiac output thereafter must come from increases in heart rate alone.

The heart rate increase, by itself, is not enough to meet the demands of the muscles for oxygen. Figure 73 shows the changes in A-V oxygen difference with increased utilization of the maximum intake.[28] These requirements are obviously met, in part, by extracting more oxygen per unit of blood pumped. The two primary factors determining oxygen delivery during severe work are, therefore, the heart rate and the A-V difference. Stroke volume changes apparently contribute only during exercises of low intensity.

The main factors responsible for the increase in stroke volume are the increase in arterial blood pressure and a better diastolic blood filling of the heart. When arterial blood pressure rises, the heart must work against greater resistance and, therefore, must use more power. In order to make contractions more forceful, the muscle fibers of the heart lengthen and cause an increase in the capacity of the heart. Meantime, the same increase in blood pressure, plus a milking action of contracting muscles, facilitates the return of venous blood to the heart. More blood enters the stretched heart and, therefore, more is ejected with each beat.

The ventricles of the heart are probably never entirely emptied at the end of a systole: as much as a third to a half of their blood content may remain undischarged. Hence it might be that the maximum stroke volume is never quite equal to the maximal capacity of a ventricle.

Rapid muscular movements result in a much greater increase of cardiac output than slow movements. Thus, flexing the right thigh once a second raises the oxygen consumption from 256 to 430 cc. and the minute-volume output of the heart from 4.2 to 7.7 liters; while alternately flexing both thighs, each every other second, raises the oxygen consumption the same as the first form of exercise, that is, to 428 cc., but raises the output of the heart only to 5.0 liters a minute.

The explanation of the difference in heart output per minute for different forms of exercise is found in the variations in pumping action of the muscles which drive venous blood on toward the heart. Such muscular activity as shivering in response to cold increases oxygen

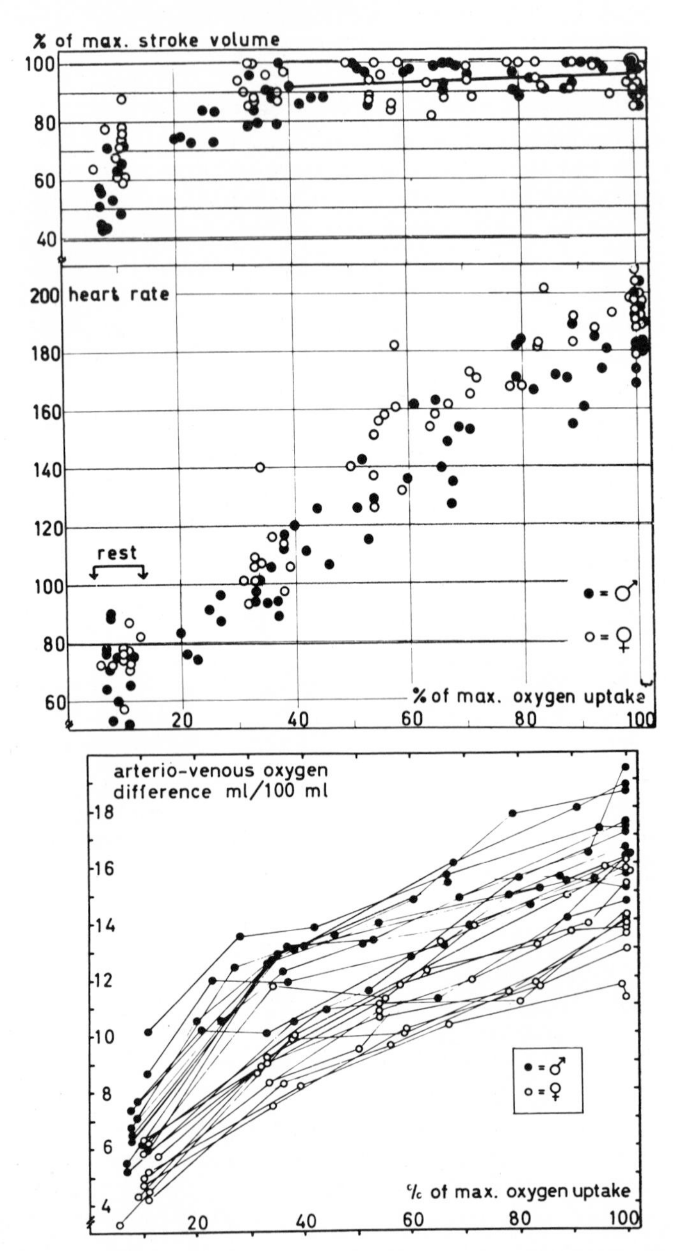

Figure 73. Changes in stroke volume, heart rate and arterio-venous oxygen difference during bicycle ergometer exercises of different intensities. The three measurements were taken simultaneously. Subjects were 11 women and 12 men ranging in age from 20 to 31 years. (Åstrand, P.-O., Saltin, B., and Stenberg, J.: J. Appl. Physiol. *19*:268, 1964.)

consumption, but has only a slight influence on cardiac output, because it does not aid the return of blood to the heart. Rapid movements, on the other hand, produce a marked increase in heart output. This explanation also accounts for the fact that, for a given expenditure of energy, cardiac output during swimming is greater than during work on a bicycle ergometer.

BLOOD CIRCULATION TIME

As has been stated, the minute-volume of blood during exercise may increase as much as ten times. It is obvious that, in order to deliver a greater amount of blood through the arteries, the blood must be moved much faster. That such is true may be demonstrated by injecting various substances into the veins and measuring the time needed for these substances to reach certain places. The average figures for circulation time at rest are: arm-to-tongue, fifteen seconds; arm-to-arm, twenty-one seconds; and arm-to-foot, one to two minutes.

Exercise increases velocity of blood flow. As one may expect, the greater the intensity of exercise, the faster the blood flow and, there-

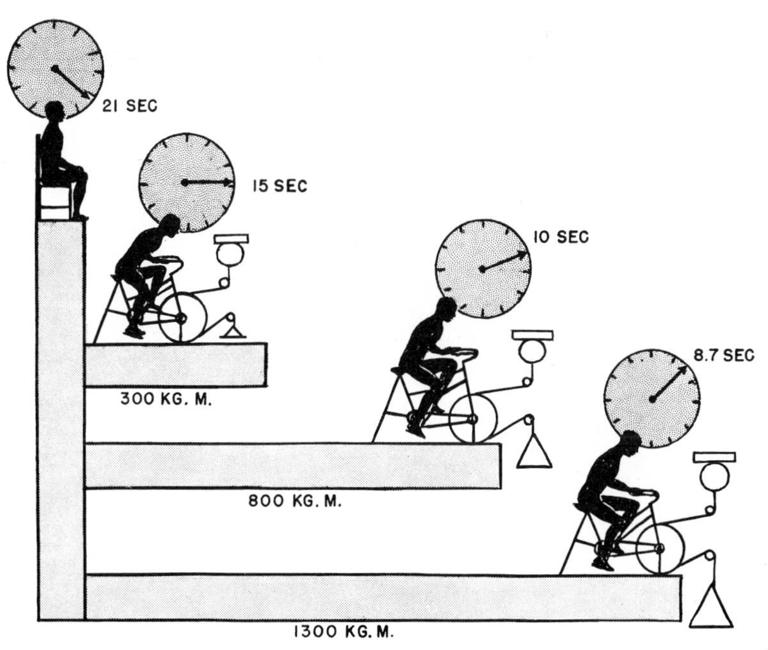

Figure 74. Relation between blood circulation time and intensity of exercise. A dye, fluorescein, was injected into a vein of one arm, and the lapse of time before it appeared in the veins of the other arm was measured.

fore, the shorter the circulating time. This relation is shown in Figure 74.

EFFECT OF TRAINING ON THE STROKE VOLUME

One of the most important differences between the trained athlete and the sedentary person is the size of the stroke volume. The studies of Clarence DeMar in the 1920's give an excellent example. DeMar maintained an excellent state of training for over 40 years. He had a stroke volume-body weight ratio of 2.98, whereas other subjects tested at the same time had ratios of less than 1. DeMar could run for two and one-half hours at a pace which required 3.5 liters of oxygen per minute; laboratory observers could not maintain it for five minutes.

Figure 75 shows a comparison of stroke volume of athletes and sedentary subjects. These graphs, as prepared by Rowell,[441] summarize the different adjustments to exercise: the decreased stroke volume on going from a supine position to standing, the increase during moderate exercise, and the plateau shown previously to occur at about 40 per cent of the maximum aerobic power. The important point is that the dynamics of adjustment are the same in the athlete and sedentary person. The athlete, however, operates at a higher level.

Similar findings are evident for the arterio-venous oxygen difference (Fig. 75*B*). Athletes have a greater capacity for unloading oxygen from the blood than non-athletes.[54] The effects of training and de-

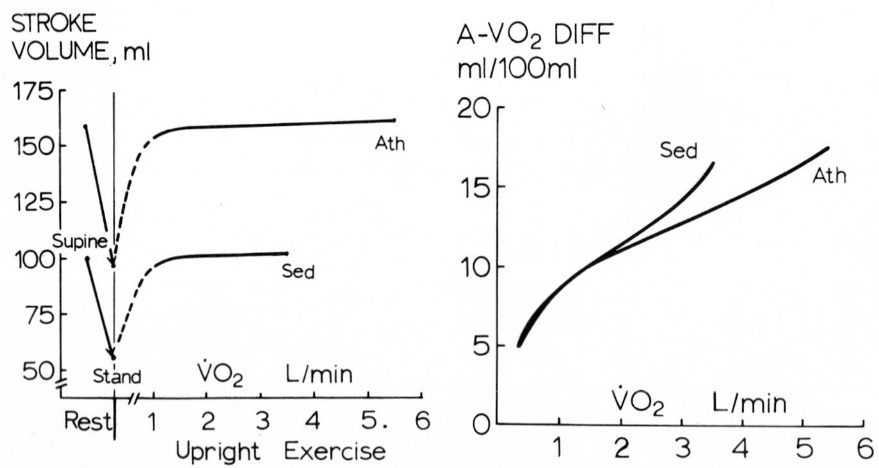

Figure 75. Comparison between athletes and nonathletes of stroke volume (*A*) and A-V oxygen difference (*B*). Compiled from a number of sources. (Rowell, L. B.: Medicine and Science in Sports *1*:15, 1969.)

training were well demonstrated by Bailie and co-workers,[34] who found that in active dogs the stroke volume during hardest work increased 82 per cent over that at rest; while a dog whose activity was restricted for three months had only a 62 per cent increase. Ekblom and co-workers[155] have demonstrated similar changes in young male students after sixteen weeks of endurance training.

EFFECT OF TRAINING UPON THE HEART

It often seems to be forgotten that the heart is essentially an endurance muscle. Since muscles of endurance increase in bulk only moderately, one should not expect any large increase in the size of the heart that happens to belong to a well-trained athlete. Moreover, even hypertrophy may not change markedly the apparent size of the heart, because the heart is a hollow organ and its walls project not only out but also in as they thicken. Therefore, x-rays cannot be depended upon to reveal the exact amount of hypertrophy. For this reason, little is known about the effect of training upon the heart of a man. When a cardiologist says the x-ray pictures indicate that the size of the heart of a trained athlete lies within the range of normal hearts, he means only that, and nothing else.

We should, therefore, draw our conclusions about the effect of training upon the heart from animal experiments. In these experiments, half the animals are trained and the other half are compelled to lead "sedentary" lives. At certain periods of time, some animals in both groups are killed and their hearts are measured. All investigators agree that the hearts of trained animals are larger and heavier. Observations of this kind leave no room for doubt that the human heart is also affected in the same way.

There is also an overwhelming amount of data from comparative anatomy studies showing that intensive and long-continued physical activity causes hypertrophy of the heart muscle, as can be judged by finding an index: $\frac{\text{heart weight}}{\text{body weight}}$. Thus wild animals have a greater index than domesticated animals of the same species. On the other hand, a caged wild animal will gradually show a decrease in this index. It has been found that the heart-body ratio in mongrel dogs is considerably smaller than that in greyhounds. Migratory birds have enormous hearts, whereas chickens' hearts, as probably everybody knows, are small.

It seems, therefore, to be clear that the heart muscle, like any other muscle, reacts by hypertrophy to the greater demand physical activity imposes upon the circulatory system. In this connection, it seems appropriate to quote Harvey, the great discoverer of blood circulation, who said in 1628: "The more muscular and powerful men are, the firmer their flesh; the stronger, thicker, denser, and more fibrous their hearts, the thicker, closer, and stronger are the auricles and arteries."

That the human heart will hypertrophy when it has to perform an excessive amount of work is well demonstrated in valvular diseases. We may mention two conditions: (1) aortic stenosis, in which the heart must overcome added resistance, and (2) leaking of aortic valves. In the latter condition, because during diastole blood regurgitates into the heart, the stroke volume has to be greatly increased. This requires more power from the heart. As a result of this defect, the hypertrophy of the heart may be very considerable.

It is obvious that extra work imposed on the heart because of this valvular defect is much greater than extra work in athletics, and, therefore, there is a striking difference in the results.

The word "athlete's heart," or rather its equivalent, *Sportherz*, was apparently introduced in 1899 by a Swedish clinician, Henschen, who, by using percussion alone, claimed to diagnose greatly enlarged hearts in skiers. He, by the way, referred to this enlargement as physiological.

The term "athlete's heart" gradually acquired, however, a connotation meaning pathological enlargement of the heart induced by athletics. This, indeed, is unfortunate, because hypertrophy of the athlete's heart is a physiological hypertrophy. The term "athlete's heart" is just as unfortunate as "athlete's foot," and should be dropped. One should speak of the heart of an athlete, conveying an idea akin to that expressed in Harvey's statement. Wolffe,[547] who has studied the hearts of many athletes, comes to the same conclusion: "As a result of training, the heart of the athlete becomes more developed, heavier and somewhat larger, as compared to the heart of the inactive individual, upon which, unfortunately, we base our concepts of norms."

Through the courtesy of Dr. Albert S. Hyman, of New York City, we reproduce here an illustration showing four athletes and their hearts. With the aid of a fluoroscope, he makes an outline of the heart, and then cuts its silhouette from black cardboard paper. He places this silhouette on the patient's chest, stands the patient in front of a special grid, and photographs him. This grid has, in large figures and letters, the written information regarding the patient's various test scores and measurements. Details of his method may be found in reference 259. In Figure 76 we may see that the volume of a normal heart (*A*, *B*, *C*) depends on the body weight: a larger man has a larger heart. Subject *D*, whose size of heart is out of proportion with his body size, has a pathological enlargement of the heart. These hearts respond to exercise differently. In the first three men, the size of the heart decreases immediately after exercise; in Subject *D* it increases. The Cardio-Body Index (the ratio of the surface area of the heart to the surface area of the body) indicates the normalcy of the first three hearts and a 35 per cent pathological enlargement of the fourth heart.

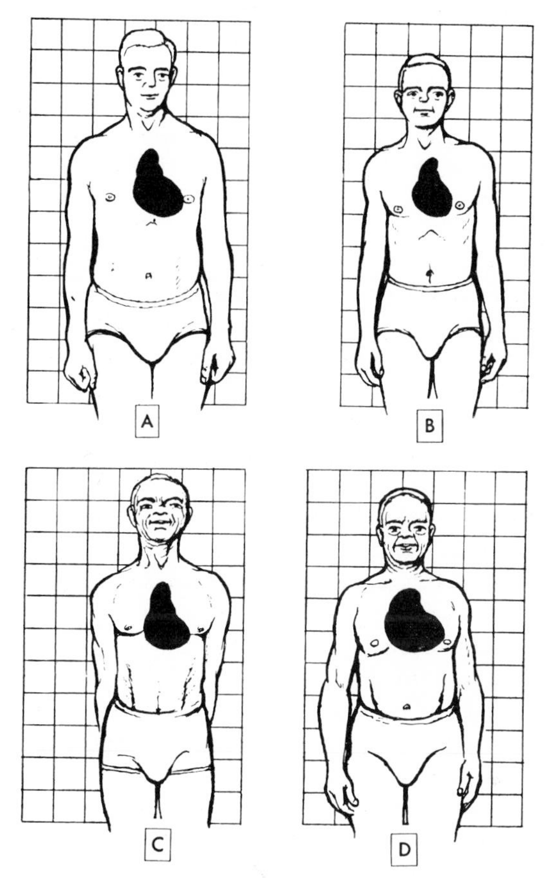

Figure 76. Four men with athletic history: first three are still active, the fourth is not. The first three hearts are normal, the fourth is enlarged and pathological. After exercise the first three hearts become smaller. The fourth—larger.

	Age	*Weight (lbs.)*	*Height (in.)*	*Heart Volume (cc.) Before—After Exercise*		*C-B Index**
A	18	222	75	942	874	950
B	21	145	72.5	865	636	708
C	65	127	67	730	684	715
D	58	162	66	1520	1675	990

*Note: C-B or Cardio-Body Index is found by dividing the heart's frontal surface area in sq. cm. by the body surface area in sq.m.: Normal range 450–550. (Courtesy, Dr. A. S. Hyman, New York City.)

Clarence DeMar, who kept on running almost to the time he died at the age of 70, had a healthy and normal heart, as revealed by autopsy. However, this does not mean that it was the same size when he was at his peak.

Heart Strain. Cases of heart strain resulting in dilatation of the heart have been reported, but in each case one suspects that the heart may have been weakened by some previous disease. The predominant opinion at this time is that a *sound heart cannot be injured by physical training.*

Parsonnet and Bernstein[404] reviewed the question of heart strain during physical activity. Their conclusion was that there is no scientific proof of a pathological *chronic heart strain* resulting from the cumulative effects of hard muscular work. Hypertrophy merely indicates a better development of the heart muscle—work hypertrophy.

These investigators, however, believe that there is such a condition as *acute heart strain,* which may develop as a result of a sudden great muscular exertion.

As has been mentioned, fear has existed and even now exists that physical activities may overstrain the heart and cause myocardial damage. This fear is justifiable only when the heart is already damaged. Then a violent exertion may even be fatal. But a person with a sound heart would probably damage his heart more with inactivity and might prevent this damage with physical activity.

During the past decade, numerous investigators in this country and abroad have pointed out that heart disease occurs more often in people doing light physical work than in those who are engaged in more intensive work. Just to cite a few examples: Morris found that in England fatal heart attacks occur more often in bus drivers than in bus conductors, who walk more than the drivers. The same is true of postal clerks, as compared to more active mailmen; and of railroad clerks compared to more active switchmen, maintenance men, and men on track-laying crews. An excellent review of this topic may be found in reference 177.

The greater vulnerability of a "sedentary" heart may lie in a higher level of plasma cholesterol, which leads to the accumulation of this substance under the intimae of the blood vessels. Plaques are formed which bulge into the lumen of the vessels and cause partial obstruction of the blood flow. This condition is called atherosclerosis. This may lead to formation of a blood clot in the constricted area and total obstruction by a thrombus.

One should not conclude, however, that lack of physical activities is the only cause of atherosclerosis. Women, who generally do less physical work than men, are also affected by atherosclerosis but less frequently than men.

The cause of atherosclerosis is not known. At one time it was believed that, by lowering the intake of cholesterol and by using vegetable polyunsaturated fats instead of solid animal fats, one could prevent or stop further development of atherosclerosis. It sounded like a logical approach to the problem. Then it was found that emotional strain can change the body chemical processes so much that excess cholesterol is produced in spite of controlled diet.

In this book we have to limit our attention to just one factor: namely, is there evidence that physical activities reduce the cholesterol level in the blood, thus either preventing or delaying the development of atherosclerosis and some types of cardiac disease caused by this condition? To this end, eminent cardiologists such as Paul Dudley White, Samuel A. Levine, and A. Salisbury Hyman, just to name a few, recommend regular physical activity for therapeutic and prophylactic purposes.

There is enough evidence that training causes lowering of the plasma cholesterol level. Golding[185] showed this on four swimmers during a 25-week training in swimming. Rochelle[432] found that in six subjects who ran two miles for time, five days a week for five weeks, the plasma cholesterol became significantly lower – 203 mg. before to 179.5 mg. after – but that it returned to "normal" during the de-training period of four to six weeks. It should be noted that immediately after the two-mile run the cholesterol level was higher than before the run. This could be explained as fat mobilization for the activity.

CARDIOVASCULAR DISEASE AND EXERCISE

Hyman[260] calls attention to the fact that there "may be little or no correlation between objective evidence of heart disease and functional capacity of the same heart." Some cardiac patients disregard the objections of their physicians and participate in strenuous sports with outstanding success. Dr. Joseph Wolffe, who examined the hearts of many athletes, observed that, on the basis of electrocardiograms and other cardiac examinations, some of these athletes should have been pronounced to be cardiac patients; yet they successfully participated in marathon running and swimming. Of course, athletes like these might, without any warning, have heart attacks or even drop dead.

THE HEART IN THE PREPUBESCENT CHILD

Occasionally even a careful investigator with a reputation for dependability commits an error in interpreting his findings. Because of his reputation, other investigators accept the findings as gospel and perpetuate this error, lending support to it by "profound" "scientific" reasoning.

A good illustration of this concerns the development of the heart and arteries in the child and its influence upon physical activities. In 1879, Beneke studied the relationship between the growth of the heart and the aorta and the pulmonary artery. As a result of numerous observations on cadavers he decided that, whereas the volume of the heart increases in proportion to the body weight, the circumference of the aorta and of the pulmonary artery increases in proportion to the body length. Thus he concluded that the arteries develop more slowly than the heart. Many noted investigators, among them Lesgaft, F. Schmidt, and Kohlrausch, just to name a few, accepted Beneke's deduction and applied it to practical life. Their interpretation found its way into textbooks on pediatrics and physical education.

Thus the idea was perpetuated that there is a discrepancy between the development of the heart and the large arteries of a child, and that the age of seven, for example, is a critical period, at which time the child's vigor begins to diminish as the result of this discrepancy.[551]

Although Beneke's observations were absolutely correct, his comparison of the heart volume with the circumference of the vessels was a grave mistake. If, for instance, the circumference of an artery increases twice, the volume of blood going through it will increase not twice but *four times*, or in proportion to the square of the radius or diameter; in other words, in proportion to the area of cross section and not to the circumference.

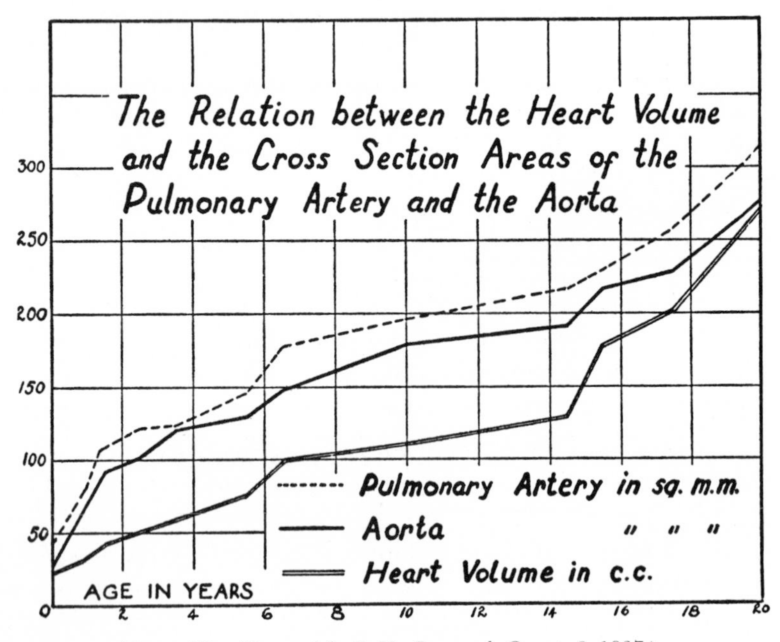

Figure 77. (Karpovich, P. V.: Research Quart. *8*, 1937.)

Karpovich[284] recalculated Beneke's original data, and compared the rate of the development of the heart with that of cross sections of the aorta and pulmonary artery. As one may see from Figure 77, all three variables follow each other closely. Thus physiological considerations regarding restriction of physical activities of children, especially at the age of seven, have been proved to be unfounded. Yet even now, 35 years later, there are still books that perpetuate the old fallacy. It is fortunate that children do not read these books, and just keep on playing.

QUESTIONS

1. Describe a laboratory method of measuring cardiac minute-volume.
2. What is stroke volume, and how is it related to minute-volume?
3. Discuss the effect of posture on heart output. What mechanism is responsible for this?
4. Discuss the effect of exercise upon the stroke volume and give some figures.
5. Discuss the effect of slow and rapid movements on the cardiac output.
6. How can blood supply more O_2 during work than at rest? Name three or four factors.
7. What is the relation between the pulse rate and stroke volume?
8. How is blood circulation time measured? Give figures for arm-to-arm and total circulation time.
9. What is the effect of training upon the stroke volume? Discuss the significance of the change.
10. What is the general effect of training upon the heart? Do the size, weight and thickness of the walls change?
11. Is the heart a muscle of strength or endurance?
12. Is there any relation between the size of the heart and body weight?
13. Discuss the size of the heart in wild and domestic animals.
14. Judging by x-ray pictures, does the size of a normal heart increase or decrease immediately after exercise?
15. How did the term "athlete's heart" originate?
16. Is there a discrepancy between the development of the heart and the largest arteries? Discuss.
17. What effect does training have upon the stroke volume?
18. (Consult Chapters 7 and 9 and solve this problem. If you can do it without help, you understand the physiology of muscular activity.) A man weighing 200 pounds climbed 100 feet in 5 minutes. His heart rate during the climb was 100/min.; his efficiency of climbing was 15 per cent; and his oxygen unloading from the blood was 60 per cent. Find the following: (a) work done; (b) amount of energy used in ft. lb., calories and horsepower; (c) amount of O_2 used (1L. O_2 = 5 cal.); (d) oxygen pulse; (e) stroke volume; (f) blood minute-volume; (g) approximate pulmonary ventilation.

Chapter Twelve

THE PULSE RATE

In considering the pulse rate, the student must understand the difference between the average and the normal. Numerous tests have shown that young men in a lying position before breakfast have an average pulse rate of 64 beats per minute. However, an examination of individual records shows that the range is from 38 to 110 (Fig. 78). This means that, for a person who is in good health and yet who, on repeated examinations, has a pulse rate of, say, 110, the rate of 110 may be considered normal. The same rate for a person whose usual rate is 50, of course, will not be normal. The American Heart Association accepts as normal a range of 50 to 100 beats per minute. The round figures, limiting this range, are arbitrary. However, the great majority of normal pulse rates will fall within these limits. Rates outside this range should be carefully re-examined before they can be accepted

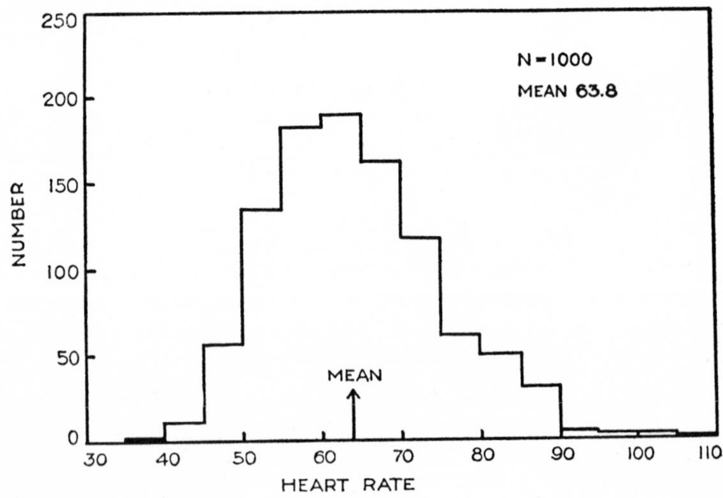

Figure 78. Mean heart rates of young men under basal conditions in lying position. (Graybiel et al.: Am. Heart J. *27*, 1944.)

as normal and not exceptional. Pulse rates for women, in basal condition, are from 7 to 8 beats higher than for men.

The pulse rate in well persons is affected by age, body size, body position, food intake, time of day, emotions, and physical activity.

PULSE RATE AND AGE

At birth the resting pulse rate may be as high as 130. Then it gradually lowers to typical adult values, until in old age it again increases slightly.

The maximum rate is of more concern than the resting rate in the study of exercise physiology. Figure 79 shows that as age increases the maximum decreases. It is obvious that there is considerable variation within each age group. It is easy to see from this how a man of 60 years could have a maximum greater than a man of 30 even though the trend would indicate the opposite. Åstrand[23] gives the following average maximums as a basis for adjusting maximum oxygen intakes predicted from submaximal heart rates (see Chapter 18): 15 years, 210; 25 years, 200; 35 years, 190; 40 years, 180; 45 years, 170; 50 years, 160; and 55 years, 150. These values tend to be higher than indicated on the graph at lower ages, while the converse is true at higher ages.

This disagreement illustrates the difficulty of setting absolute values unless a very large sample is available. Either scale does, however, show the decrease in adaptive capacity with age. It gives information essential to setting guidelines on performance capacity for cardiovascular testing and exercise programs for various age groups.

POSTURAL PULSE RATE CHANGE

Most observations have shown that the pulse rate is definitely affected by body position. The rate is lowest in lying, higher in sitting, and highest in standing. The extent of variation, however, differs with the subject. Schneider and Truesdell[463] found that in 2000 healthy young men the differences between pulse rates in standing and in lying varied from +57 to −15. About 1 per cent of the men had lower pulse rates standing than reclining, and 2.4 per cent showed no difference at all. Undoubtedly, large differences were caused by nervousness.

It was found that men who have a slow pulse rate (60 to 70) while standing will experience little or no decrease when reclining. On the other hand, if the standing rate is high, it is likely to decrease decidedly when the man lies down. Thus men with a rate of 99 beats standing show an average decrease of 18 to 21 beats when reclining.

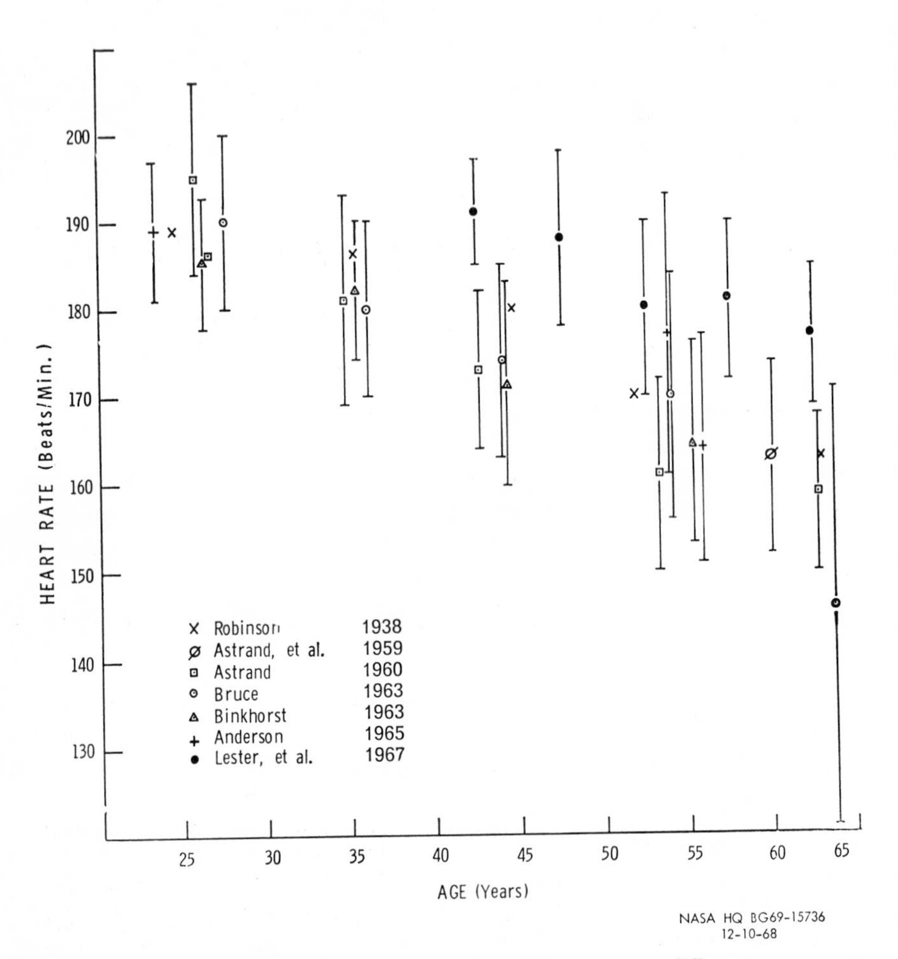

Figure 79. A summary of findings on maximal heart rates at different ages as prepared by Haskell and Fox. Studies included Robinson: Arbeitsphysiol. *10*:251, 1938; Åstrand et al.: J. Appl. Physiol. *14*:562, 1959; Åstrand, I.: Acta Physiol. Scandinav. *49*:suppl. 169, 1960; Bruce et al.: Pediatrics *32*:suppl. 742, 1963; Binkhorst and Van Leeuwen: Int. Z. Angew. Physiol. *19*:459, 1963; Anderson and Hermansen: J. Appl. Physiol. *20*:432, 1965; Lester et al.: Circulation *36*:5, 1967. (From Haskell, W. L., and Fox, S. M.: Mal. Cardiovasc. *10*:189, 1969.)

Men who are physically fit show a smaller difference between reclining and standing than do men in general. Pulse rates tend to be slower in athletes than in non-athletes. A slow pulse rate in the reclining and standing positions, with a small difference between the two, is usually regarded as a sign of excellent physical condition.

Turner, who studied the effects of prolonged standing on the pulse rate in college girls,[514] determined that the most physically fit girls had a reclining pulse rate of 50 to 60 beats per minute. On standing, this rate was increased by 10 beats. On the other hand, girls whose reclining pulse rate was high and who had a large increase in the standing position often felt dizzy and even fainted during a fifteen-minute standing test.

FOOD INTAKE AND TIME OF DAY, AND PULSE RATE

The digestion of food invariably accelerates the heart rate for two or three hours. This effect should be taken into consideration when diurnal variations in pulse rate are studied. In order to detect the effect of the time of day upon the pulse rate, the subject should abstain from eating. Observations on fasting men have shown that the time of day seems to have no effect on the reclining pulse rate, whereas the standing pulse is higher by about 6 beats per minute in the late afternoon.[464]

EMOTIONS AND PULSE RATE

Because emotions accelerate the pulse rate, it is sometimes very difficult and even impossible to obtain a normal resting pulse rate. The subject may appear relaxed, while his pulse rate tells a different story. Variations in the emotional state affect the pulse rate much more than postural changes. One is forced to believe that some unusual pulse rates credited to postural changes have resulted from emotional factors, which either were not even suspected or could not be eliminated.

During World War II, many volunteer applicants for the Air Force frequently had higher pulse rates at the beginning of a medical examination than at the end when they had regained composure.

A seemingly simple thing like waiting for a test may greatly affect the rate of the pulse. This fact was well demonstrated by Brouha on a group of college students in one of his experiments. During a medical examination this group had a mean rate of 82 (range from 50 to 130). While the students waited for their turn to run on a treadmill, their mean rate rose to 125 and their range became 79 to 170. During the treadmill run, their mean was 193 and their range 160 to 220. From this observation, it is also possible to conclude that the pre-exercise average rate of 82 beats per minute was not a true average because of excitement caused by the medical examination.

Obtaining a reliable resting pulse in children is much more difficult than obtaining a reliable count in adults. Besides being more excitable, children are usually restless while waiting for examinations. Often they begin to play or fight with each other. Muscular movements involved in restlessness, play, and fighting, of course, raise the pulse rate.

PULSE RATE BEFORE EXERCISE

In taking a resting pulse rate before an exercise involving an element of contest, one should remember that instead of a "resting" rate there might be a "start" pulse, accelerated by the excitement of anticipation. For example, in a group of well-trained weight lifters whose usual resting pulse was 72 beats per minute, the pulse before a contest was 135 to 160 beats per minute. Spurious resting rates have even found their way into the reports of reputable clinicians who have failed to ascertain the true resting rate. One should not forget that a pulse rate obtained during a period of apparent rest may not necessarily be a resting pulse.

PULSE RATE DURING EXERCISE

At the beginning of muscular exercise, the pulse rate increases rapidly. The greatest rise takes place within one minute. Sometimes half of this increase occurs within fifteen seconds. Gradually a plateau is reached. If the exercise is intensive, a secondary rise may occasionally be observed (see Figure 80). As one may expect, the change in pulse rate depends on the individual. For equal intensity of work, one subject may have a pulse rate of 160, while another's may reach 220. For example, while in adults the pulse rate immediately after a 5-kilometer ski run is under 200, in girls between fifteen and seventeen it is 250. One girl had a rate of 270.[95]

Excitement may also affect the work pulse rate. In one experiment a subject had to walk on a treadmill for six 10-minute periods; during the third period, his pulse rate was 120 beats/min. While getting off the treadmill he stumbled. When he resumed walking, his pulse, during the remaining periods, rose to 160 beats/min. The following day his work pulse rate was still high, although it gradually declined and became, during the fifth period, 130 beats/min.[503]

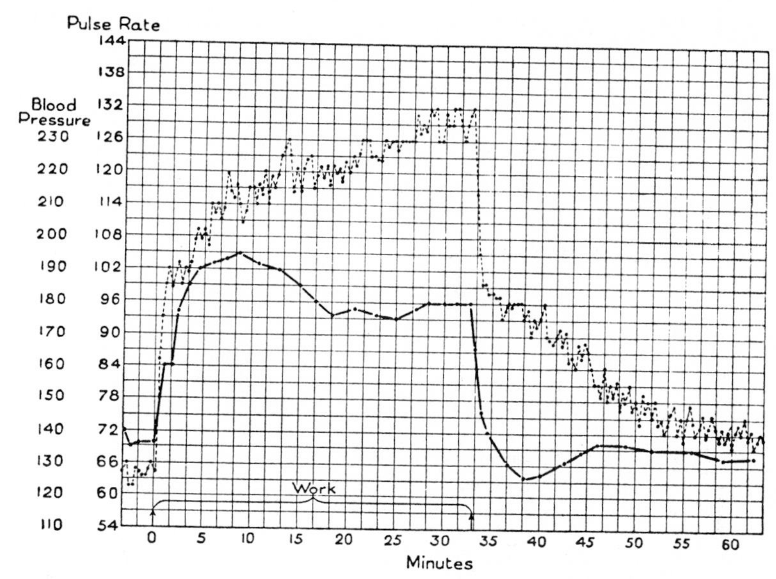

Figure 80. Curves showing relative changes in the pulse rate and the systolic blood pressure during thirty-two minutes of work and the following thirty minutes of rest. Note that the pulse rate (.------.) has not returned to normal in twenty minutes and that the systolic pressure (—·—·—) is subnormal for a time after exercise. (Bowen: Am. J. Physiol. *11,* 1904.)

When a catheter is introduced into the superior vena cava and another into the brachial artery, subjects are in a state of apprehension and one may expect their work pulse rates to be increased. Yet a group of seven men with catheters inserted, walking on a treadmill, had identical pulse rates for work bouts of equal strenuousness. After three months of intensive conditioning, the tests were repeated. This time the submaximal pulse rates were higher with the catheters than without. The maximal heart rates, however, were equal.[503]

The frequency of the heart rate during a period of exercise, particularly if a steady state is established, is in linear relationship with the load of work. This linear relationship remains true up to a certain limit, depending on the individual.

For the same intensity of work, as judged by the amount of oxygen used, more physically fit men have lower pulse rates. This is well illustrated in Figure 81. DeMar was a trained runner, and his pulse rate at rest and at work was much slower than those of the other three subjects. C.V.C. was a man of no athletic experience whatever. Notice that, with a load of work requiring 2 liters of oxygen per minute, the heart frequency of DeMar was 118 and C.V.C. 160 beats per minute. The demand made by this load on DeMar's heart was much less than that on C.V.C.'s.

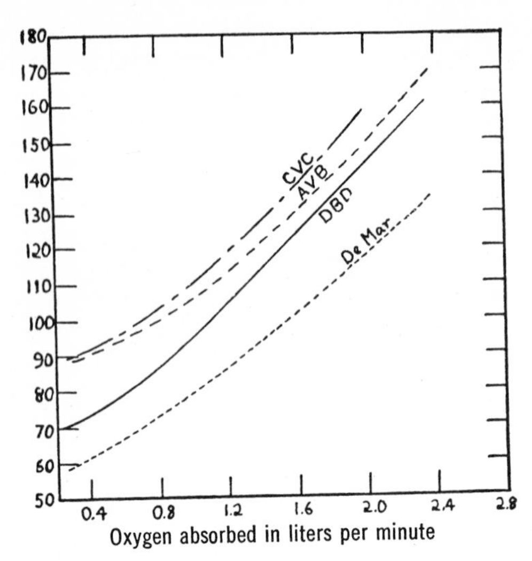

Figure 81. Plotted data of pulse rate obtained from four men while working on a bicycle ergometer. Note the linear relationship between the load of work (oxygen absorbed per minute) and the pulse rate. (Bock and co-workers: J. Physiol. *66*, 1929.)

PULSE RATE AND PARTICIPATION IN SPECIAL PHYSICAL ACTIVITIES

Diving. Craig[109A] found that in good and poor swimmers (both adults and children) during diving while holding the breath, the pulse rate slowed down to 40 to 55 beats per minute. Before submersion the pulse rate was over 120. Even during underwater swimming, the pulse rate was much lower than during swimming while breathing.

Weight Lifting. As has been previously stated, the pulse rate changes in the same direction as the work load. Tests conducted on a group of the best weight lifters in Soviet Russia showed that when these men lifted their maximal weights the pulse rate increased each time.[323] Expressed in per cent of resting rate, this increase was: 99 for the *snatch*, 89 for the *clean and jerk*, and 76 for the *press*. It is interesting to note that the anticipatory acceleration of the heart rate also increased in the same order: snatch, clean and jerk, and press.

The reason why, in lifting, the press has less effect than the other two lifts may be found in the fact that it is the slowest of the three movements.

Track. The use of radio-telemetry equipment has contributed considerably to our knowledge of heart rate changes during track events. McArdle, Foglia, and Patti[370] studied responses in eighteen varsity trackmen and four untrained subjects. Trackmen ran only in their regular events, while untrained subjects ran in all events: 60-, 220-, 440-, and 880-yard as well as 1- and 2-mile distances. In the 60-yard dash the anticipatory heart rate was equal to 74 per cent of the total adjustment, but decreased with length of race until it accounted for only 33 per cent before the 2-mile run. Skubic and Hilgendorf[483] found anticipatory increases accounted for 59 per cent of the total in girls who ran distances from 220 yards to a mile.

Heart rate increases rapidly at the beginning of each race. In the study by McArdle, Foglia, and Patti it reached about 180 beats per minute within 28 seconds in the mile and 2-mile events, and within 10 seconds in the "220," with higher peak rates being attained in longer events. Bowles and Sigerseth[66] similarly found that the response was almost complete within the first 220 yards when experienced distance men ran set paces.

The time it takes for the heart rate to return to normal after running does not markedly differ for the various distances. The recovery following the 60-yard dash has been found to be significantly faster than for the other events, but no differences have been shown among the recovery rates for longer distances.[370, 483]

The studies reviewed dealt with sprints and middle-distance events. In a study of highly-trained distance runners, heart rates over 180 beats per minute were recorded during the final one-third of a run in which the metabolic demands exceeded 85 per cent of the maximum oxygen intake.[104]

Heavy Apparatus: Gymnastics. Kozar[324] recorded heart rate data from a highly-skilled gymnast performing on the parallel bars, highbar, side horse, and still rings. The heart rate response is apparently related to the difficulty of the routine. All routines tend to cause a rapid increase during the first few seconds which may or may not plateau. The routines recorded lasted no more than 35 seconds, and values ranging from 140 beats per minute in a highbar warmup to 169 beats per minute in a "tough" parallel bars routine were recorded.

HEART RATE AND STEP-UP EXERCISE

Because of the ever-increasing use of various types of step-up exercises in testing physical fitness, it is of practical interest to discuss here one effect of stepping-up upon the heart rate. As in every repetitive exercise, pulse rate in stepping-up reflects the intensity of the effort. Pulse rate naturally is affected by the height of the bench used, and by the number of steps per minute (cadence).

Elbel and Green[156] tested seventy-two aviation students on benches of five different heights: 12, 14, 16, 18 and 20 inches. Each subject was tested twice on each bench: once for half a minute and again for one minute. The stepping rate in the experimental series was constant—24 complete steps per minute. A complete step consisted of four counts. On count one, the left foot was placed on the bench; on count two the right foot was placed on the bench; on counts three and four, the feet

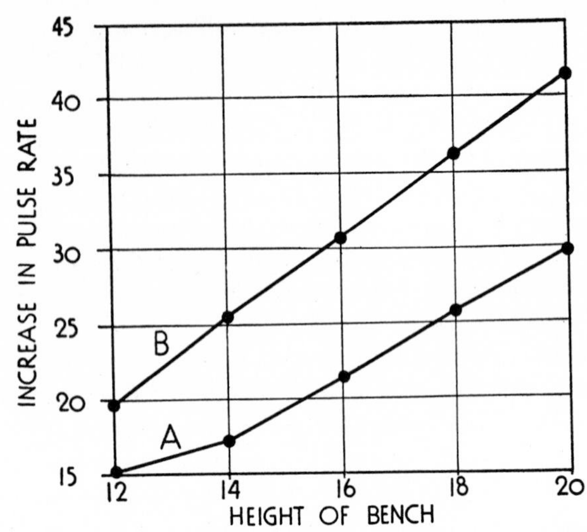

Figure 82. Relationship between the height of a bench, used in stepping up and down, and the pulse rate immediately after exercise. *A*, After thirty seconds of exercise; *B*, after sixty seconds of exercise. (Courtesy E. R. Elbel.)

were placed on the floor. The pulse rate immediately after the thirty-second exercise was, on the average, 3.7 beats per minute faster for each additional 2-inch increase in the height of the bench (Fig. 82*A*). After the sixty-second exercise, the increment in pulse rate was 5.6 beats per minute for each additional 2-inch increase (Fig. 82*B*). Thus,

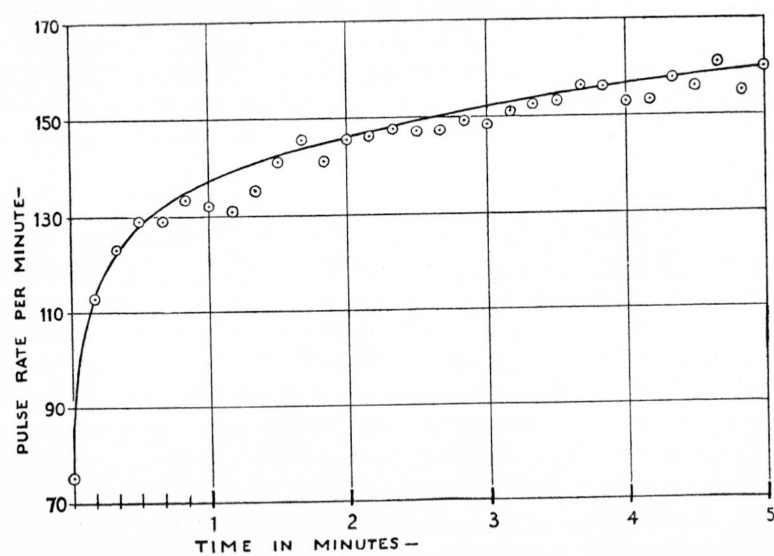

Figure 83. Pulse rate during five minutes of vigorous exercise, which consisted of stepping up and down on a 20-inch bench at a rate of 24 complete steps per minute. The curve represents the average data for ten subjects. Heart beats were recorded by an electrocardiotachometer at ten-second intervals.

longer exercise on the same height bench caused a greater increase in the pulse rate.

The pulse rates taken from sixty to ninety seconds after exercise indicated that one minute is a sufficiently long period of time for the return of the pulse to normal, after all intensities of exercise used in this series.

The change in pulse rate during a five-minute stepping-up exercise is shown in Figure 83. Ten healthy young men performed the exercise, at a rate of 24 complete steps per minute, on a bench 20 inches high, for a period of five minutes. The pulse rate was automatically recorded by an electrocardiotachometer every ten seconds. The curve represents the average pulse rate for the ten subjects. It may be observed that the pulse rate increased rather rapidly at first and then continued to increase slowly until the end of the exercise. The total rise in pulse rate was 85 beats. Approximately 45 per cent was gained during the first ten-second period of exercise, and an additional 28 per cent was gained during the following fifty seconds.

RETURN OF PULSE RATE TO NORMAL

The time required for the pulse rate to return to normal after exercise depends upon the intensity of the exercise and upon the condition of the individual. Increasing the intensity of exercise increases the time required for recovery. On the other hand, better physical condition tends to shorten the period of recovery.

By recording the pulse rate continuously by means of a cardiotachometer, Cotton and Dill[106] showed that, in ten seconds immediately after the cessation of strenuous exercise, the heart rate decreases on an average of about 1 beat per minute. After that the decline is more rapid. Bowen[65] found that a sudden and rapid primary fall of pulse rate may at times be followed by a plateau or constant rate with a subsequent slower secondary fall. The pulse rate occasionally may fall below the pre-exercise level. This happens even in those whose resting pulse rates have been obtained under carefully controlled conditions.

A drop below normal may, obviously, be expected in subjects whose pre-exercise pulse rates have been elevated on account of various factors, some of which may be psychological. For this reason, some investigators think that the pulse rate after a standard exercise is more reliable than the pre-exercise resting pulse rate, which may be affected temporarily by various complicating influences. The importance of painstaking precautions for obtaining reliable normal resting pulse records cannot be overemphasized.

Studies conducted at the Harvard Fatigue Laboratory on the pulse rates of many subjects after strenuous exercise showed that the post-exercise pulse follows exponential curves, and suitable formulas have been evolved. The trained subjects recovered faster than untrained ones.

The time necessary for the pulse rate to return to normal has a wide range. After a half minute of stepping-up on benches 12 to 20 inches high, the rate should be back to normal within a minute. After exhaustive exercise, it may not be back to normal for several hours.

From Figure 84 it may be seen that the pulse rate does not return to normal as fast as does the oxygen intake after an oxygen debt has been incurred. It has been suggested that the immediate fall in oxygen intake is determined not by the requirement of the body for oxygen, but rather by an alteration in the mechanisms by which it is supplied. The immediate fall in oxygen intake is attributed largely to a change in the circulation of the blood, which is a result of the cessation of bodily movements. As a result of the discontinuation of muscular contraction, the rate of return of venous blood to the heart is diminished; hence heart output and the amount of blood passing through the lungs are decreased. Thus, less oxygen is absorbed from the lungs. This can be demonstrated from a study of the oxygen pulse, shown in Figure 85. Oxygen intake may reach the pre-exercise rate long before the heart rate does.

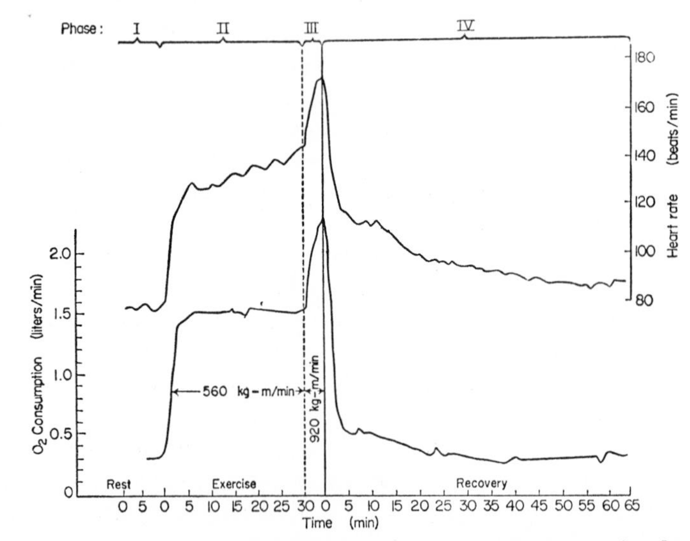

Figure 84. Effects of exercise on heart rate and oxygen consumption for six male subjects at moderate temperature and humidity (72° F. and 50 per cent R.H.). The exercise consisted of pedaling a bicycle ergometer for 30 minutes at submaximum work, followed by a 4-minute period of maximum work. Although a steady state of oxygen consumption was reached at the lower load, the heart rate continued to increase. Oxygen consumption returned rapidly after work to the resting level, but even after an hour the heart rate did not return to normal. (Brouha, L.: Physiology in Industry. Pergamon Press, 1960.)

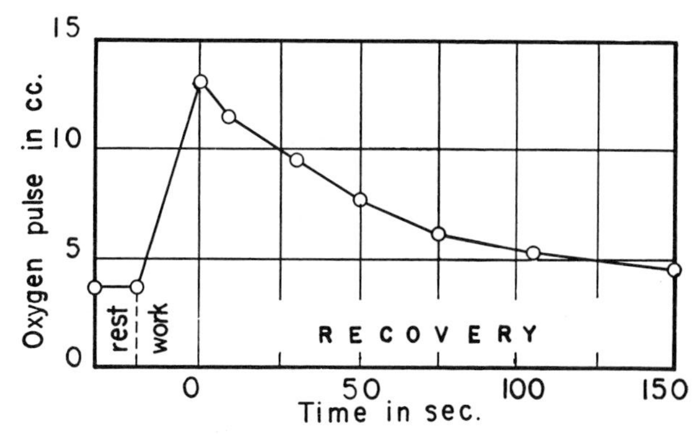

Figure 85. Changes in the oxygen pulse after strenuous exercise. (Lythgoe and Pereira: Proc. Roy. Soc., London, s. B. *98*, 1925.)

RELATION BETWEEN RESTING AND POST-EXERCISE PULSE RATES

It is a common belief that, in a group of subjects after a standard exercise, pulse rates will be higher in those individuals whose resting pulse rates are also higher. This belief was expressed by Robinson[430] in reporting his tests on subjects of various ages. Cogswell and co-workers[102] reported a statistically justified relationship between the resting and post-exercise pulse rates after strenuous work on a treadmill or a bicycle ergometer, or after the Harvard step-up test. Coefficients of correlation ranged from 0.63 to 0.88.

Tuttle and Salit,[518] experimenting on young men and women who exercised on a bicycle ergometer, came to the conclusion that the relationship between resting and post-exercise pulse rates depends on the strenuousness of the exercise. After mild or moderate exercise, coefficients of correlation between the resting pulse rates and the increase in pulse rates caused by exercise were either positive or negative, but too low to be statistically significant. After strenuous exercise, however, the coefficients of correlation became statistically significant, being −0.46 for women and −0.731 for men. The negative sign may be explained in the following manner. Strenuous exercise causes the heart rate to become maximal, which is approximately the same for all individuals. The higher the resting pulse, the smaller the difference between it and the maximal pulse. Thus, with an increase in resting pulse, there is a corresponding decrease in a possible exercise rise in the pulse rate.

Taylor[497] and Knehr and his co-workers,[319] however, observed no relationship between basal pulse rate and post-exercise pulse rate.

Unpublished experiments by Karpovich on thirty aviation students produced rather conflicting results. He used two tests, one consisting of making 12 steps up-and-down in thirty seconds on a bench 12 inches high; and the other requiring 12 steps in thirty seconds on a bench 20 inches high. There was a low, but significant, correlation (0.47) in the test with the 12-inch bench, and no statistical significance (0.14) in the test with the 20-inch bench. These findings substantiate those of Tuttle and Salit.[518]

The chief complicating factor in studying the relationship between resting and post-exercise pulse rates is the difficulty of obtaining a true resting pulse. It takes so much time and precaution that often the acceptable resting pulse is that which is obtained after an insufficient period of rest from all disturbing influences, at which time two consecutive readings happen to check.

The authors are of the opinion that it is quite probable that the post-exercise rate is related to the level of the basal or even resting pulse rate, and hope that new investigators dealing with this problem will specify in their reports each time whether they used true or "conventional" resting pulses.

REGULATION OF THE FREQUENCY OF THE HEART BEAT

The heart is governed by two sets of nerves from the autonomic nervous system: the sympathetic and the vagus. The effects are accelerative and inhibitory, respectively.

There is normally a balance between the two systems, with the vagus influence slightly stronger. An increase in the heart rate is produced by inhibition of the cardio-inhibitory center, by an increase in the tone of the cardio-accelerator center, or by both actions at the same time.

Exercise Acceleration of the Heart. If there is no anticipatory acceleration of the heart (start pulse), then, at the beginning of exercise, there occurs an immediate increase in the frequency of the heart beat. The first heart cycle after work begins is always shorter than those that immediately precede the work period. Each succeeding heart cycle for several beats continues to decrease. This first period of acceleration is wholly due to a shortening of the diastole of the heart. It has been found, by electrocardiographic methods of recording, that an exercise which consisted of clenching the fist shortened the first heart cycle 9 per cent and the second 25 per cent.[181]

Such a rapid response of the heart to beginning an exercise eliminates the possibility of any chemical change in the blood as a factor in this acceleration. Only the nervous mechanism could act within such time limits.

The initial increase in frequency of the heart beat is first caused by a depression of the tone of the cardio-inhibitory center and, some seconds later, by an augmentation of activity of the cardio-accelerator center. The other factors responsible for an increase in heart rate during exercise are secretion of epinephrine, a rise in body temperature, and better return of the venous blood to the heart because of the milking action of contracting muscles.

That epinephrine does appear in the blood during physical exertion was demonstrated by Hartman and co-workers[212] through experiments on cats. Their test for the presence of epinephrine made use of the fact that the denervated pupil of an eye dilates when blood containing this hormone flows through it. The authors found that a good dilatation of the denervated pupil accompanied work on a treadmill. The dilatation begins within a few minutes after the start of work and increases as work progresses. Spurts of work are accompanied by greater increases in the dilatation. The importance of epinephrine for physical work may be seen further from the fact that intensive physical exercise causes enlargement of the adrenal glands.

Effect of Temperature. The quickening effect on the pulse rate of a rise in body temperature during exercise cannot be large. Ordinarily the body temperature during exertion does not rise above 102° F. The increased temperature of the blood acts directly on that area in the right auricle of the heart, the pacemaker (sinoauricular node), responsible for controlling the rate of the heart. Under usual conditions of work it appears that in man the rise in body temperature during exertion can at best be responsible for only a moderate quickening of the heart rate.

A high environmental temperature may greatly increase the frequency of heart beat. It stands to reason that a person with a weakened heart takes an unnecessary risk when indulging in vigorous activity on a hot day. Besides the increased work of the heart to provide a sufficient amount of blood for the active muscles, an additional strain will be imposed because of an augmented peripheral (skin) circulation for the purpose of heat dissipation. This double work may sometimes be fatal.

REFLEX FROM WORKING MUSCLES

It may seem strange, but, as in respiratory acceleration, the effect of stimuli arising from working muscles upon the heart rate has been either neglected or minimized.

Alan and Smirk[5] showed that if tourniquets are placed on the arm or leg, so that blood circulation is stopped, contraction of the muscles of these limbs causes acceleration of the heart rate. Asmussen, Christensen, and Nielsen,[14] in similar experiments, excluded 50 per cent of working muscles from circulation by means of pneumatic cuffs, and found that cardiac output increased in proportion to the intensity of work, although the amount of oxygen consumed was reduced to one-half. This study led to the conclusion that the presence in the blood of metabolites produced in working muscles is of little importance for control of the heart output, and that nerve control is the decisive factor. However, the experiment still left undecided whether this control was operated by motor impulses originating in the cerebral cortex or by reflexes arising in the working muscles. To solve this problem, Asmussen, Nielsen, and Wieth-Pedersen[19] carried out a study in which the increase in heart rate was studied during voluntary work of the legs, and during similar work induced by electric stimuli. The apparatus used for this study may be seen in Figure 64, Chapter 10, and the description of the use of electrodes in the corresponding text.

These investigators observed that voluntary work, as well as electrically induced work, had the same effect upon the pulse rate and the minute-volume of blood. This leads to the conclusion that, during a steady state of muscular activity, the pulse rate and the rate of blood circulation are controlled by impulses arising from working muscles, and not from the motor area of the cerebral cortex.

Thus, many factors separately affect the rate of work of the heart. Their combined effect one day will be expressed in a formula showing some sort of algebraic summation of partial effects. Of these, during exercise, stimuli arising from working muscles will be shown to be the most important.

EFFECT OF TRAINING UPON PULSE RATE

Pulse Rate during Rest. Some idea of the influence of various kinds of training may be secured from a study by Bramwell and Ellis[67] of 202 Olympic athletes. Most of these men were examined for a period of ten days preceding the various contests. Table 13 gives a summary of the observations on the pulse frequency.

The four classes of runners, who showed a decreasing frequency of pulse rate as the length of their run increased, are of special interest, for in the first three groups there was considerable similarity in age, though the men who ran the longer distances were slightly older. The average ages of the first three groups were twenty-one, twenty-three, and twenty-four years respectively. The marathon runners, however, were older men, in their late twenties and early thirties. In Table 13 the runners are, in fact, arranged according to the duration of their experience in athletic exercise. The question naturally arises, what part do quantitative differences in training and age play in the retardation of pulse frequency.

Submaximal exercises of the "endurance" type seem to have more

TABLE 13. *Pulse Rate in Olympic Athletes*

	Average Resting Pulse Rate	*Typical Range of Pulse Rate*
Sprinters (100–200 meters)	65	58–76
Middle distance runners (400–800 meters)	63	49–76
Long distance runners (1500–10,000 meters)	61	46–64
Marathon runners	58	50–67
Cyclists, sprinters	67	53–76
Cyclists, long-distance	64.5	51–73
Weight lifters	80	55–106

effect on slowing down the resting pulse than maximal ones of the "sprint" type.[326]

Pulse Rate during Exertion. In the performance of the same task, the trained man's heart has the advantage of starting at a slower rate of beating; but, on the whole, it accelerates as many beats in response to the task as does the heart of the untrained subject.[223A] Table 14 shows changes in the pulse rate during work in three men. DeMar was in excellent condition, A.V.B. in fair condition, and C.V.C. in poor condition.

These data show clearly that, for any given task, the pulse rate during work is slower in an athlete than in an untrained man, but the relative acceleration (expressed in per cent of resting pulse) is greater in the athlete.

TABLE 14. *Pulse Rates and Load of Work*

Oxygen Used, in cc. per Minute	*DeMar*		*A. V. B.*		*C. V. C.*	
	Pulse Rate					
	Absolute	*Relative in Per Cent*	*Absolute*	*Relative in Per Cent*	*Absolute*	*Relative in Per Cent*
250	58	100	88	100	90	100
500	64	110	93	106	96	107
750	71	123	99	115	102	113
1000	79	136	107	121	111	123
1250	89	153	115	130	122	135
1500	98	170	126	143	134	150
1750	108	186	138	157	146	162
2000	118	203	150	170	160	177

The effect of training on the heart rate may be observed during physical reconditioning of convalescents. With the regaining of physical fitness, their pulse rates in response to a standard exercise gradually begin to decrease. Absence of this decrease may be interpreted as a lack of improvement or as an indication that the exercise is too strenuous.

QUESTIONS

1. What is the difference between the terms "normal" and "average" pulse rate?
2. What is the range of the normal pulse?
3. Discuss the relation between the heart rate and: body position; food intake; emotions; changes in metabolism.
4. Discuss changes in the pulse rate during and after work.
5. What is the effect upon the heart rate of anticipation before an exercise?
6. Is there any relation between pulse rate and oxygen consumption? How can this relationship be utilized in physiology?
7. Does weight lifting affect pulse rate?
8. Does straining such as in the Valsalva phenomenon affect the heart rate?
9. How is heart rate affected in a step-up test by (1) height of the bench, (2) cadence, (3) weight of the subject?
10. Which returns to normal faster after exercise, oxygen consumption or pulse rate?
11. Is the pulse rate's return to normal after exercise of any importance in testing physical fitness? Why?
12. What is the oxygen pulse, and how is it affected by exercise?
13. Discuss the relation between pre- and post-exercise pulse rates.
14. Discuss the mechanisms regulating the pulse rate.
15. What effect does diving have upon the heart rate?
16. Discuss the importance of reflexes from working muscles in regulating the heart rate.
17. What effect does training have upon the heart rate?

Chapter Thirteen

ARTERIAL AND VENOUS BLOOD PRESSURE

FUNCTION OF THE ARTERIES

The quantity of blood passing through different organs is not constant, but alters with variations in activity. It is the function of the smallest arteries and the arterioles, which immediately supply the delicate capillaries, to vary the amount of flow according to the needs of the tissues. The arterioles, which are the vessels opening directly into the capillaries, are composed of only endothelial cells and spindle-shaped muscle cells. The muscle cells are supplied by two types of nerve fibers—vasoconstrictors and vasodilators—capable of varying the caliber of these small vessels by increasing or decreasing their contraction. The arterioles act as stopcocks in control of the stream of blood that flows from the arteries to the capillaries, and, furthermore, constitute the chief physiological mechanism to determine the peripheral resistance to the onward flow of blood.

The vasoconstrictor nervous mechanism is continually active and is responsible for the tone of the arterial walls. This tone is brought about through a partial contraction of the smooth muscles in the arterial walls.

The vasodilator nervous mechanism does not act continuously, but is called into action on special occasions, particularly when the diameter of the arterioles must be enlarged.

The distribution of the two kinds of vasomotor nerves is unequal. The vasoconstrictor nerve fibers are most abundant in the arteries of the skin and abdominal viscera, while the vasodilator fibers are most abundant in the glands and muscles.

Control of the arterial diameter is maintained by the vasomotor center or centers in the medulla of the brain. There is a distinct vasoconstrictor center, which, by its tonic activity, constantly sends nerve impulses to the arteriole muscles along the vasoconstrictor nerve fibers.

The vasoconstrictor center is kept active in part by chemical action of the gases and other compounds in the blood and in part by being continuously bombarded by afferent nerve impulses coming to it from all parts of the body. While there is also a vasodilator center, it is not so definitely localized as the vasoconstrictor center. The vasodilator center acts only intermittently, probably by reflex. A coordinated action of both centers is indispensable for the control of blood flow.

The blood vessels of the body, if fully relaxed, would have a capacity far greater than the entire blood volume. It is evident, therefore, that, if the blood is to circulate, the capacity of the vessels must be reduced to a size sufficiently small to allow the blood to fill the entire stream bed. The tonus of the arteries, which is only one of the means that provide for this narrowing of the circulatory bed, also plays an important part in a reciprocal manner. Thus, for example, when more blood is required for the muscles, the arterioles reduce the blood supply to the abdomen and skin by vasoconstriction, and increase the flow into the muscles by vasodilatation. Further, if the arms are being used and the legs are at rest, the arm muscles are especially in need of blood. The arterioles of the arms then become dilated, while those of the legs remain unchanged; or, if more blood is needed in the arms, the legs become more constricted.

ARTERIAL BLOOD PRESSURE

The factors upon which blood pressure depends are the pumping action of the heart, the peripheral resistance offered to the outflow of blood from the arteries, which varies with elasticity and vasoconstriction, and the volume of the circulating blood. Only the first two are important variables during physical activity of the body.

The blood pressure is varied during physical activity to provide an adequate blood supply. The variations are brought about by the regulatory activity of the vasomotor and cardiac centers in the brain.

Arterial blood pressure of man is usually measured in the brachial artery of the arm. It is considered indicative of the pressure in the arteries generally, although pressure varies from artery to artery. The maximum pressure caused by the systole of the heart is spoken of as the systolic pressure; the minimum pressure in the artery between heart beats, that is, the pressure at the end of the diastole of the heart, is known as the diastolic pressure. The difference between systolic and diastolic pressure is designated as the pulse pressure. The systolic pressure is considered an index of heart energy expended, and indicates the strain to which the arteries are subjected. The diastolic pressure is generally considered a measure of the peripheral resistance to the circulation of the blood and, therefore, an index of vasomotor tone.

Changes in blood pressure from birth to twenty years of age are of special interest to physical educators. One group of investigators reported an abrupt rise in blood pressure at the age of six; another group reported that the rise is steady until eleven years, when an abrupt rise takes place. Since an abrupt rise in blood pressure may be interpreted as a warning that children at the time of its occurrence should not indulge in strenuous exercise, one may ask, "When should a child 'take it easy'—at six or at eleven?" The authors of this book are of the opinion that this warning is a theoretical scarecrow. As has been shown, in Chapter 11, a rigid control of activities was once insisted upon because of a mythical discrepancy between the development of the heart and the large arteries. A debatable abrupt rise in blood pressure can hardly be considered a sound basis for modifying activities.

It appears to be the general opinion that the limits in normal individuals at rest range, for systolic pressure, from 110 to 135 mm. Hg; for diastolic pressure, 60 to 99 mm. Hg; and for pulse pressure, 30 to 55 mm. Hg. These pressures are slightly lower in women than in men.

From an intensive study, Alvarez and Stanley[7] concluded that blood pressure during health varies but little from youth to old age. The mean pressure does not increase until after the age of forty years. In the years from twenty to forty, the percentage of people with pressures above 140 mm. Hg remains almost constant. This suggests that all those who have a high pressure at the age of forty had it at the age of twenty. After middle age, fatness tends to increase and thinness to decrease the blood pressure. Persons subjected to a life of worry and fatigue have pressures somewhat above the normal. Some claim that, beginning with the menopause in women, and at the age of fifty in men, there is a steady increase in systolic and diastolic pressure, resulting in a higher pulse pressure.

POSTURAL BLOOD PRESSURE CHANGES

In spite of all the work that has been done on this subject, there is no general agreement about the relative values of the brachial arterial blood pressures in the recumbent and erect postures of the body. The preponderance of evidence is that a healthy man in the standing position may have a blood pressure either higher, lower, or the same as in recumbency. However, in the position of standing on the head, the blood pressure rises.

The daily ritual of standing on the head is practiced not only in India but also in this country. Whatever subjective satisfaction is derived goes beyond the scope of this book. *A priori* this stunt should be contraindicated in people with high blood pressure and hardening of the arteries. Rao[421] tested a number of subjects and found that, compared with the upright position, the following changes took place when the subjects stood on their heads: the pulse rate went down 15 beats per minute and the systolic blood pressure in the arm rose 18 mm., but in the tibial artery it dropped from 196 mm. to 10 mm. It should be noted that the Indian subjects had low systolic blood pressure, the average being only 90 mm.

Changes in the blood pressure observed during the first minute or two after taking the erect posture are of little value in interpreting postural circulatory condition, since such changes are often the effect of the muscular work of getting up.

ANTICIPATORY RISE IN BLOOD PRESSURE

That excitement causes a rise in the systolic blood pressure is well known. For this reason lie-detecting machines have been used to record blood pressure changes in criminal suspects.

Athletes exhibit blood pressure changes before contests and during the days of contests. In a group of weight lifters the systolic blood pressure fluctuated as follows: one day before contest, 108 to 124; first day of contest, 120 to 136; second day, 127 to 156; third day, 129 to 163; two days later, 108 to 137.[331]

ARTERIAL BLOOD PRESSURE DURING MUSCULAR EXERTION

During any exercise, the demand for oxygen increases. In order to meet this demand, the amount of blood passing through the lungs and working muscles in a unit of time must also be increased. This is achieved through a greater velocity of the blood flow. Since the velocity of blood flow depends largely on the arterial pressure, it is clear that this pressure must also rise during muscular work.

Figure 86 shows systolic and diastolic blood pressure changes in the brachial artery, pulmonary capillaries, pulmonary artery, and right ventricle in athletes and non-athletes during supine exercise. The athletes were eight top-flight cyclists at the beginning of their training season.[54]

Systolic pressure rises with increases in heart rate in both athletes and non-athletes. The latter, however, do not show as large an increase for a given heart rate. The brachial artery pressure shows rather substantial changes, up to 211.1 ± 28.1 mm. Hg during sitting exercise. This is apparently related to the higher stroke volume found in athletes.

Pressures on the pulmonary side of the circulation do not show as much change as those on the systemic side (Figure 86). This serves a useful purpose in that extreme pressures in the lungs would result in the diffusion of water from the blood to the alveoli. Guyton[200] points out that the pulmonary arterioles are very passive and that they expand with pressure, offering little resistance to the passage of blood, and thereby reduce the load on the heart.

During steady-state work, the systolic blood pressure may rise and remain more-or-less constant. In fatigue, however, the person unadapted to exertion may show a drop to below resting level.[114]

The changes in diastolic pressure tend to be quite small compared to the systolic changes. Values shown for athletes in Figure 86 for rest, low work load, and high work load are 68 ± 7.1, 73.5 ± 5.6, and 84.6 ± 7.3 mm. Hg. This observation is even more strongly supported by data

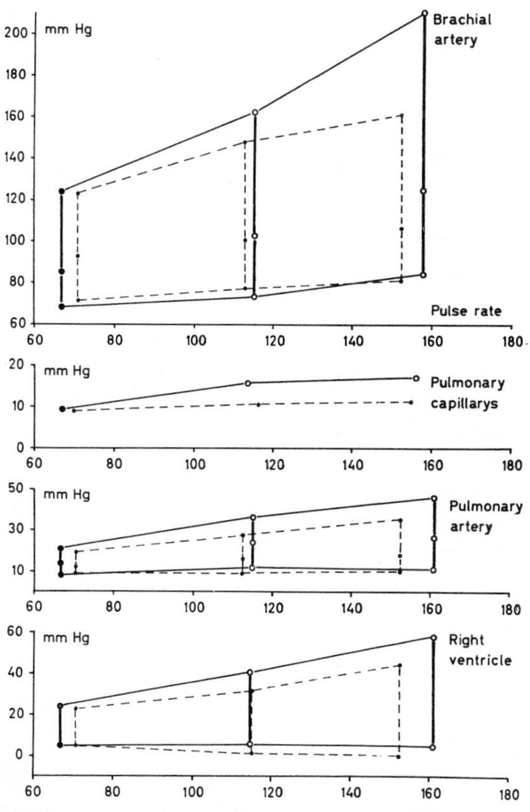

Figure 86. Blood pressure changes in response to exercise. Solid lines show data for athletes, while broken lines show data for nonathletes. Exercise was pedaling a bicycle ergometer in the supine position. (Bevegard, S., Holmgren, A., and Jonsson, B.: Acta Physiol. Scandinav. 57:26, 1963.)

taken in the sitting position, where the respective values were 81.3 ± 11.7, 83.0 ± 8, and 87 ± 11.5 mm. Hg.[54]

The small changes in diastolic pressure during exercise are related to arterial blood flow dynamics and peripheral resistance. As the blood is pumped from the heart, during systole, it also flows from the arterioles. However, the flow from the arterioles is not as fast as the flow into the aorta. This difference is absorbed by the elasticity of the aorta and arterial network; in other words, the structures stretch. When the heart goes into diastole, the force absorbed during systole is applied to the column of blood as the stretched structures return to usual dimensions. The amount of pressure remaining at the next systole is the diastolic pressure.

The rate at which the stretched arterial network returns to its regular dimensions is dependent on the rate at which the blood flows out of the arterioles into the capillaries. Most of the resistance to blood flow is found in the latter segment of the vascular system. We have noted already in this chapter that the arterioles act as "stopcocks" in the control of blood flow and that they provide peripheral resistance.

If the peripheral resistance is high, the diastolic pressure will be high. The force stored in the elasticity of the aorta will not be dissipated between systoles if the blood cannot escape into the capillaries. Obviously, if the peripheral resistance remained the same during exercise as during rest, diastolic pressure would increase with systolic pressure rather than show the relatively constant pattern shown in Figure 86.

During exercise there is a marked decrease in peripheral resistance due to vasodilatation. The arterioles of the active muscles open due to the effects of exercise allowing the blood to flow freely through the capillary bed, causing the peripheral resistance to become less.[85] The effects are shown in Figure 87. Because of this reduction, the active muscles are richly supplied with blood during exercise. Non-active muscles, however, show vasoconstriction.

Another factor affecting blood pressure during exercise is the rhythmic compression of the capillary bed during muscle contractions. This tends to act as a resistance to capillary flow as well as an aid to the return of blood to the heart, because of the massaging effect on the veins. The diastolic pressure is maintained somewhat as a result of this flow restriction in opposition to the peripheral vasodilatation.

The *pulse pressure* is the arithmetical difference between the systolic and diastolic pressures. Since, during exercise, the diastolic pressure changes little and the systolic pressure increases considerably, the pulse pressure tends to increase and decrease with the systolic pressure. It is a rough indication of the effective pumping force of the heart.

The segment of the body being exercised also has an effect on the blood pressure response. Åstrand and co-workers[29] found that blood

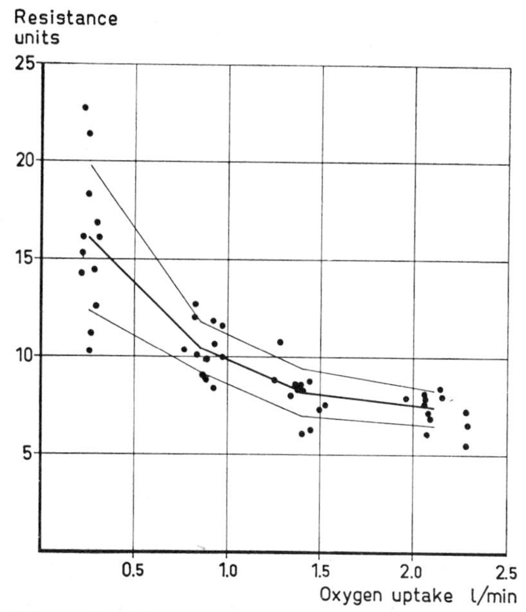

Figure 87. Changes in peripheral vascular resistance in response to exercise. Resistance units are expressed as mm. Hg/liter flow/min. Work rates ranged from 300 to 900–1000 kgm./min. (Carlsten, A., and Grimby, G.: The Circulatory Response to Muscular Exercise in Man. Charles C Thomas, 1966.)

pressure in both arm and leg cranking exercises increased linearly with oxygen intake. The increase was, however, more pronounced for arm than for leg work. At about 30 per cent of maximum oxygen intake, mean femoral catheter values were 152/86 for arm work and 129/70 for leg work. At 67 per cent the pressures were, respectively, 198/100 and 156/72. The higher pressures for arm work were considered to be due to higher peripheral resistance and greater static effort.

FACTORS INFLUENCING ARTERIAL BLOOD PRESSURE DURING EXERCISE

The anticipatory and the initial rises of blood pressure before and during exercise are brought about by a flow of nerve impulses from the cortex of the brain to the cardiac and the vasoconstricting centers in the medulla. Stimulation of these centers causes an acceleration of the heart rate and constriction of the blood vessels in the splanchnic area. The combined effects result in an increased arterial blood pressure.

Once exercise is started, other factors add their influence to increase arterial blood pressure; for example, the simultaneous and considerable degree of vasoconstriction in the skin, as evidenced by slight pallor. This vasoconstriction is of nervous origin and is caused by cortical impulses or reflexes from the skin. Later in the exercise, however, as the body becomes heated, the skin vessels again dilate; so their constriction is not a continuous aid in augmenting blood pressure. We know that the arterioles and capillaries in the active muscles dilate and can only be adequately filled with blood if, at the same time, there is compensatory constriction of blood vessels in the inactive parts of the body. That such compensatory constriction occurs is undoubted. The splanchnic vessels in the abdomen offer the most favorable field for this compensation. It therefore seems likely that, as long as there is a need, the vessels of that area remain constricted.

POSTEXERCISE BLOOD PRESSURE

Systolic pressure values tend to be high immediately after exercise, while diastolic pressures are typically below resting values. During recovery, vasodilatation remains until the vasomotor centers can restore normal tone in the arterioles. With the loss of resistance caused by the cessation of skeletal muscle contraction, the blood flows freely into the capillaries and the diastolic pressure drops. In ice skating and middle-distance racing diastolic pressure has been shown to be subnormal, but the amount is apparently not related to the distance covered.[331] Schellong[454] observed a drop in subjects after they had completed 2 runs on a 23-step stairway in 45 seconds.

If exercise is very severe, the loss of the massaging action of the muscles on the veins and the high rate of blood flow into the area due to the decreased peripheral resistance after exercise may lead to excessive pooling of blood in the muscles if vasomotor tone is not quickly restored. This in turn will lead to a reduction in the volume of circulating blood, reducing heart output, and a decrease in systolic as well as diastolic pressure. Average systolic and diastolic blood pressure, taken within 15 seconds after all-out bicycle ergometer tests in a group of 21 subjects before and after training, were 160/64 and 162/63, the diastolic pressure showing the expected low values. Individuals within the group, however, showed marked reductions in both phases: the two most extreme cases were 100/48 and 90/45. After five minutes these values increased to 114/68 and 118/60 respectively (unpublished data). Both subjects experienced syncope after exercise. This problem can generally be avoided if one does not stop exercise abruptly but continues to move, to keep the muscles massaging the blood back into circulation.

VALSALVA PHENOMENON

It is a fact that a combination of static muscular contraction and breath holding may be taken advantage of with occasional success in wrestling. If, while the opponent is in a bridge position, a powerful pressure is applied to his chest and abdomen during an expiration, in order to interfere with his inspiration, it may cause dizziness. A somewhat similar phenomenon is observed during a stunt used by boys. The "victim" stands still, holding his breath after deep inspiration. The "lifter" stands behind the victim and, after placing his arms around the victim's chest, lifts him up by squeezing the chest as hard as possible. The combined effect of the squeezing and the pull on the victim's body may cause a considerable fall in blood pressure and fainting.

These observations may be explained by the *Valsalva phenomenon*, which occurs in exercises of strain performed with the glottis closed and breathing suspended. This tends to compress the chest and thereby cause a rise in intrathoracic pressure. The systolic pressure first rapidly rises to 180 or 200 mm. Hg, and then just as rapidly falls to below 60 mm. Hg. The fall in pressure is caused by the fact that intrathoracic pressure rises so high that it prevents the return of venous blood to the heart, with the result that the heart receives little or no blood to pump. A vigorous compression of the chest, when the breath is held, produces the Valsalva phenomenon. Fox and co-workers[176] showed a reduction of 35 per cent in the thoracic blood flow with increases to 19 per cent over control values immediately after voluntary inducement.

WEIGHT LIFTING AND ISOMETRIC CONTRACTIONS

An increase in the systolic blood pressure has been observed after weight lifting.

A feature in the after-effects of exercise on the systolic blood pressure that is not easily observed has been brought out by Cotton, Lewis, and Rapport.[107] They had their subjects lift 20-pound dumbbells from the floor to the full stretch of the arms above the head, swinging them up in one motion and down in another, seven to sixty times. Immediately after the exercise the systolic pressure showed a marked fall, being little, if at all, above the normal resting level. Within about ten seconds the pressure began to mount, to reach a maximum in twenty to sixty seconds. The fact that the muscles are relaxed and empty at the end of such exercise explains the changes during the first half minute or so. It takes a few seconds for the blood to fill veins in the muscles, and during this interval the systolic pressure falls. It is suggested that for this period the supply of blood to the heart is reduced and its output, therefore, decreased. As soon as the veins are again filled the heart resumes its large output and returns the systolic pressure toward that of the exercise period. It has been suggested that the sudden drop in blood pressure immediately at the close of exercise is the result of the sudden stoppage of the muscular pump when exercise ceases; this brings about a momentary stagnation of blood in the capillaries, which is rapidly succeeded by restoration of the venous inflow to the heart as the capillaries begin gradually to empty themselves into the great veins.

Lind and co-workers[345] have studied extensively the effects of isometric contractions on cardiovascular responses. Figure 88 shows typical changes during an isometric handgrip contraction and an exhausting progressive treadmill exercise. Note the excessive increase in blood pressure compared to heart rate response, as well as the parallel increase in systolic and diastolic pressures during isometric exertion.

There is a small increase in heart rate, cardiac output, blood pressure, and muscle blood flow to a steady state, when isometric contractions are less than 15 per cent of the maximum force. Exercises of this kind can be maintained indefinitely. During isometric contractions greater than 15 per cent of the maximum, however, heart rate, cardiac output, blood pressure, and muscle blood flow increase continu-

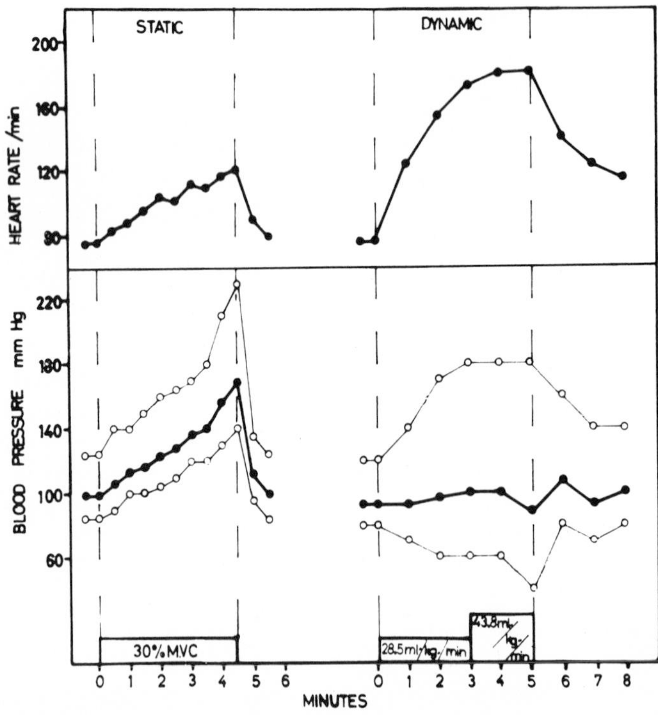

Figure 88. Cardiovascular responses to static and dynamic exercise. Static exercise was a sustained hand-grip contraction at a resistance equal to 30 per cent of the maximum voluntary force. Dynamic exercise was a progressive treadmill test to exhaustion. (Lind, A. R.: Canad. Med. Ass. J. *96*:706, 1967.)

ously until fatigue occurs. The heart rate and blood flow increases tend to be small, while the pressure increase is large.[345]

Fatigue sets in earlier as the amount of exertion increases. Periods of 10 to 13, 4 to 6, and 1 to 2 minutes are reported for forces equal to 20, 30, and 50 per cent of the maximum voluntary contraction. Muscle blood flow remains high or even increases after exercise, while heart rate and blood pressure quickly return to normal. After strenuous exertions, muscle fatigue, as shown by lack of ability to apply force, has been shown to last up to 40 minutes.[345]

These observations are related to the compression of the vascular bed of the contracting muscle. When muscles contract, a reflex apparently originating in the muscle accelerates the heart.[345] The vascular compression causes a high peripheral resistance. The cardiac output remains high also—the additional force of the pumping action being absorbed by the arterial side of the vascular system, since the blood has no place to go. As a result, both systolic and diastolic pressure increase.

Lind[344] noted that the unusual pressure increase indicates a need for caution in the use of isometric exercise for training programs. The body tolerates the increased blood flow of isotonic type activities better than increased blood pressure. It is advised that persons with aneurysms, incompetent valves, or angina pectoris not use isometric exercises or isotonic exercises such as push-ups that have strong static contractions accompanying them.[344]

EFFECT OF TRAINING ON ARTERIAL BLOOD PRESSURE

Perusal of the literature pertaining to this topic reveals that training may cause an increase, a decrease, or no change in the systolic blood pressure. Although a report by Cogswell and co-workers[102] indicated a tendency for the systolic pressure to decrease, the statement made by Dawson more than 50 years ago—that the effects of training on the resting blood pressure are neither striking nor constant—still seems to hold true.

In this connection, of interest is a study on 202 Olympic athletes by Bramwell and Ellis,[67] which showed that Olympic athletes have systolic, diastolic, and pulse pressures within the range common to people of similar ages (Table 15).

EFFECT OF MUSCULAR ACTIVITY UPON VENOUS PRESSURE

The rise in venous pressure during a steady type of work may begin immediately or may be delayed several minutes. If the load of work is not too heavy, the pressure gradually continues to rise and

TABLE 15. *Arterial Blood Pressure in 202 Olympic Athletes*

	Systolic		*Diastolic*		*Pulse Pressure*	
	Average	*Typical Range*	*Average*	*Typical Range*	*Average*	*Typical Range*
Sprinters................	116	105–120	77	70–80	39	35–45
Middle distance runners	119	105–130	81	75–90	38	30–55
Long distance runners	116	110–125	76	70–85	40	35–45
Marathon runners...	123	110–145	78	65–80	45	35–55
Cyclists, sprinters.....	124	....	86	....	38	....
Cyclists, long distance	123	......	76	....	47	
Weight lifters	134	110–150	90	70–120	44	25–60

eventually enters a steady state. If the load is too heavy the pressure rises steadily until fatigue occurs.[459] There is a rough linear relationship between venous pressure and load. This may be obscured by the heavy breathing of exertion, since during expiration the venous pressure may be as much as 2 cm. of water pressure higher than during inspiration. Ordinarily, after exertion the venous pressure returns slowly to the pre-exercise level. It does not as a rule fall below that level. The time required for recovery varies, being longer after severe exertions, when it may extend through a period of twenty-five or thirty minutes. The rise during dynamic muscular contractions is much greater than during static.[536, 537, 538]

The effect of the "milking action" exerted by working muscles upon the blood in the veins has been well demonstrated by continuous measurements of pressure in the great saphenous vein at the ankle level during a walk on a treadmill. Venous pressure falls considerably but continues to rise and fall synchronously with contraction and relaxation of the calf muscles.

EFFECT OF RESPIRATION UPON VENOUS PRESSURE

It is known that changes in the intrathoracic pressure during respiration affect venous blood pressure and flow. During inspiration the blood pressure in the central veins falls, and during expiration it rises. The total effect of deep respiration results in a diastolic increase in the heart volume. Prolonged observations of heart output, however, have shown that a deeper respiration does not cause any increase in the heart output.[161] Although venous pressure during exercise has a positive linear relationship with pulmonary ventilation, this relation is coincidental and not interdependent.

QUESTIONS

1. Describe the mechanism controlling the size of the blood vessels.
2. What is the average normal blood pressure? How does it change with age?
3. What is pulse pressure? What does the name indicate?
4. Discuss the effect of posture upon the blood pressure and pulse rate.
5. What is the effect of exercise upon the blood pressure?
6. Explain the effect of straining upon the blood pressure.
7. What is the Valsalva phenomenon?
8. Discuss the post-exercise systolic and diastolic blood pressure.
9. What is the immediate effect of weight lifting upon the blood pressure? Of isometric exercise?
10. What is anticipatory rise in blood pressure?
11. What is the effect of training upon blood pressure?
12. Discuss the effect of respiration on venous pressure.

Chapter Fourteen

COORDINATION OF FUNCTIONS OF VARIOUS ORGANS FOR MUSCULAR WORK

Before one engages in physical activity, such as an important contest involving emotions, there is an anticipatory mobilization of various functions—respiration, blood circulation, and secretion of epinephrine—which raises the sugar level of the blood. But when a non-emotional activity is involved, mobilization of bodily functions takes place during the activity.

LOCAL CONTROL

An excess amount of metabolites resulting from an inadequate oxygen supply produces several changes, some local and others remote. Local changes consist of the direct action of metabolites upon blood vessels. Specifically, the arterioles dilate, allowing blood to flow through the capillary bed in greater amounts.

The dilation of the capillaries and of the arterioles favors a better blood supply to the muscle. There are also two more factors which help the muscle to secure more oxygen. The increased production of carbon dioxide facilitates dissociation of oxygen from hemoglobin. Thus more oxygen is taken from the same amount of blood or, in other words, the coefficient of utilization of oxygen is increased.

REMOTE CONTROL

When muscles commence contracting, their sensory nerve receptors are stimulated, and volleys of nerve impulses are sent to the cerebrospinal system. Some of these impulses reach the level of consciousness, but most of them do not, and remain at a reflex level. The latter messages are relayed to the cerebellum, and to the medulla. In the cerebellum these messages are indispensable for the proper execution of movements. For example, a person suffering from tabes dorsalis, a condition in which the dorsal columns of the spinal cord have lost their ability to transmit messages to the brain, cannot perform well even such a relatively simple activity as walking.

Impulses reaching the medulla stimulate sympathetic centers, among which are the respiratory, cardiac, and vasomotor. These centers in their turn stimulate the subordinate organs. For example, as a result of the stimulation of the respiratory center, the minute-volume of breathing is stepped up, the bronchioles of the lungs are expanded, and the infundibula are opened. The sum total of these responses provides a better aeration of the blood, more oxygen, and a greater elimination of carbon dioxide.

Through the action on the vasomotor center, there is constriction of blood vessels in unused regions of the body, notably in the splanchnic area and in the skin. (The action on the skin vessels lasts only until the rise of body temperature requires increased elimination of heat.) As a result of vasoconstriction a larger amount of blood becomes available for the muscles. Confining the blood to a smaller stream bed raises the arterial and the venous blood pressure. The rise in arterial pressure favors the flow of blood to the muscles; and the rise in venous pressure favors the return of blood to the heart.

When nerve impulses from working muscles reach the cardiac inhibitory center in the medulla, the action of this center is limited. In consequence, the heart rate accelerates. Moreover, the cardiac inhibitory center becomes less sensitive to stimuli coming from the carotid sinuses and the aortic arch. During exercise, arterial blood pressure causes distention and, therefore, stimulation of the carotid sinuses and the aortic arch which, under normal conditions, stimulate the cardiac inhibitory center to slow the heart rate. Because of a lowered sensitivity of the cardiac inhibitory center, the heart rate, during muscular work, is allowed to accelerate.

When work is prolonged, then, a further acceleration of the heart rate is effected also through a reflex from the muscles. This time the cardio-accelerator center is stimulated.

EPINEPHRINE

Epinephrine, produced by the adrenal glands, acts directly on organs to bring about the same action brought about by the sympathetic nerves. Therefore, in vigorous physical activity the sympathetic system and the adrenal gland system reinforce each other. For this reason, the two systems are frequently referred to as one sympathetico-adrenal system. Through the action of this system the heart rate is accelerated; the arterial blood pressure is raised; the liver is stimulated and glycogen is converted into sugar and sent to the muscles.

EFFECT OF HEAT

During exercise, excess heat is produced. If no means is provided for heat dissipation, the body temperature will rise to such an extent that either further work will become impossible or life itself will be endangered. Thus it is essential to prevent an undue rise in temperature. This is most effectively achieved by getting rid of heat through evaporation of sweat.* It is true that much more sweat is produced than can be evaporated, but this is characteristic of most bodily reactions—to provide safety measures in excess.

Increased sweating during muscular work is effected through activation of the hypothalamus. When blood heated in the muscles reaches the hypothalamus, the latter stimulates activity in the sweat glands and vasodilatation in the skin, which will neutralize the original vasoconstriction caused at the beginning of muscular work. In spite of an increased heat dissipation during muscular work, body temperature as high as 105° F. has been recorded after running.

"MILKING" ACTION OF MUSCLES

The muscles also aid mechanically in bringing about an increase in the minute-volume and output of the heart per beat. The intermittent contraction of muscles engaged in performance of work squeezes the blood on into the veins. The valves in the veins prevent the blood from surging back, and favor the inflow of blood from arteries to capillaries as a muscle relaxes. The pump-like action of the muscles increases the return of venous blood to the heart and raises the venous pressure. The rise in venous pressure causes a more rapid filling of the heart in its diastole and, provided the heart has not already been fully distended, stretches the heart muscle. According to Starling's "law of the heart," this causes a more forcible contraction and a larger output of blood per beat. As a consequence, the volume of blood pumped by the heart per minute is also increased.

---

*Approximately 0.58 Cal. (large) is used to evaporate 1 gm. of sweat.

EFFECTIVENESS OF REFLEX VERSUS CHEMICAL CONTROL

It has been shown that a small oxygen debt is acquired during the first minute or so of exercise, before a steady state has been reached. This indicates that neither the stimulation of the adaptive mechanisms by the psychic or higher brain centers during the anticipatory phase nor the reflexes from the muscles is adequate. Evidently the local changes in the muscles, such as dilatation of capillaries and small arterioles caused by a momentary increase in metabolites, are necessary.

When the limit of the transportation capacity of the blood has been exceeded, further muscular activity is possible only to the extent of a maximal oxygen debt. When this limit is reached, the excess of metabolites, chief among which is lactic acid, sets in operation a process which may be considered as a safety device. An excessive amount of lactic acid causes a toxic effect upon the motor area of the cerebral cortex. Muscular movements, therefore, become slower, uncoordinated, and sometimes entirely impossible. A further production of lactic acid is thus prevented, and recovery begins.

FACTORS LIMITING ATHLETIC PERFORMANCE

Why some athletes, at the peak of their training, cannot lift as much, jump as high or run as fast as others who may be built very much like them is a real question. What limits them? The explanation may be found in the inferior quality of their muscles and in a more limited capacity for work mobilization of their physiological systems and functions (such as respiration, circulation, and coefficient of utilization of oxygen). In athletic events of the sprint type, muscle quality plays the most important part. In events of the endurance type, the capacity for work mobilization comes to the fore.

It is very rarely that endurance is limited by a lack of fuel for muscular contraction. One has to run a marathon race or swim across the English Channel to experience such a condition. When an athlete gives up a race, he does so because he is out of breath and his muscles seem to have lost their power.

From this common observation one may conclude, and rightly so, that the athlete in question has stopped, or has been forced to slow down, because of a lack of oxygen. The lack of oxygen has caused an increase in the amount of lactic acid in the blood, which has deleteriously affected the muscles. A question may be asked: What was responsible for this lack of oxygen?

RESPIRATION

Since this lack of oxygen coincides with a great difficulty in respiration, termed "intolerable breathing," one might conclude that it is inadequate respiration that causes the lack of oxygen.

This conclusion is wrong. Even during intolerable breathing, there is a sufficient amount of oxygen present in the lungs. At rest, the partial pressure of oxygen in the alveolar air is about 100 mm. of Hg. During exercise, because of a greatly increased pulmonary ventilation, the partial pressure is at least as high as or even higher than during rest. Therefore, it is not the fault of respiration that athletes suffer from a lack of oxygen.

If this be true, why then does an athlete develop a state of "intolerable respiration"? The explanation is rather simple. Because of an insufficient amount of oxygen available to the tissues, an excess of lactic acid is present. Lactic acid stimulates the respiratory center, and respiration is augmented. The more acid produced, the greater the augmentation. Finally, a state is reached which overtaxes the respiratory muscles, and an accumulation of waste products creates a feeling of "intolerable respiration."[101]

TRANSPORTATION OF OXYGEN

Since, at all times during exercise, there is an abundance of oxygen in the lungs but not enough in the muscles, the fault then must lie in an inadequacy of transportation. Let us consider various physiological features involved in the transportation of oxygen:

1. *Blood Saturation with Oxygen in the Lungs.* During work blood moves faster through the lungs than at rest. May the speed increase so much that the blood cannot pick up its normal quota of oxygen? This does happen in some cases. At rest, blood may be saturated from 95.6 to 100 per cent. During exhaustive work, saturations of 91 per cent[213] and even 85.2 per cent[444] have been reported. Persons with a markedly lowered saturation probably have thicker alveolar epithelium, or probably there has not been a sufficient mobilization of red blood corpuscles. In any case, these men cannot be champion distance runners; and they will never make good high-mountain climbers, because at high altitude the partial pressure is lower than at sea level and, therefore, absorption of oxygen from the lungs will become more inadequate.

2. *Unloading of Oxygen in the Muscles.* During muscular work the per cent of oxygen unloaded from arterial blood is increased 2 to 2.5 times. It is obvious that the lower this limit, the less oxygen will be available, and, therefore, endurance will be less.

3. *The Heart Output per Minute.* The greater the heart output, the

greater may be the amount of oxygen transported. The heart output may be increased in two ways: by increasing the stroke volume, and by increasing the heart rate.

As has been shown, better athletes have a larger stroke volume, and in this lies their superiority. Of course, the heart rate of a non-athlete may reach a higher level than that of an athlete, but after it has reached a certain frequency, the stroke volume may become smaller, because the diastolic filling is reduced.

4. *Removal of Carbon Dioxide.* When lactic acid is produced in excess, it appears in the blood. Being stronger than carbonic acid, it will combine with sodium bicarbonate and other substances used for carrying carbon dioxide, thus reducing the ability of the blood to transport this gas to the lungs for elimination. A person who has more buffers in the blood, or who can mobilize an additional amount of buffers, is able to eliminate more carbon dioxide in spite of lactic acid interference and, therefore, has greater endurance than a man lacking in these characteristics.[101]

EFFECT OF COMPENSATORY ADJUSTMENTS ON DIGESTION

The compensatory adjustments made for the benefit of working muscles are made at the expense of other organs. This is notably true for the organs of digestion, particularly the stomach. The effect depends on the extent to which the blood is diverted from these organs. Light exercise, such as moderate walking, probably does not divert any blood, and, therefore, digestion and the rate of the emptying of the stomach are not affected. Moderate exercise, such as running slowly 2 or 3 miles, may retard digestion and emptying of the stomach.[80] The extent of the interference of physical work with gastric function depends on the degree of physical fitness. A seasoned lumberjack will work comfortably all morning after a hearty breakfast and will be hungry as a wolf at lunch time.

EFFECT ON THE KIDNEYS

Every normal athlete has experienced certain changes in excretory function during the anticipatory period preceding an important contest. After a customary emptying of the bladder he may feel an urge to urinate again, only to find that but a few drops of urine may be excreted. His excitement has lowered the threshold of sensitivity of the bladder.

During strenuous exercise, the amount of urine produced is greatly diminished or its production entirely discontinued. Three factors are responsible for this: (1) movement of water from blood plasma into tissues; (2) diversion of blood from the kidneys to active muscles; and (3) increased sweating. The effect of the last factor is debatable. While it is true that the kidney function is inhibited *before sweating begins*, a copious sweating during work also reduces the amount of urine produced. It is probable that the purpose of inhibition of kidney function is, mainly, to save water which can be used for heat dissipation in sweating.

The urine is strongly colored, after exercise, and has an acid reaction. Lactic acid appears in the urine, its amount roughly indicating the intensity of the exertion.

After long and strenuous physical activity, most athletes have protein (proteinuria), and some may also have erythrocytes, in the urine. In this connection observations made in Sweden on skiers are of interest. After a 5-km. run, 92 per cent of the well-trained boys and 80 per cent of the trained girls had proteinuria. Among the less trained participants, 92 per cent of the boys and 93 per cent of the girls had proteinuria. In well-trained adults, after 10- to 50-km. runs, the incidence of proteinuria was 92 per cent for men and 100 per cent for women. Erythrocytes were found in the urine of 28 per cent of the men and 53 per cent of the women.[95]

All these chemical and microscopic changes in the urine seem to indicate pathological conditions of the kidneys. This point of view, however, can be debated. For instance, vigorous exercise produces morphological changes in the blood similar to those observed during appendicitis, but this similarity does not necessarily mean that such changes indicate pathological conditions.

On the same basis we cannot say that changes in the urine after excessive exertion represent pathological conditions of the kidneys. These changes may be regarded as an adaptation of renal functions to a high metabolic level. It is possible, however, that a considerable change in the composition of the urine may indicate that the limit of physiological adjustment has been reached.

SUMMARY

Voluntary work is initiated by the cerebrum. The proper coordination is maintained by the cerebellum. During work, there is a constant flow of sensory stimuli, from working muscles. These stimuli are utilized by the central nervous system for the proper execution of contractions. Well-learned activity is performed mostly in a reflex manner. The brain also prepares the body for exertion during the anticipatory period by augmenting respiration and heart rate, and by constricting the blood vessels of the splanchnic area and skin. Working muscles exert a control on the other organs: by reflex, chemically, through a rise in temperature, and by "milking action" upon the blood in the veins. Local control consists of opening muscle capillaries and small arterioles. Remote control consists of stimulation of the respiratory, cardiac, and vasomotor centers in the medulla oblongata. During a steady state, reflexes from the muscles suffice to maintain functions of the other organs at a desired level. When the intensity of work increases, excess metabolites produced in the muscles act as additional stimulants to the central nervous system. Increased temperature of the blood stimulates the hypothalamus, which in its turn causes vasodilation in the skin and an increased activity of sweat glands.

Factors which limit athletic performance are individual differences and the capacity of the blood to transport oxygen. There is always enough oxygen in the lungs, and the blood has enough time to become saturated with oxygen, but the supply may still be inadequate. A greater heart output per minute, a greater amount of hemoglobin, and buffers in the blood increase the capacity of the blood to transport oxygen and carbon dioxide.

Digestion and time needed for emptying of the stomach may be impaired by strenuous exercise, especially when a person is not used to this type of work.

Production of urine during exercise is inhibited. When intensity of work reaches the limit of man's tolerance, not only protein but also erythrocytes may appear in the urine. The incidence of proteinuria after distance runs on skis may be as high as 100 per cent for women and 92 per cent for men. Erythrocytes may appear in the urine of 28 per cent of the men and 53 per cent of the women.

QUESTIONS

1. Discuss the local control of constriction and dilatation of the capillaries and arterioles.
2. Which area of the brain maintains proper coordination?
3. Discuss the remote control of blood vessels.
4. What effect has epinephrine upon the blood vessels, spleen, heart, blood pressure?
5. What is the sympathetico-adrenal system?
6. How is an undue rise in body temperature prevented during exercise?
7. What is the "milking" action of muscles?
8. Discuss the effectiveness of reflex control of organs during exercise as compared with chemical control.
9. Discuss factors limiting athletic performance.
10. What is "intolerable breathing"? Is oxygen content of the lungs during this state very low?
11. What per cent of oxygen is present in the blood at rest? During exercise?
12. What effect does exercise have upon the unloading of oxygen in the muscles, heart minute-volume, and digestion?
13. What effect does vigorous exercise have upon the function of the kidneys?

Chapter Fifteen

FATIGUE AND STALENESS

Fatigue may be defined as *decrease in work capacity caused by work itself.* It is important to state that fatigue is caused by work, because work capacity may also be lowered by other causes: by drugs, illness, or lack of incentive. In every case there is a sensation of fatigue, although no work may have been done.

TYPES OF FATIGUE

Fatigue is often divided into two types: mental and physical. The first type implies a state of fatigue which results from mental work. It is often due to boredom because of a lack of interest, and is a problem for the psychologist, the psychiatrist, and the sociologist, as much as for the physiologist.

Physical fatigue is caused by physical or muscular work, and should be of great interest to the physiologist. It is true that the physiologist has all too frequently emphasized the effects of fatigue on the working muscles, but he ordinarily recognizes a twofold nature of fatigue. This he does when he speaks of "neuromuscular fatigue," which at once indicates the respective share of muscular and nervous elements.

It is difficult, if not impossible, to separate wholly these two types of fatigue. Mental concentration and emotions are factors in much of the fatigue associated with work, and enter also into athletic activity. It is generally recognized that excessive muscular work may cause mental weariness. But that excessive mental work may cause muscular weariness seems to be a new thought to many people.

It must be admitted that the term "fatigue" is indefinite and inexact. The objective changes can be definitely measured only when it has developed to a relatively high degree. Dill[125] makes a distinction between the fatigue of moderate work, hard work, and maximal work. He believes that the first of these is of rather remote interest to the physiologist since it includes the type described as boredom. Here the expenditure of energy throughout an eight-hour work day is relatively small: so much so that, when the day's work is done, the worker enjoys such physical activities as gardening, strenuous games, and dancing. Dill points out that, with ideal social conditions, outside as well as inside a school or factory, work is carried on happily and at a uniform rate without the appearance of fatigue and boredom.

The distinction between moderate and hard work is made on the basis of metabolism, which, of course, relates them to the individual's capacity for supplying oxygen to his tissues. Moderate work is defined as that amount of activity which uses energy at a rate of three times or less that of the basal metabolic rate. With hard work, the use of energy ranges between three and eight times the basal rate. It is said that a mean metabolic rate of about eight times the basal rate is as much as can be maintained for eight hours. Up to this rate the circulatory and respiratory systems effectively provide the body with the necessary oxygen.

In moderate and hard work, only minor blood changes occur, lactic acid concentration and alkaline reserve are unchanged, and the heart rate, respiratory volume, and circulation rate remain in a linear relationship with the metabolic rate. In maximal work, Dill's third type, the worker enters the "over load" zone, in which a steady state cannot be maintained and breakdown is not far off.

SYMPTOMS OF FATIGUE

Some of the manifestations of fatigue are subjective; others are objective in character. If it be accepted that the outward sign of fatigue is a diminished capacity for doing effective work, then it must be recognized that the subjective sense of fatigue is often a fallacious index, since one frequently feels quite tired and yet finds, if he goes to work, that his capacity for performance is large and that the tired feeling disappears as he warms up to his task.

The subjective feeling of physical fatigue is really a great complex of sensations, differing in some degree according to the kind of work. There may be a feeling of local tiredness in the active muscles, a general bodily sensation of tiredness, or a feeling of sleepiness. There may be a tired feeling in the head, obscure and poorly localized pains in the back of the head, pain and soreness in the muscles, stiffness in the joints, and swelling of the hands and feet.

Symptoms of mental fatigue are sometimes clearly recognized by the subject. He complains of inability to keep attention fixed, of impaired memory, of failure to grasp new ideas, and of difficulty and slowness in reasoning. Arithmetical calculations and the like are slow and inaccurate.

CAUSES OF FATIGUE

The primary cause of fatigue, both mental and physical, must be activity involving the expenditure of energy by the body, as there is no fatigue when all expenditure is excluded. Such a state is rest. Fatigue of either type is chemical in character. It may be the result of (1) a depletion or non-availability of stores of energy in the body; or (2) the accumulation of end products of metabolism which become a hindrance to vital exchanges of the body; or (3) an alteration of the physiochemical state, a breakdown of homeostasis.

1. The fact that fatigue can be delayed by administering sugar to men during hard and long physical labor is sufficient evidence that a reduction in the store of energy-producing substances is a causative factor.

2. The end product or waste product theory of fatigue was suggested by the nineteenth-century German physiologist Ranke, when he found that certain substances formed during contraction depress or inhibit the power of muscle contraction. Among these products are lactic acid, carbon dioxide, and acid phosphates. It should be noted that the extent of the occurrence of some of these substances depends in part on the inadequacy of the oxygen supply to the muscles during their activity. Oxygen is required for the chemical processes within an organ. There is no simpler way of hastening fatigue than to subject the individual to a diminished oxygen supply.

3. When the average man finishes his day's work, his fatigue cannot be ascribed to a specific fatigue substance, to hypoglycemia or to anoxemia. We must fall back on some other sort of explanation.

4. Changes in the internal environment, the physiochemical state of the blood and lymph, may also cause fatigue. A large number of delicately interrelated substances cooperate in maintaining the balanced condition of these fluids. A marked increase or decrease in any one of the substances may modify the fluids sufficiently to affect adversely the living cells of the body. Fatigue owing to chloride losses illustrates one type of cause. McCord and Ferenbaugh have studied it among soldiers; and the Harvard Fatigue Laboratory group studied it among workers in "hot" industries where a natural loss of sodium chloride may cause worker fatigue ranging from a mild form to total incapacitation. These conditions are not alleviated by water alone and, in fact, are in some measure aggravated by water intake. Salinized

drinking water (sodium chloride from 0.04 to 0.14 per cent) goes far to prevent the form of fatigue or exhaustion occasioned by mineral losses.

The work of Campos, Cannon, and others[81] showed that dogs driven to exhaustion on a treadmill will begin to run again after an injection of epinephrine. The animals did not stop because of a failure of sugar supply or because of a high concentration of lactic acid in the blood. A preliminary injection of epinephrine, and occasional injections throughout the period of running, did not increase an animal's capacity for running. Epinephrine was helpful only to the fatigued animal. Why it is helpful then and how it acts have not been determined. In this connection it is of interest to note the findings of Dill, Edwards, and de Meio,[130] in a study of moderate work. In the early stages of moderate work, about one-half the energy was derived from carbohydrates, and, in the last two hours, less than one-tenth was so derived. The experimenters found that injections of epinephrine had no effect on protein metabolism, but increased the carbohydrate utilization more than one-half over the corresponding period in a control experiment.

PROBABLE SEATS OF FATIGUE

The places where fatigue may be located are more numerous than would at first be expected. Considering now only the neuromuscular apparatus, it appears that there are six possible seats of fatigue: (1) the muscle fiber; (2) the motor nerve end plate in the muscle fiber; (3) the motor nerve fiber; (4) the synapses within the nerve ganglia and the central nervous system; (5) the nerve cell body; and (6) the end organs of sense in the muscle and elsewhere in the body (see Fig. 89).

Muscle Fiber and Motor End Plates. If we take a nerve-muscle preparation and stimulate the nerve once or twice per second, the muscle, after a period of time, will show signs of fatigue and eventually will not be able to contract. If, now, we stimulate this muscle directly by applying electrodes to its surface, the muscle will again respond with the same vigor as when it was originally stimulated through the nerve. This observation proves that the seat of fatigue is *not in the muscle tissue.* It, therefore, must be either in the motor nerve or in the motor end plates.

Motor Nerve Fiber. This fiber is relatively indefatigable. This may be demonstrated by an old laboratory experiment which consists of blocking the transmission of nerve impulses to a muscle either by passing a galvanic current through a small segment of the nerve or by placing a piece of ice on the nerve near the muscle. The nerve is then stimulated continuously for hours at some point on the side beyond the muscle and the block. If, now, while the nerve is still being stimulated, the block is lifted, the muscle is found at once to be responsive to the

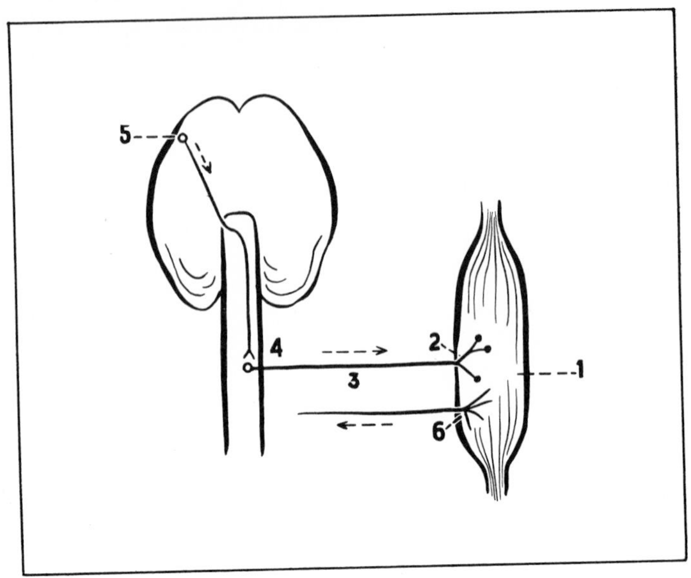

Figure 89. The possible seats of fatigue. 1, Muscle fiber; 2, motor nerve end plate; 3, motor nerve fibers; 4, synapse; 5, nerve cell body; 6, sensory nerve endings.

nerve, and immediately proceeds to produce a typical performance curve. Thus, the seat of fatigue in our nerve-muscle preparation was neither in the muscle tissue itself *nor in the motor nerve.* It must, therefore, be in the *motor end plate.*

Synapses. Stimulation of any spot on the saddle-shaped part of the back of a dog will produce a scratch reflex on the same side. If this stimulation is continued, eventually there will be no reflex response. If, now, another spot on the back, close to the first one, is stimulated, the scratch reflex is obtained again.[472] It is also possible to cause muscular contractions by direct stimulation of sensory nerves. When, after repetitive stimulation of a certain sensory nerve, a muscle begins to show signs of fatigue, stimulation of other sensory nerves will again produce contractions, equal to those obtained at the beginning of the experiment.[174] These two observations can be explained only by assuming that the seat of fatigue in each case was a synapse between a sensory and a motor neuron.

Nerve Cell Body. Although there is abundant evidence that structural changes occur within the cell bodies in extreme fatigue it is probable that moderate fatigue can be explained in a similar manner. There is not doubt that the functioning of the cells of the cerebral cortex is modified by fatigue. Proof of this may be found in the effect of fatigue upon conditioned reflexes. If a dog, with several well-established conditioned reflexes, is made to pull a cart until he is tired, it will be observed that fatigue affects the functioning of these reflexes. The reflexes most recently established may completely disappear, and the old ones may decrease by 50 per cent. Since conditioned reflexes involve activity of cerebral nerve cells, it is natural to conclude that fatigue products, although developed elsewhere, cause a state of fatigue in these nerve cells. Such experiments explain the value of a change of interest during a working day and explain why a tired soldier may march with renewed vigor when a band begins to play.

Sensory End Organs. A local feeling of fatigue is familiar to everyone. It arises from the sensory end organs located in muscles, and gives a subjective measure of the working condition of the neuromuscular system. Because of fluctuations in the threshold of sensitivity to fatigue, this measure is not very reliable. When the threshold is raised, one may overwork without knowing it. If the threshold is lowered, one may feel extreme fatigue without any appreciable decrement in work capacity within the neuromuscular system. Although no sensation of muscular fatigue is possible without the sensory end organs, and although muscular work is often discontinued because of these sensations, the seat of fatigue is not in the sensory end organs. The ringing of a fire alarm is not the cause of a fire.

FACTORS CONTRIBUTING TO INEFFICIENCY AND FATIGUE IN INDUSTRY

Modern industry does not consider a man merely an animated gadget capable of doing work. At all times, the worker remains a human being; and, no matter how some of his reactions may be controlled or suppressed, they have a cumulative effect which will become evident sooner or later. For these reasons, modern industries have established special departments whose aim is to help in ameliorating personal problems of the worker.

It is beyond the scope of this book to go into a detailed discussion of this topic; therefore, only an outline of conditions which affect industrial workers will be presented here. It is believed that the implications arising from these conditions are obvious.

1. *Physical and Social Conditions at Work; Nature of Work.*
   a. Time and place of work, amount of space, temperature, conditions of air, light, and so forth.
   b. Bosses, co-workers, degree of security, incentives, and the like.
   c. Nature of work. Is work well planned? Is there adequate equipment and time allowed for the completion of assignments?

2. *The Worker Himself.*
   a. Health.
   b. Fitness for the work he is doing.
   c. Interest in the work.
   d. Ambitions and aspirations.
   e. Use of alcohol.
   f. Gambling and such.
3. *Home and Community Conditions.*
   a. Kind of wife and relations with her.
   b. Children, in-laws.
   c. Diet.
   d. Amount of recreation and sleep.
   e. Adequacy of living quarters.
   f. Participation in community life.

BOREDOM

When a person *has* to participate in an activity, physical, mental, or social, without adequate motivation and, therefore, without any interest, he usually experiences a feeling of disinclination to continue this activity. This feeling is referred to as boredom. Boredom sometimes may simulate fatigue, because the person may feel tired and the work output may be diminished. Closer observation indicates that both the feeling of fatigue and the work output are too irregular for true fatigue. If the person is made interested in his work, symptoms of fatigue disappear and output increases. For this reason, fatigue based on boredom is called "pseudo-fatigue." There are two methods for minimizing boredom: to develop interest, or to do work automatically while thinking of something more interesting.

STALENESS

Sometimes an athlete eager to excel in some sport begins to train frequently and intensely. At first he may improve somewhat, but finally his record becomes stationary and much below what he set for his goal. Anxious to pass this dead point he begins to train incessantly. Instead of improving, his performance becomes worse. With this result comes a feeling of personal inadequacy and frustration. Besides a decline in performance, some changes in personality and behavior may also be detected. He has developed a state of *staleness.*

The subjective symptoms of staleness are numerous, and are somewhat as follows: One is likely to notice first that he is beginning to feel generally tired and that he has lost some of his original keenness.

His sleep does not refresh him. He gets occasional headaches. Later he does not go off to sleep quite so well as he did, or he may go off fairly soon, and yet wake up early in the morning. He may lose his appetite. His digestion may trouble him, and he may often suffer from constipation. His sleep may be troubled by dreams about his daily work and by nightmares of all kinds. He may find that, while he has to force himself to go to his work, he nevertheless feels quite fit after he gets into it; but, after it is over, he may feel shaky and utterly exhausted. He probably finds that he is getting irritable; that he cannot enjoy his friends; and he prefers to go off by himself. Although he feels tired, fits of restlessness overcome him; he cannot sit down quietly, but must be puttering about at something.

These conditions may become even more aggravated if an unsympathetic coach tells him that "he has not got the stuff"—that he will never become any good.

To recover from staleness, training should be temporarily suspended. The boy should be told why he has developed this state, and should be advised not to attempt to reach a high goal in a hurry. A sympathetic coach may also show the boy, without hurting him too much, that every person has his limitations, and that one should do the best he can with what he has, without hoping for the impossible. Usually athletes recover from this staleness. Sometimes they do not, and may even leave the college or the club, give up the sport they loved so much, and nurse a grudge against the world indefinitely. These men probably had a seed of neurosis, which sprouted under the impact of frustration.

During World War I, the term staleness was used as a polite substitute for psychoneurosis developed among flyers. During World War II, the word staleness was not abused in this way. Instead, other polite terms were invented: flyer's fatigue; and, if the person did not fly, battle fatigue, although the person might never have got closer to the battlefield than several thousand miles.

PREVENTION OF NERVOUS BREAKDOWN

What may start at first as boredom may lead to a great degree of dissatisfaction, frustration, and even to a nervous breakdown. On the other hand, even work which is essentially interesting, if it must be pursued under constantly unsatisfying conditions, such as lack of appreciation or insecurity, may at first lower efficiency and be called "staleness," and then it may also lead to nervous breakdown. Hard but enjoyable work in an atmosphere of appreciation and security, with sufficient periods of physical and mental relief or rest, cannot cause any untoward effects. Since in life one is often compelled to work under adverse conditions, the practice of hygiene, especially of mental hygiene, may be of great help. Enjoyable physical games in which one excels, or hobbies, may be of great help. If one finds it difficult to relax at will, and so to recuperate from fatigue and staleness, one can turn to such sources of help as techniques of relaxation advocated by Jacobson,[270] and Rathbone.[424]

QUESTIONS

1. Define fatigue.
2. Can a state of fatigue be induced by other causes than work?
3. How many types of fatigue are known?
4. What is boredom?
5. What is the physiological basis for grading the severity of fatigue?
6. Discuss objective and subjective symptoms of fatigue.
7. Discuss physiological causes of fatigue.
8. Discuss probable seats of fatigue.
9. Do muscles actually become tired?
10. Discuss some of the factors contributing to inefficiency and fatigue in industry.
11. What are the usual causes of nervous breakdown?

Chapter Sixteen

EXERCISE UNDER UNUSUAL ENVIRONMENTAL CONDITIONS

It is man's nature to explore and expand his environment as well as to test his ability to perform within it. Mountain climbing and free diving have historically been accepted challenges, while work in varying temperatures has been a necessity. Man's ability to adapt to such conditions has long been of interest to experimental physiologists, and the knowledge they have gained has helped us to prevent unfortunate accidents and to understand the reasons for observed performance limitations. Our purpose here is to discuss the effects of three environmental conditions which are frequently of interest to workers and athletes: temperature, the low atmospheric pressure of high altitude, and the high atmospheric pressure during underwater swimming.

TEMPERATURE

Temperature variations affect nearly every physical, chemical, and biological process. We are all familiar with the ill effects on mankind of the disturbance of the heat equilibrium of the body seen in heat-stroke or heat exhaustion, but we are unfamiliar with the effects of moderate heat. In the industrial worker these results may be cumulative. For more information the reader is referred to an excellent book by Brouha.[61]

Environmental and Body Temperatures

Normally, man's internal temperature in health and during rest remains fairly constant wherever he may be, with a diurnal range of from 1° to 2½° F. Man readily adapts himself to extremes of temperature through responses made by his vasomotor system and sweat glands. He is constantly and necessarily eliminating heat. The loss of heat results from radiation, conduction, and the evaporation of water. The amount of heat lost by radiation and conduction depends largely upon the temperature of the surrounding air, while the amount lost by evaporation depends upon the relative humidity of the body's immediate environment. Some conditions permit loss of heat by radiation and conduction only. In a hot, dry environment, loss of heat by evaporation is at its maximum. When air temperature increases above that of the body, most heat is lost by evaporation. An enormous quantity of water may be lost in this manner, which makes it possible for a man to withstand exceedingly high temperatures. A man may remain for eight minutes in an oven at 260° F., a temperature high enough for cooking. On the other hand, if air is saturated with water vapor, a temperature of 118 to 122° F. cannot be tolerated for more than a few minutes. Even at lower temperatures, the heat-dissipating mechanism may fail and the body temperature rise. Thus the body temperature in a hot mine may rise.

Cooling Power of the Environment

The environmental factors of importance in the control of the body temperature are the atmospheric temperature, humidity, and air movement. The cooling effect may be obtained even without lowering temperature by a decrease in humidity and increase in air movement.

It is of interest that cooling of a small part of the skin may produce a sensation of comfort under adverse temperature and humidity conditions. For example, at 105° F. and 75 per cent humidity, a pleasant

TABLE 16. *Heat Loss in Man in Twenty-Four Hours at Ordinary Room Temperature*

Manner of Loss	*Cal.*	*Per Cent*
Radiation	1650	55.0
Convection and conduction	450	15.0
Evaporation of water (skin and lungs)	780	26.0
Warming inspired air	75	2.5
Urine and feces	45	1.5
	3000	100.0

relief will be obtained by immersing an arm in cool water.[77] The same effect may be obtained by immersing the feet in cool water or running tap water over the hands, especially the wrists.

Physiologic Responses to Heat

Metabolism. When environmental temperature rises about 80° F., body temperature and metabolism also rise because of intensified chemical reactions, and not as a result of lessened heat dissipation.

Respiration. The frequency of breathing increases by some 5 to 6 breaths per minute with 1.8° F. increase in rectal temperature. If the rise in temperature is rapid, the breathing not only accelerates, but also deepens; the movements appear to be designed to produce a maximum alveolar ventilation. The augmentation in breathing causes a marked fall in the alveolar carbon dioxide tension, in some instances to as low as 25 or even 23 mm. Hg instead of the normal 40 mm. The net result of the loss of carbon dioxide is a marked alkalinity of the blood.

Pulmonary ventilation and oxygen consumption during the same intensity of work are higher at hot (90° F.) and humid conditions, but are *lower* at hot (108° F.) and dry conditions than at room temperature.[71]

Circulation. The circulatory system is also affected by high temperatures. As the external temperature is raised, dilatation of the cutaneous vessels occurs; therefore, a greater volume of blood is exposed to the outside air. Loss of heat by conduction and radiation is thus facilitated. At the same time, the volume of blood of the abdominal organs is diminished. The blood leaving the skin is highly saturated with oxygen, thus indicating that the increased blood supply to the skin is determined by requirements of temperature regulation and not by those of tissue activity.

The pulse rate increases concomitantly with body temperature. An increase in pulse rate of 37 beats per minute has been recorded by Bazett[46] for a rise of 3.6° F. in rectal temperature. There is rarely an exact parallel between the rise in rectal temperature and pulse rate, although the statement has been made that there is an increase of 15 beats per minute during reclining and 20 during standing for each rise of 1° F. in rectal temperature.

An increase in the minute-volume output of the heart occurs with a rise in temperature. This is caused mainly by the increase in pulse rate. As to the blood pressure, the systolic pressure may either rise or fall, but the diastolic pressure constantly shows a fall.

When the peripheral blood vessels are profoundly dilated, there may be difficulty in maintaining the standing posture. This explains why some persons experience dizziness on standing after a long hot bath. With more blood going to the skin during exposure to heat, it may be that the internal organs have to function with a greatly decreased amount of blood.

Blood Volume. A prolonged exposure to high temperature and high humidity leads to a slow rise in blood volume which may be ascertained in a day or two. The source of extra fluid is not well known. It is suggested that the intestinal tract contains a considerable reservoir of fluid and salts which may be drawn upon, that a part may be drawn from the tissues of the body, and that the kidneys conserve fluid by secreting a more concentrated urine.

During exposure to dry heat a considerable amount of water may be lost by sweating. Though no liquid is drunk, the percentage of hemoglobin may be the same before and after an experiment, thus indicating that no concentration of the blood has occurred. In this instance it is believed that the tissues supply the fluid eliminated in the sweat.

Effects of External Heat During Muscular Activity

Rectal Temperature. Since the mouth temperature is subject to considerable variation, the rectal temperature is usually taken in studies of exercise. With heavy muscular exercise, a rise of rectal temperature is common. Even 105° F. is sometimes reached. It has, however, been shown that the rise may not be detectable for some minutes. The rectal temperature may be taken as evidence of the direction of change in the temperature of the arterial blood.

According to Dill and co-workers,[129] the body temperature ordinarily increases steadily for the first minutes at a given rate of work and then, under favorable conditions for heat dissipation, becomes constant. When, however, the conditions were such that the body temperature did not reach a level, the temperature continued to rise until exhaustion intervened.

Loss of Water by Evaporation. According to Campbell and Angus, the loss of water is much increased by moderate muscular work in still air at temperatures of 50° to 77° F.; the increase in one case was as much as 146 gm. per hour, the figures for rest and for work being 30 gm. and 176 gm., respectively. There is considerable variation among individuals. The wearing of clothes, the lack of wind during work, and high humidity retard evaporation from the skin.[79]

Under certain conditions of temperature, humidity, and muscular exertion, as much as 6 to 13 pounds of water may be lost from the body in a day. During work at high temperature and humidity, 50 per cent of the sweat is lost through the skin of the trunk, 25 per cent from the head and upper limbs, and 25 per cent from the lower limbs.

Blyth and Burt[58] experimented with the effects of dehydration and superhydration upon the duration of all-out runs on the treadmill, with an ambient temperature of 120° F. They tested 11 athletes and 7 non-athletes under three conditions of water intake: (1) normal (control); (2) superhydration, by drinking 2 liters 30 minutes before the test; and (3) dehydration, by abstaining from water for 24 hours and sweating in a hot room, so that 3 per cent of the body weight was lost. Dehydration lowered endurance in both groups. Superhydration increased endurance in athletes and decreased it in non-athletes.

While sweating induces thirst, excessive drinking of water on a hot day increases sweating. An observer working in India, in a hot, dry climate, noted that he had used 3.6 gallons of water in one day (about 30 pounds) without gaining weight. Most of the water was eliminated through perspiration, because the amount of urine was scant.[126]

It has been observed that the profuse sweating in industries where men have to work at high temperatures may lead to severe cramps in the muscles of the limbs and of the abdominal wall. Significantly, these *heat cramps* are known as stoker's or miner's cramps. Studies of individuals suffering from such cramps have revealed that, besides a dehydration of body tissues, there is a lowered concentration of sodium and chloride ions in the blood plasma and either an absence or a drastic reduction of sodium and chloride in the urine. According to several reports, the addition of 0.1 per cent of salt to drinking water or the use of salt tablets has considerably reduced the incidence of heat cramps among workers.

The amount of sodium chloride in sweat varies from 0.2 to 0.5 per cent. Therefore, theoretically it is possible to lose from 16 to 40 gm. of salt a day. The latter figure, however, is too high, because, during profuse sweating, the amount of sodium and chloride in the sweat is diminished. It has been estimated that, under the most severe conditions of the tropics, 13 to 17 gm. intake of salt a day is sufficient. Since the average daily intake of salt is about 10 to 12 gm., it is obvious that an extra supply of salt may be necessary for people working at conditions of high external temperature.

During World War II, men in the armed forces stationed in subtropical and tropical areas were supplied with salt tablets. They soon learned that these tablets had to be taken with a liberal amount of water. Just washing them down with a couple of swallows of water resulted in a feeling of discomfort in the stomach.

Attention should be drawn, however, to observations by Taylor and his associates[502] on a group of men who worked to exhaustion at a temperature of 120° F. (wet thermometer bulb being 85° F.) and sweated to the extent of 5 to 8 liters a day. They showed that cramps do not necessarily follow a drop in the sodium chloride content of the blood. Of eight men, five developed heat prostration, but no cramps.

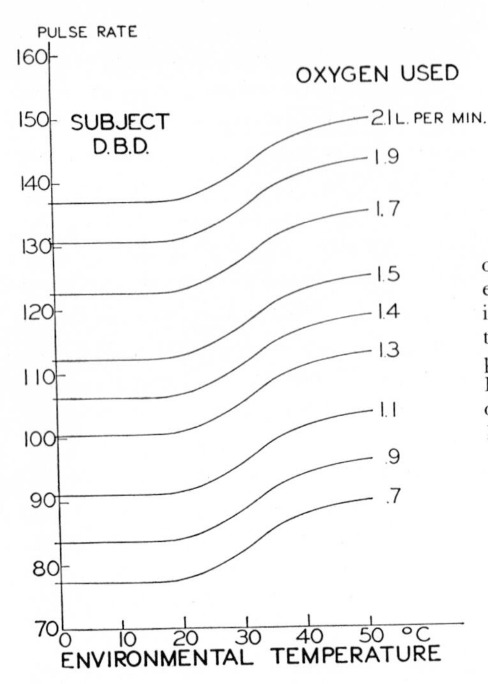

Figure 90. Effects of temperature on pulse rate during exercise on bicycle ergometer. Nine exercises of different intensities were performed at different temperatures. The higher was the temperature, the higher was the pulse rate. Intensity of exercise was assessed by oxygen used. (Dill: Am. Heart J. *23,* 1942.)

Even though depletion in sodium and chloride ions may not be the sole cause of heat cramps, an abolition of the prophylactic use of salt in occupations causing profuse sweating is not warranted at the present time.

Pulse Rate. For the same intensity of work, the pulse rate is higher with an increase in the air temperature. Figure 90 illustrates the effect of temperature on the pulse rate during work of various intensities. No wonder that people find that it is harder to do the same work on a hot day than on a cool one. On a hot day the heart has to do more work because the amount of blood circulating through the skin may be greatly increased. As a result of this, less oxygen is supplied to working muscles and lactic acid begins to rise in the blood at a much lower intensity of work.[530]

Rowell and co-workers[444] found that because the maximal pulse rate is reached at a lower intensity of work at high temperature than at room temperature, the current procedures of predicting maximal oxygen intake become invalidated.

Blood Output of the Heart. In four out of five subjects, who did the same amount of work in a hot and in a cold room, the heart output per minute increased from 1 to 4 liters more in the hot room than in the cold.[129] Since the pulse rate is augmented proportionately more than the minute-volume, it is evident that the output of blood by the heart per beat is diminished with increasing external temperature.

Respiration and the Respiratory Exchange. Observations made by Dill and his co-workers[129] on men who exercised in a hot room are of special interest. The minute-volume of air breathed throughout the period of work in the hot room increased rapidly for the first ten minutes and then more slowly to the end of the period. As a result, the alveolar air carbon dioxide pressure fell steadily, to a greater degree than it would have during rest at the same temperature. It is evident that carbon dioxide was eliminated more rapidly than it was produced by oxidation. The consumption of oxygen was only slightly higher, not more than 5 per cent, in the warm room than in the cold. The fuels used were not significantly different at the two temperatures.

Influence on Working Capacity. It has been demonstrated in industry that, in heavy work, especially work that involves exposure to high temperature, output undergoes a seasonal variation. It is, as a rule, greatest in winter, least in summer, and intermediate between the two in spring and autumn. The influence of high temperature is especially evident when the work is strenuous. It has been found that no man, under such circumstances, is able to work continuously, but must take short rests from time to time. These periods of rest are more effective if they are taken at comfortable temperatures. Industries find it profitable to install air conditioned rest rooms because they provide the body temperature a chance to return to normal (see Fig. 91).

The effect of combined high temperature and high humidity upon

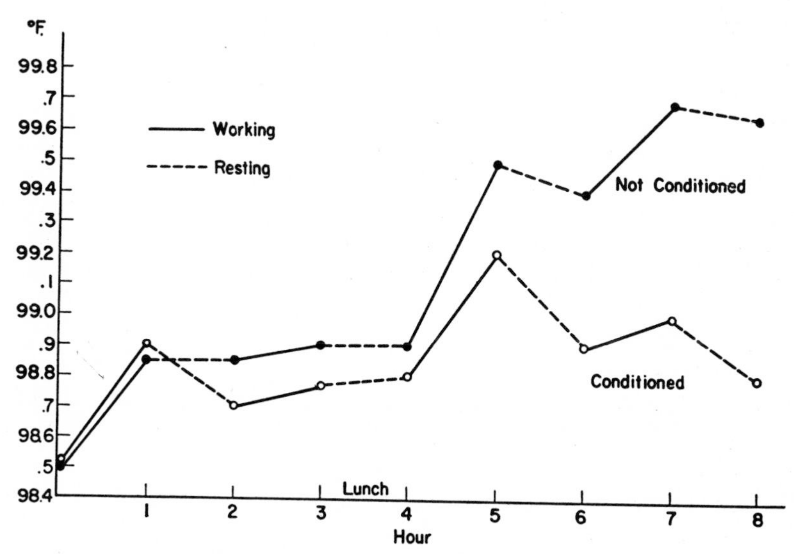

Figure 91. Effect of air conditioned rest rooms upon the body temperature in men on an 8-hour shift (Brouha, L., reference 60).

heart rate may be observed from Figure 92. At temperatures of 90° to 95° F., and relative humidity between 65 and 90 per cent, the heart rate after exercise does not return to normal even in 45 minutes.

Some observations made by Vernon[520] on 138 miners are of practical importance. Under the best conditions of temperature and humidity, they rested only 7.3 minutes per hour; but, when temperature and humidity rose to a definitely uncomfortable degree, they rested for 22.4 minutes per hour.

Yaglou[550] found that, between temperatures of 40° and 75° F., men work practically steadily. Above 75° F. their output falls off gradually until 80° F. is exceeded, and from there on the fall is rapid. The output of work at an effective temperature of 93° F. was only one-half that at 70° F. The movement of the air is an important factor in its cooling power.

A study of geography and history indicates a definite effect of climate upon human capacity for work. While men in the Arctic and equatorial zones seem to have spent all their energy on surviving, men in temperate zones have had plenty of energy for satisfying their urge to produce, explore, and conquer. Although the importance of climate for personal comfort has long been realized and utilized, it is only recently that an artificial climate has been introduced in industries. Probably the line of reasoning which has kept industry from introducing improvements sooner was influenced by work conditions in the coal and ore mines. One might say that if a man can work under-

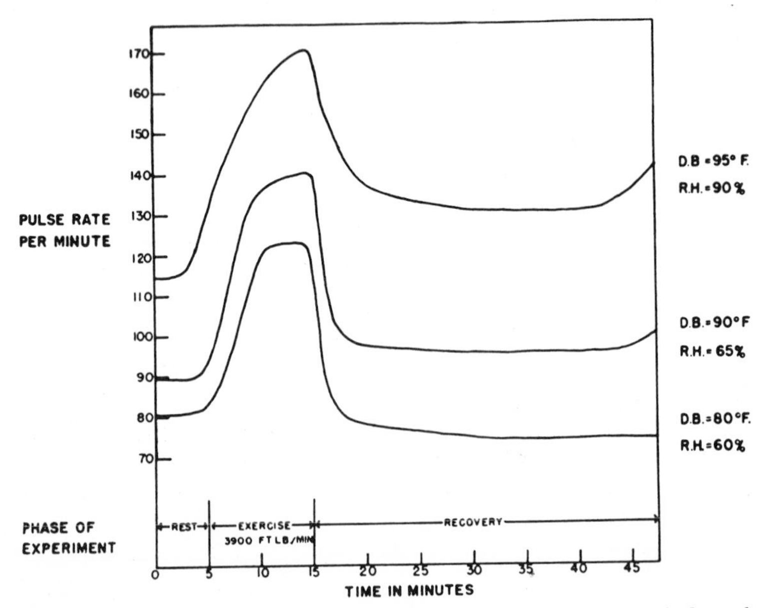

Figure 92. Effect of temperature and humidity upon the heart rate before, during and after standard exercise (Brouha, L., reference 60).

ground, where there is not enough light and ventilation and where the temperature may be high, why should one worry about men working in dingy sweatshops above ground? Modern industrial experience has shown, however, that it pays to improve conditions under which any work is done, because such improvements increase the efficiency of the worker and reduce the cost of production. The most favorable conditions for work are: 68° F. and 50 per cent relative humidity.

Exhaustion at High Temperatures. It will be recalled that Dill and co-workers had their men, while working in the hot room, ride the bicycle ergometer until the men became exhausted. Of the five subjects, one had to give up in thirty-seven minutes, two in forty, and a fourth in forty-nine minutes; while the fifth man, after working sixty minutes, was not exhausted. On analyzing the data obtained from these exhausted men, the investigators could find no considerable accumulation of lactic acid, no exhaustion of fuel reserves, and no excessive inroad on the respiratory capacities. Furthermore, the exhaustion in the hot room was not caused by an uncomfortably high temperature, because three of the exhausted men had a body temperature only 0.9° F. higher in the hot room than in the cold room, where they had carried the same load of work easily and comfortably for an hour. As judged by the output of blood per minute, and by the systolic arterial blood pressure, the mechanical work done by the heart was about the same in the hot room as in the cold. It was significant, however, that the final heart rate was very high, ranging between 162 and 180. By a process of elimination they concluded that the cause of exhaustion was most likely in the *heart muscle.* The evidence is, of course, indirect and, therefore, inconclusive; but the important fact is that, in every case of exhaustion, the heart had reached its upper limit of response when exhaustion occurred, while every other function which could be measured still had ample reserve.

The Bureau of Mines workers have repeatedly affirmed that the pulse rate rather than the rise in body temperature apparently determines the extent of discomfort experienced in hot environments. Subjects become uncomfortable after the pulse rate exceeds 135 per minute. In South African mines, a rectal temperature of 103° F. has been accepted as the upper limit of safety. It has been suggested that an oral temperature of 101° F. may be used as a similar indicator.[493]

Investigations by Taylor and his co-workers,[501] who lived from two to eight days under simulated desert conditions, showed that it takes from four to five days to become acclimatized to high external temperature. Acclimatization in some people occurs in five to eight days and is characterized by cardiovascular stability and better thermo-regulation, leading to a decrease in skin and rectal temperatures and to a reduction in cardiovascular strain.[433]

Strydom et al.[493] found that rectal temperature and pulse rate of men working in mines five hours daily at 93° F. decreased within four to five days, but sweat rate reached maximum value only on the tenth day. They also found that although a better physical condition of men caused by preliminary work at a cool temperature facilitated performance in a hot environment, it could not replace acclimatization.

Effect of Cold on Capacity for Work

The problem of cold is much simpler than that of heat, because, usually, it can be solved by wearing heavier clothes. While a nude man working in a tropical climate is at the mercy of his own physiological adaptation, a man in the Arctic puts on a parka, extra-heavy boots, gloves, and other warm clothing. This solution, however, brings another problem—the effect of heavy clothing on efficiency. The extra weight of clothes and the resulting interference with body movement increase the amount of energy used and limit the degree of skill. A man whose body is warm can take off his gloves when the temperature is as low as −30° F. and his hands will still be warm. In one such experiment, performed on two men (see Fig. 93), the temperature of the hands remained sufficiently high for work.[422] The only precaution that must be used under these conditions is to avoid contact between the bare skin and cold metal.

When people who are inadequately clothed are exposed to cold, they start to shiver. This phenomenon is more vigorous when wind is

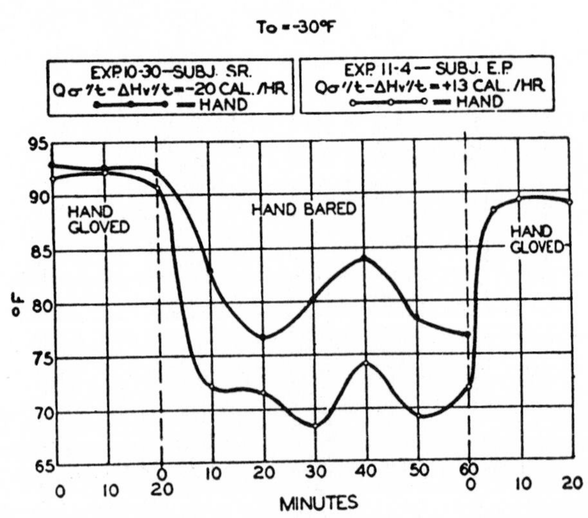

Figure 93. Effect of body warmth on hand temperature. The hand temperature of two warmly dressed men remained sufficiently high at −30° F. (Rapaport et al.: J. Applied Physiol, 2, 1949.)

present, because its purpose is to raise body metabolism. Even when no visible shivering is noticed, there is an increase in heat production. At maximum shivering, about 425 Cal./hr. may be produced, almost seven times greater than man's resting metabolism at room temperature.[261]

People who live in cold regions have increased metabolism; therefore, they consume more food and voluntarily increase the amount of fat in the diet.

EFFECTS OF HIGH ALTITUDE

The effect of high altitude on performance has long been of concern to mountain climbers and aviators. Considerable research on high-altitude physiology was completed during World War II, when it became desirable to fly airplanes as high as possible. Most of the earlier research involved altitudes over 10,000 feet. Holding the Summer Olympic Games at the 7,350-foot (2,300-meter) elevation of Mexico City, in 1968, stimulated considerable research on performance at moderate altitudes, and excellent reports have resulted from symposia on this topic. (See references 274 and 361.)

The effects of altitude result from the "thinning-out" of the air as one ascends to higher elevations. There is a progressive decrease in the barometric pressure, as shown in Table 17. The percentage composition of the gases in air is the same at all elevations (for example, oxygen is always 20.94 per cent of a dry sample), while the decrease in pressure is due to the reduced number of gas molecules per unit of volume. Several bodily responses occur to counteract the effects of the "thin" air, some on immediate exposure (which we refer to as *acute*) and some on prolonged exposure (which we refer to as *chronic*).

Oxygen Intake and Oxygen Debt. The oxygen requirement for a given submaximal work load is the same at all altitudes,[210, 413] but the maximum oxygen intake progressively decreases as the altitude increases. The amount of decrease shown has varied. Faulkner and associates[163] found a loss of 13, 20, and 29 per cent of near-sea-level values at 2300, 3100, and 4300 meters, respectively. Other values reported were 17 per cent at 4300 meters,[210] 13 per cent at 3800 meters,[317] and 14.6 per cent at 2270 meters.[412] Because of this decrease, a given submaximal exercise intensity at a high altitude utilizes a greater proportion of the aerobic capacity, and the performer will be working closer to his physiological potential than he would be at sea level.

The dynamics of anaerobic reactions at high altitude have not been studied as intensively as those of aerobic reactions. It is generally conceded, however, that the anaerobic release of energy is not affected.[92] For example, in trained distance runners, 40-minute recovery oxygen intakes following 5 minutes of intense exercise were almost identical at sea level and at 2300 meters.[412] There have been indications, however, that the end products of anaerobic metabolism, pyruvate and lactate, are more apt to appear in the blood during rest and submaximal exercise on ascent to higher elevations, but tend to decrease toward normal values with acclimatization.[209]

Respiration. Since the saturation of hemoglobin depends upon the partial pressure of oxygen in the alveoli, the lower pressure at higher altitudes leads to a reduction in the amount of blood oxygen (Table 17). This effect becomes more pronounced as elevation increases, since the descending segment of the oxygen dissociation curve operates (see Fig. 65, Chapter 10). The eventual result is *hypoxia,* a deficiency of oxygen in the tissues.

The most apparent compensation to hypoxia is an increase in pulmonary ventilation. Respiratory minute-volume, during rest and submaximal exercise, is greater at higher altitude when corrected to BTPS; but when corrected to STPD it, like oxygen consumption, is dependent on work load and is independent of altitude.[316, 412] The BTPS correction shows the volume of the low density air breathed while the STPD correction indicates the actual number of molecules in the air if its volume were to be measured at sea level. The body compensates for the lower oxygen partial pressure by moving more of the thinner air through the lungs.

The increased ventilation also reduces the alveolar carbon dioxide

TABLE 17. *Gas Tensions at Various Altitudes (mm. Hg)*

Meters	*Feet*	*BP*	PIO_2	PaO_2[1]	$PaCO_2$	SaO_2(%)
0	0	760	149	94	41	97
1,500	5,000	630	122	66	39	92
2,500	8,000	564	108	60	37	89
3,000	10,000	523	100	53	36	85
3,600	12,000	483	91	52	35	83
4,600	15,000	412	76	44	32	75
5,500	18,000	379	69	40	29	71
6,100	20,000	349	63	38	21	65
7,300	24,000	280	52	34	16	50
8,500	28,000	250	42			
9,100	30,000	226	37			

The data represent average values, without consideration to individual degrees of hyperventilation and variations in barometric pressure.

[1]A small alveolar-arterial gradient (5–10 mm. Hg) is assumed. It does not change significantly with altitude.

BP is barometric pressure; other symbols indicating inspired (*PI*) and arterial oxygen tension, arterial carbon dioxide tension, and oxygen saturation. Mount Everest figures (above 8,000 m.) are incomplete.

(Reproduced from Hecht, H. H.: Certain Vascular Adjustments and Maladjustments at Altitudes. *In* Jokl, E., and Jokl, P. (editors): Exercise and Altitude. S. Karger, 1968.)

partial pressure, leading to an excessive diffusion of that gas from the blood, lowering its arterial partial pressure[209] (Table 17). The loss of carbon dioxide from the blood also lowers the hydrogen ion concentration (that is, increases the pH) leading to *respiratory alkalosis.* The kidneys thereafter remove excess bicarbonate to restore a more normal acid-base balance during adaptation to high altitude.[316]

The water vapor tension of alveolar air also presents a problem, since it depends only on temperature and is independent of altitude.[348] Since the body temperature is constant, the alveolar partial pressure of water vapor is 47 mm. Hg no matter what altitude. As the total alveolar gas pressure decreases with ascent, the relative contribution of water vapor increases, until there is no room for other gases. This can be somewhat compensated for by hyperventilation, but eventually even this plus oxygen inhalation is inadequate.

The dynamics of the breathing response to exercise on acute exposure are shown in Figure 94. The overall pattern is similar at sea level and higher altitudes, in that at all elevations there is almost a linear relationship between oxygen consumption and ventilation until maximum oxygen intake values are approached. Then the increase in ventilation is disproportionately greater due to the accumulation of the acid end products of anaerobic metabolism and the role of the respiratory system in buffering them.[348] It should be noted that the oxygen intakes charted in Figure 94 were always comparable at a given work load.[24] There is apparently very little effect on maximal ventilation on ascending to high altitudes, but both increases[163, 414] and decreases[93] have been reported.

Cardiovascular Responses. Just as the respiratory system attempts to overcome the deficiency of oxygen in the air during submaximal exercise by increasing the volume of air moved, the cardiovascular system moves more blood to make up for the decreased oxygen saturation. Asmussen and Nielsen[16] reported a 10 to 20 per cent increase in cardiac output when the arterial blood saturation was 65 to 70 per cent of its maximum at 3800 meters. Vogel, Hansen, and Harris[522] have similarly shown increases in cardiac output: 12 per cent during rest, 16 to 18 per cent during exercise, and 20 per cent during recovery, at 4300 meters.

The increase in cardiac output is apparently primarily dependent on an increase in heart rate. Higher heart rates are commonly reported for submaximal exercise at higher altitudes.[416, 414] Findings on stroke volume have been variable, with studies showing either a decrease[16] or a slight increase[522] of such small magnitude that it could not account for the major portion of the increased output.

The cardiac output potential is limited at peak work, since the maximum heart rate is less than at sea level.[413, 522, 93, 28] The amount of decrease may be quite impressive: Pugh[413] reported maximum values

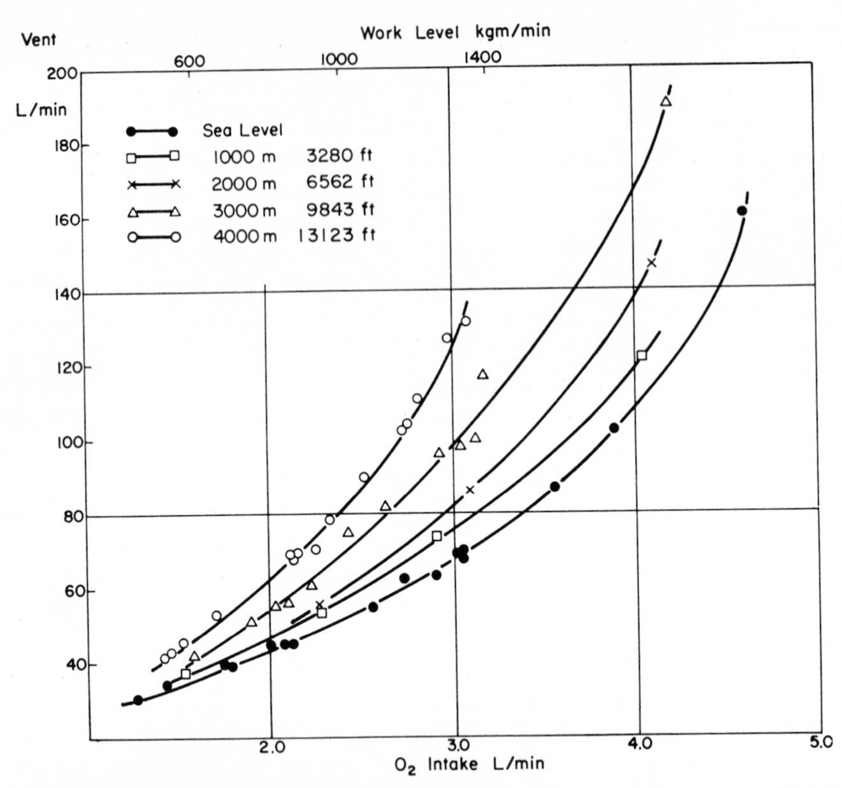

Figure 94. Pulmonary ventilation at different altitudes during exercise. Highest values indicate maximum oxygen intakes at each elevation. (From Åstrand, P.-O.: Acta Physiol. Scandinav. *30*:343, 1954, as adapted by Luft, U. C.: Aviation Physiology—The Effects of Altitude. *In* Handbook of Physiology, Section 3; Respiration, Volume II. Edited by Fenn, W. O., and Rahn, H. American Physiological Society, 1965.)

in subjects of 180 to 196 and 130 to 150 beats per minute at sea level and 5800 meters, respectively; while Vogel, Hansen, and Harris[522] reported an average maximum of 187 at sea level, with decreases to 181 and 178 beats per minute after up to four and eighteen days at 4300 meters. The effects at moderate altitude have been shown to be very minor.[163] The hypoxic conditions apparently have a direct effect on the heart, since administering oxygen to acclimatized subjects during maximal work will be accompanied by an increase in the maximum heart rate. The opposite effect is expected at sea level.[25, 413]

Peripheral vascular adjustments are also apparent. A slight elevation in arterial blood pressure, with a decrease in total peripheral resistance, has been reported.[522] Such an adjustment indicates a better distribution of blood to the tissues, which would be of benefit in the exchange of the limited oxygen supply between the tissues and the blood.

Sensations Accompanying Acute Exposure. In addition to a general malaise, the unacclimatized person may experience many unpleasant sensations on arriving at high altitude. There is considerable variability in individual responses, but typical patterns have been observed.

Observations on British distance runners who participated in controlled experiments in Mexico City prior to the 1968 Olympic Games serve as an example.[416] Disturbed sleep patterns were common to everyone on arrival. Some experienced shortness of breath on mild exertion, while all suffered during strenuous exercise. Dry throats, dry lips, and resultant coughing were also common, since at higher elevations the water vapor pressure in the air is quite low, causing a constant evaporation from the respiratory passages. Other sensations reported during the first few days were transient headaches and some dizziness on sudden movements. Head colds were common. Such symptoms are gradually alleviated as one becomes acclimatized.

At even higher elevations such sensations are intensified. Sleep is sometimes interrupted by Cheyne-Stokes respiration, where there are cycles of increasing intensity of breathing movements to the point of dyspnea, followed by a total cessation of breathing lasting the better part of a minute. Some may be affected by a condition of nausea and vomiting, called "Mountain Sickness." Sunburn is also common, since the thinner air does not effectively filter out ultraviolet rays.

Performance Potential. Studies of athletic potential at or near the altitude of Mexico City prior to the 1968 Summer Olympics showed that sprints and field events requiring only anaerobic energy sources were not affected.[39] Decreased marks were shown for events relying on the aerobic power of the performer. Faulkner and co-workers[163] found times for 1- to 3-mile runs to be from 2 to 13 per cent slower. In general, runs lasting over two minutes were generally 5 to 6 per cent slower than at sea level. Pugh[412] reported increases in 3-mile run times of 8.5 per cent on the fourth day at altitude, but this was reduced to 5 to 7 per cent by the twenty-ninth day. Similarly, 1-mile run times were reduced 3.6 per cent the first week and 1.5 per cent the fourth. Acclimatization helps approximate, but does not completely restore, sea-level capacity.

In a "Post-Mortem" on the 1968 Olympics, Craig[111] evaluated the winning performances relative to past results and expectations based on research completed prior to the Olympics. He found that winning performances averaged 0.9 per cent below world marks in 1968, while in previous Olympics the average was 2.9 per cent. Also, in 1968, 29 per cent of the winners equaled or exceeded world marks, compared to 15 per cent in the past. Such evidence would seem to contradict the implication that performance is impeded at high altitudes. However, winning times in running and men's swimming followed the pattern shown in previous Olympics and the implications of research: the longer the event, the further the winning time was from the world mark. The exceptional marks came in men's field events, which were not expected to be affected, and in women's running and swimming. The results of predictive studies were within 1 to 2 per cent of the actual results.

There is, of course, a progressive deterioration in performance as altitude increases. Pugh,[414] in his report of studies on Mount Everest, vividly described the plight of men working at high altitudes without supplementary oxygen. Above 27,000 feet sustained effort is apparently impossible, and one must proceed by following the principles of intermittent work. Even then, the climber suffers unusual respiratory distress. His calculations indicate that it is highly improbable that man can reach the peak of Mount Everest without oxygen.

Acclimatization and Training. After a stay at altitude, adaptive changes occur, restoring comfort and productive activity. The nature of such changes and the time required to produce them is important to both coach and athlete. Some of these changes have been reviewed previously.

The findings regarding acclimatization are difficult to interpret, since there has been considerable variation in the altitudes studied and the physical condition of the subjects. Subjects are frequently more active during their stay at altitude, and it is questionable whether reported changes are due to acclimatization or training.

Perhaps the most universally documented acclimatization response is the increased oxygen carrying capacity of the blood. Increases in hemoglobin and hematocrit have been frequently reported after adaptation to both moderate and extreme elevations.[416, 93, 210, 317] This compensates for the decreased oxygen saturation of each oxygen carrying unit by increasing the number of units available.

Most studies show an increase in ventilation over that found on acute exposure during both maximal and submaximal exercise. A 6 per cent increase in sea-level maximum ventilation was found at 2300 meters,[163] while a 14 per cent increase in STPD and a 22 per cent increase in BTPS values were shown after five weeks at 3800 meters.[317] Ventilation has also been shown to increase over acute exposure values after five days in a pressure chamber.[24] However, after sixty days acclimatization at 7760 meters, Ceretelli[93] found that maximum exercise ventilation was only 78 per cent of sea-level values. This might be explained by the overall physical deterioration common during long residency at such heights.

The maximum oxygen intake tends to increase, but does not reach sea-level values. Klausen and co-workers[317] found a 4 per cent increase in maximum oxygen intake despite a further depression of the maximum heart rate during a 5-week acclimatization period. Cardiac output and stroke volume decreased during submaximal work as a result of acclimatization.[316, 317] Highly-trained athletes who are already near their potential physical condition on ascent to moderate altitude may not show any improvement in the maximum oxygen intake during their stay. This has been found true for both runners and swimmers.[162, 163]

It has been pointed out repeatedly that complete acclimatization is a long process. British athletes exposed to the 2300-meter altitude of Mexico City were still showing signs of adaptation after four weeks.[412] However, the major adaptation is completed within this time, and only relatively small changes may occur thereafter. Age has been shown to be no barrier to acclimatization.[134]

Man has demonstrated his ability to adapt to extremely high elevations. Both in Asia and in South America, large populations carry out normal life activities at elevations approximating 4000 meters. The short-term acclimatization that occurs on arrival at high altitude and the long-term acclimatization acquired over many years are different. Lahiri and his co-workers[335] studied Sherpas (highlanders) and lowlanders, over an 8-month period, at an elevation of 4880 meters in the Himalaya Mountains. Steady state exercise oxygen intakes were similar in both groups, but the Sherpas showed a smaller ventilation and a higher arterial carbon dioxide tension. They also showed a lower pH. The lowlanders had maximum heart rates of 146 to 165 beats per minute, while the Sherpas' were as high as 196, values usually expected when testing at sea-level. Similar results were reported by Mazess[369] on Peruvian Indians. He concluded that their superior performance was due to acclimatization and training and that no racial or genetic factors were responsible.

Training for athletic competition at a high altitude is something of a problem. Pugh and Owen[416] feel that training activity should be reduced during the first week at altitude; while Balke, Daniels, and Faulkner[37] recommend starting immediately, at a high intensity, so that athletes quickly become accustomed to the changed conditions. There are no apparent dangers to the athlete with this procedure.

Another problem discussed by the latter authors is the possibility of lower intensity workouts at high altitudes. One must have maximal muscle power to perform successfully, but the lowered overall work capacity at altitude makes it difficult to adequately challenge the muscles, leading to a gradual detraining. This may eventually lead to a decrease in the overall cardiovascular-respiratory response owing to a reduced ability to challenge these systems as the muscle power deteriorates.

Two proposals are made[37] to overcome this difficulty. First, speed work must be emphasized to maintain or increase muscle power; but this routine is difficult, since at high altitude, periods of exertion must be shorter and periods of recovery longer. The second suggestion is intermittent training at low and high altitude. Such a program would enable the runner to maintain his muscle power during sea-level exposures and to adapt to the thinner atmosphere during the high-altitude exposures. An example of this, as used in the Balke, Daniels, and Faulkner study at a 2300 meter elevation, follows: 9 days at altitude and 5 days at sea level; 9 days at altitude and 7 days at sea level; 16 days at altitude and final return to sea level.

UNDERWATER SWIMMING

As one descends into water, the pressure increases, because the density of the water is greater than that of air. Since the body is compressible, the pressure of the water is transferred to the internal structures and fluids. The pressure not only has a direct mechanical effect, but also causes more of the respiratory gases to be forced into solution, leading to disturbances of normal gas diffusion and transportation. Some of the more important consequences are described below.

Snorkels. A snorkel is a short tube which allows the swimmer to breathe outside air while his face is below the surface of the water. Most people cannot make a single inhalation when their chests are about 4.5 feet below the surface, because water pressure becomes greater than the force which inspiratory muscles can develop. Therefore this device can operate only close to the surface.

SCUBA (***Self-Contained Underwater Breathing Apparatus***). There are two main types: *the open circuit* and *the closed circuit.* In the open circuit the swimmer inhales air from a cylinder or a bottle containing compressed air, and exhales directly into the water. From a military standpoint this method is not always acceptable because of the tell-tale bubbles of expired air. The closed circuit consists of a cylinder with compressed oxygen, a breathing bag, and a canister containing an absorbent for carbon dioxide. The tube leading from the swimmer's mouth has two one-way valves which control the direction of air flow so that inspired "air" comes directly from the bag and expired air must go through the canister before it returns to the bag. Oxygen in the bag is replenished from the cylinder by a valve that is either automatically or manually controlled. Care should be taken to assure complete absorption of carbon dioxide, for an accumulation of this gas may be dangerous. From this standpoint, an open circuit is safer.

The cylinders in such an apparatus are filled under pressure of 1800 to 2000 lb. per square inch, and contain from 38 cu. ft. (1076 L.) to 70 cu. ft. (1983 L.) of air. The first is rated as a half-hour unit, and the second as a one-hour unit. In rating these cylinders, a rather liberal allowance of 30 L. per minute was made.

Among various physiological dangers associated with underwater swimming, we might mention *nitrogen narcosis*, *decompression sickness* (bends and embolism), and *oxygen poisoning*.

Nitrogen Narcosis. When air is breathed under pressure, the amount of nitrogen dissolved in the body fluids is increased, and a state of general anesthesia may be reached. French divers coined a name for this state, "rapture of the deep." The diver's judgment is distorted, and he acts as if he were drunk. For this reason, a "martini" rule has been used. At the depth of 100 feet, one feels as if he had had one martini on an empty stomach. At 200 feet, it feels like two or three martinis; at 300 feet, like four; and at 400 feet, like too many. Since tolerance to martinis is different in different men, some feel the effect sooner and at a lesser depth. The diver, if he is connected by a telephone with the surface, talks like an intoxicated man. One has even been known to remove his mask and try putting it on a fish.[144] The outcome may be fatal.

Decompression Sickness (Bends). On ascent, nitrogen begins to escape from the body fluids. If "decompression" is sufficiently rapid, the escaping nitrogen cannot be brought by the blood to the lungs in an orderly manner and eliminated from there. It may accumulate in various tissues, and cause either discomfort or pain. It may compress or even damage nerve tissue, and cause a temporary or permanent paralysis. Air bubbles in blood vessels may even cause *air embolism* and stop circulation through those vessels. To avoid decompression sickness, the diver should adjust his speed of surfacing to the depth and the length of his dive. For depths up to 90 feet, one can stay down for 30 minutes without danger of decompression sickness, if one surfaces at the rate of *60 feet per minute*. For other depths and durations of dives, see Table 18.

TABLE 18. *Relation Between the Depth and Safe Duration of a Dive When Air and Oxygen are Used*

Air		*Oxygen*	
DEPTH FT.	TIME MIN.	DEPTH FT.	TIME MIN.
40	120	10	240
50	78	15	150
60	55	20	110
70	43	25	75
80	35	30	45
90	30	35	25
100	25	40	10
110	20		
120	18		
130	15		

When symptoms of decompression sickness appear, serious consequences may be avoided by placing the victim in a decompression chamber where pressure is rapidly brought to the level of his dive, and then slowly bringing him back to normal. It may take from a few to many hours to complete decompression.

Oxygen Poisoning. Under pressure, oxygen may become harmful. The symptoms are convulsions, followed by unconsciousness. Although oxygen poisoning may occur without any warning, frequently it is preceded by signs. The most common signs are twitching, nausea, dizziness, ringing in the ears, and visual disturbance.

Of 50 trials in underwater swimming conducted by Schaefer,[453] 14 had to be terminated because of symptoms of oxygen poisoning. Analysis of the gases breathed indicated that poisoning was caused by the high oxygen content (79.15 per cent) at 20 to 40 feet, and was not related to the carbon dioxide content. Oxygen under pressure leads to "dyspnea characterized by rapid, shallow breathing and apparent inspiratory inhibition." During underwater swimming with oxygen, the pulse is relatively slow (90 beats per minute). When poisoning begins, the pulse suddenly rises to over 150 beats per minute. A diver using oxygen should periodically count his pulse. If there is a marked rise, he should surface before it is too late.

There is no treatment other than reducing oxygen pressure to normal. Oxygen poisoning may occur when the closed circuit Scuba is used. For safety reasons, it is advisable not to use pure oxygen below 25 feet. If it is necessary to go deeper, a special table should be consulted for safe time limit (see Table 18).

Danger from Hyperventilating before Free Dives. Recently, newspapers have carried a number of stories concerning near-fatal or fatal results when people either became unconscious or drowned while trying to break their records swimming under water, depending only on their ability to hold their breath. On land, people practicing breath holding are more fortunate—they just faint and recover.

Since some instructors of swimming insist on long dives, it is only appropriate to know what may happen. With this purpose in mind, Craig[109] undertook two experiments, one on land and the other in water. His subjects held their breath as long as they could while riding ergocycles or while swimming under water.

The alveolar air was analyzed for oxygen and carbon dioxide. It was observed that the lowest concentrations of oxygen or carbon dioxide were found when the subjects hyperventilated their lungs before the exercise. The oxygen pressure was 34 mm. or lower in several subjects. Such a low oxygen content often leads to unconsciousness. Hyperventilation makes this stunt more dangerous, because it lowers the carbon dioxide content of the blood, which normally acts as a stimulant for the respiratory center.

Thus, fatal and near-fatal results during breath holding while exercising may be attributed to hypoxia, aggravated by a low carbon dioxide content.

QUESTIONS

1. How much does a man's normal temperature vary during the day?
2. When is a high ambient temperature tolerated better, in humid or in dry air? Why?
3. How much, and in what manner, is heat dissipated by the body?
4. How does high ambient temperature affect respiration, circulation, and blood volume?
5. Discuss the relation between external heat and muscular activity.
6. How much water may be lost through perspiration?
7. Why is taking salt tablets recommended to people working under conditions of high temperature?
8. How does work at high temperature affect pulse rate, blood output of the heart, and respiration?
9. What are the safe upper limits or rectal and oral temperature?
10. What is the advantage if rest periods during intensive work at high temperature are taken in air-conditioned rooms?
11. What is the effect of cold upon work capacity?
12. Will the hands feel colder on a cold day when the person is scantily or when he is heavily dressed?
13. What effect does high altitude have on sprint and endurance events?
14. List the physiological changes that occur during the acute adaptation to altitude.
15. How would you train an athlete for competition at high altitude?
16. High-altitude dwellers have special physiological adaptive powers that enable them to live there: true or false? Defend your answer.
17. Why might employing the principles of intermittent work be of benefit at very great altitudes?
18. Why might high hemoglobin concentrations be of value at high altitudes?
19. What does *SCUBA* mean?
20. What is nitrogen narcosis? What do French divers call it?
21. What are the bends?
22. What is air embolism?
23. How deep can one dive with compressed air, stay there for thirty minutes, and come back in one and one-half minutes?
24. Describe oxygen poisoning.
25. Why is hyperventilating before a long dive dangerous?

Chapter Seventeen

HEALTH, PHYSICAL FITNESS AND AGE

An attempt to define physical fitness is a most provocative task. It seems almost, if not entirely, impossible to find a definition which will satisfy everybody. There are two main stumbling blocks: one, the relation of physical fitness to health; and the other, the consideration of what constitutes a physical fitness test.

An instructor using this book might find it very profitable to give his students an assignment to collect as many definitions as possible of health and physical fitness. After collecting about two dozen of these definitions they should be grouped and analyzed. The result of this assignment will be very interesting, and will reveal how much confusion may be caused when wishful thinking is substituted for logic and facts and how even meaningless statements may find their way into print if they are high sounding.

DEFINITION OF HEALTH

One will find in medical dictionaries that health means *absence of disease*. This definition is applicable only to *perfect health*. Since most people suffer from some disease which may be as trivial as a wart or as serious as the last stage of cancer, this definition does not fit everybody. We have to speak about a degree of health rather than about perfect health. The degree of health evidently depends on the seriousness of the disease, and may be measured by subtracting the degree of illness from perfect health:

Degree of health = perfect health − degree of illness

This formula is consciously or unconsciously used by everybody. It is shown in Figure 95, where the state of health is represented by a diagonal line which is the result of the interaction between perfect health and disease. Since disease may be defined as a disorder or abnormality of function of an organ or a system of the body, health may be understood as follows: *health is a state of the organism in respect to the degree of normality of its physiological processes.** Subjectively, the degree of health may be appraised through deviation from the sense of well-being. Although subjective sensations associated with a state of health are not always reliable and are sometimes misleading, nevertheless they are often the first manifestations of either decline or gain in health. For this reason, health may be defined as *the degree of freedom from disease and defects which disturb homeostasis.*

PHYSICAL FITNESS

Confusion regarding the definition of physical fitness stems from an understandable desire to make a definition apply to everything under the sun: health, economic success, and happiness. It can be said that there are three types of fitness: fitness for living, fitness for holding a job, and fitness for recreational hobbies. All of them are affected by age and sex.[291]

Strictly speaking, physical fitness means that a person possessing it meets certain physical requirements. These requirements may be anatomical (structural), physiological (functional), or both. Anatomical fitness may require a person to be of a certain height or weight, or have

*Mental health may be considered as depending upon the processes in the brain.

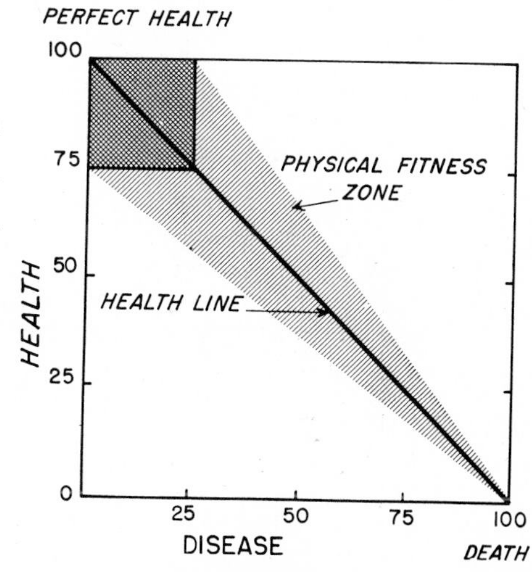

Figure 95. Relation between Health, Disease and Physical Fitness. Health line = perfect health − degree of disease. Physical fitness is shown as a zone which depends on the type of fitness tests. Lower border—difficult test and the upper—easy test. These borders, obviously, are not necessarily straight lines. In the upper left corner area the relation between the degree of health and fitness may be shown on groups but hardly on individuals.

specified dimensions of various parts of the body. Physiological fitness may require a person to be able to withstand certain temperatures or altitudes, or able to perform specific physical tasks involving muscular effort. A person may be perfectly fit to meet some of these requirements and yet be unfit for others. A person physically fit in all respects does not exist. A grown-up person who is fit to be a jockey will never become a champion heavy weight lifter.

One should not forget that, at the present time, *physical fitness measures merely the ability to pass physical fitness tests;* and, therefore, the so-called degree of fitness possessed by an individual depends on the character of the test. If an easy test is given, the score obtained will be high, and physical fitness will appear to be high. On the other hand, a difficult test may put the same person in a low category. Suppose that a group of men is given a pull-up test. Since there are men who can not make even a single pull-up, these will fail, and therefore will be considered unfit. Unfortunately, too often too much is expected from a single test, especially when the test involves little exertion. It must not be forgotten that, after all, a test determines primarily the degree of physical fitness relative to the test. Therefore, it is important that the test used provide a true evaluation of the physical fitness characteristic under study.

It is obvious that, with more numerous test items, more information can be obtained about the individual. Efforts are made, however, to select those items that can reveal the most and thus reduce the number of tests. The use of too many tests is impractical: they either tire the subject, and therefore affect the score, or demand more time for testing than is justifiable or available.

From an occupational point of view, physical fitness may be defined as *the degree of the ability to execute a specific physical task under specific ambient conditions.* Most industrial work requires only a medium degree of exertion, and therefore applicants need not be physically tested. If strenuous work is involved, a fitness test can be devised very easily by imitating the strenuous phases of the work.

Fitness for recreational hobbies, which can be expected to result in happiness, depends on what a man wants and likes to do.

HOW MUCH PHYSICAL FITNESS IS NECESSARY?

Nobody knows exactly. One answer might be: Much less than physical culturists insist, and more than some advocates of inactivity recommend. Surely, though, fitness for living requires a minimum applicable to all. Men should be able to walk, run, carry some weights, and be ready for an emergency. Even this minimum is not indispensable, as may be observed on handicapped people. The famous Pépin of Paris had no arms or legs, yet he lived to be 62 years of age. Incidentally, he was quite intelligent and could *write* in several languages.[191] Since we do not know how much fitness will be necessary, it is safer to follow the example set by Nature, and have an excess of fitness to guarantee a sufficient reserve for emergencies.

One cannot escape the idea, however, that there is a general physical fitness—present or potential—for most activities involving physical work. A good illustration of this may be seen in the processing of men called into the armed forces. Medical examiners, after weeding out those who are medically and anatomically unfit, accept those who, in their opinion, could immediately, or after a period of training, be fit to perform the occasionally arduous tasks incidental to military life. Since this degree of physical fitness has been determined without any special objective tests, it can be concluded that it is "potential" fitness that has constituted the guiding basis of this kind of judgment.

RELATION BETWEEN HEALTH AND PHYSICAL FITNESS

There is no question that there is some relationship between health and physical fitness as measured by muscular performance tests. A sick man, at a low degree of health, may serve as a clear-cut illustration of this relationship. When, however, we examine this relationship at the upper end of the health scale, the situation becomes uncertain, obscure, and contradictory. Men with impaired health have been known to have more fitness for athletic competitions than those in good health.

Most students in physical education schools are in good health and good physical fitness. In Figure 95, they will be found in the upper left corner area. This important area is very crowded, and the spread is so small that efforts to establish a correlation between health and physical performance tests give some positive results only for groups, but not for individuals. One should not be either surprised or discouraged, because these tests measure the combined effect of aptitude and training, which are independent of health above a certain level. The small area in the diagram represents this area of independence.

If, however, tests are very easy, the level of independence will be much lower. For example, one may ask a dying man to open and close his eyes. This test, certainly, will not measure the degree of health: it will merely indicate that the man is still alive.

The most reliable information regarding the relation between health and physical fitness will come from studies based on testing physical fitness before and during illness. This will require frequent testing of a large group. Then, when any member of this group becomes ill, he can be tested through the entire period of illness and convalescence.

Coaches of sports such as track, swimming, and apparatus gymnastics, in which the activities used may also serve as tests of fitness, can usually detect when a member of a team becomes ill, although the individual himself may attempt to hide his condition. Even though men have been known to compete successfully in spite of indispositions, such feats should be regarded as exceptions rather than the rule. A champion sprinter came down with tuberculosis two months after winning the championship, and died a year later. One gymnast had lung tuberculosis during the height of his career.[273]

Interpretation of results obtained even with a test of strenuous performance should be made cautiously. Undoubtedly such a test reflects the function of various physiological systems under stress. It reflects, for example, the functioning of the heart and circulatory system. On the basis of such a test, however, little can be said about the organic condition of the heart. Men with valvular defects of the heart may score excellently in a strenuous test because of compensatory mechanisms in the heart. The test will merely demonstrate the functional extent of the compensation.

It has been taken for granted that physical exercise improves health. Such an assumption appears, on the surface, to be logical: physical training increases or, as is customarily understood, improves capacities of various physiological processes. Since health represents the sum total of the states of all physiological processes, one is tempted to conclude that physical training improves health.

Let us consider a man convalescing from a serious disease. He can barely walk and gets dizzy easily upon exertion. Such a man is usually considered to be an example of both a low degree of health and a low degree of physical fitness. He practices walking, and eventually he is able to walk well, lift 200 lb., and swim 100 yards. He continues to exercise six more months. Now he can hike long distances, lift 300 lb., and swim one mile. Undoubtedly he is much better fit physically. But is his health better than it was six months before? Some would say, "Yes"; some would say, "No." This difference is caused by the vagueness in the definition of health.

It seems more logical to conclude that, beyond a certain indispensable minimum of physical fitness, an additional improvement in physical fitness has no effect upon health (no matter how one defines the word). Excessive physical activities, on the other hand, may be definitely detrimental to persons with some diseases. The rather common practice of "sweating out a cold" by playing hand ball or some other strenuous game can hardly be recommended as a cure. What may appear to be a simple cold may be the beginning of some serious disease. Furthermore, a sick player may spread the disease among the other participants.

One may now ask the question: What is the minimum degree of physical fitness indispensable for health? Although physicians, almost unconsciously, take this minimum into consideration in their daily practice, nobody can give a precise definition. Undoubtedly it is very low.

MENS SANA IN CORPORE SANO

One of the most famous historical slogans applied to physical education is: "A sound mind in a sound body." This slogan has been used so frequently that is has acquired practically the rank of an axiom.

If one assumes that an abnormal function denotes some abnormality in structure—not necessarily gross anatomical or histological, but submicroscopic—it would be difficult to refute this assumption. On the other hand, the appalling number of psychoneurotic and psychotic individuals with apparently sound bodies indicates that soundness of the body, as judged by development of the muscles, does not always guarantee soundness of the mind. Yet it is undeniable that physical activities may help to keep the mind sound. Taking part in an enjoyable game increases the zest for life, eliminates unwholesome moodiness for the time being, erases worries, and, at least to an extent, neutralizes the damaging effects of repetitive morbid ideas. A moderate degree of fatigue which follows physical activity may help to combat insomnia, and consequently the individual gets more rest.

Recently this slogan, "a sound mind in a sound body," has been misinterpreted as signifying that "better physical fitness means a better learning capacity." Since learning is a function of the nervous system, not of the muscles, we could have dismissed this interpretation as sheer nonsense. One may, however, be surprised to learn that many people in the profession of physical education *believe* in this interpretation. It is true that a belief does not have to be based on logic or facts; yet some professional people, who should have known better, have tried to muster experimental data for supporting this belief.

The "experiment" consisted of selecting a group of pupils with poor class marks, and giving them special classes in physical fitness. Results: fitness improved, and so did the marks. It should be noted that the experimenters did not confine their efforts to a mechanical administration of activities. They also tried to stimulate their pupils' interest in school life in general, which is probably the explanation of the success.

Delinquency in studies often depends on lack of interest. Raise this interest, and better marks follow. This type of experiment is similar to an "experiment" in poker with stacked cards. We are sure, however, that the experimenters did not realize that the cards were stacked. It must be mentioned here that this Latin quotation is really a misquotation and should be read "Orandum est ut sit mens sana in corpore sano"—"It is to be prayed that there would be a sound mind in a sound body."*

In this connection, an investigation, conducted by Asmussen and Heebøll-Nielsen in Denmark, is of special interest.[15]

They divided their subjects into three groups on the basis of Intelligence Quotient, the average being 120, 103, and 83; and gave them a physical fitness test which included measuring the strength of leg extensors and finger flexors, maximum expiratory and inspiratory force, and height of the vertical jump.

Although the difference between the fitness scores obtained in the first and second groups was not statistically significant, the score tended to be higher in the second group. Evidently, the "book worms" composing the first group were less motor-minded than those in the second group with lower Intelligence Quotients.

The third group, with the lowest I.Q. rating, also had the lowest physical fitness score, because its members lacked intelligence and could not properly follow directions.

PHYSICAL FITNESS AND IMMUNITY TO DISEASE

It is common to meet "strong" men who boast that no disease can get them. These men claim to have iron stomachs, to have no need to dress warmly in cold weather, or to carry umbrellas or wear rubbers.

The idea that physically strong men are less susceptible to diseases has, probably, developed by association of ideas. A strong man can take greater physical punishment; he also can endure privations better than a weaker man: therefore, a strong man is less susceptible to disease.

Histories of epidemics are full of illustrations that "strong" men are affected as well as ordinary people. The great epidemic of typhus fever in Russia immediately after World War I certainly showed no favoritism toward strong men. One of us (P.V.K.) examined about 1000 adult male patients with typhus. In their physical development they represented a typical cross section of the male population.

And, of course, the history of venereal diseases shows that no amount of physical training will make an individual immune to this type of infection. It is a common belief that a physically well-developed man is less susceptible to the upper respiratory diseases, including the common cold. Again, one cannot find sufficient proof even for this contention.

Observations on 18,823 school athletes in Indiana showed that the

---

*Thanks to Mr. Philip Smithels of New Zealand for calling attention to the full quotation: Juvenal, 10th Satire.

incidence of illness among them was the same as among non-athletes.[406]

Spaeth[488] attempted experimentally to find in rats the relationship between physical exercise and susceptibility to infection by Type I pneumococcus. His conclusions were that the resistance of rats after seventeen to twenty-six weeks of training was less than that of "sedentary" rats. Moreover, none of the sedentary rats died, whereas 12.5 per cent of the trained rats did. Spaeth also found that rats exercised in a motor-driven revolving cage until exhaustion became more resistant to infection than did resting animals.

There is not enough evidence to support a contention that physical activities make a man more immune to upper respiratory infections. The fact that a person engaged in outdoor physical work may do it with apparent impunity seems to depend on two factors: (1) while working he produces enough heat to prevent chilling; and (2) it is possible that his vasomotor reactions have been better adapted to changes in temperature than have those of a person cooped-up indoors.

While the first factor can operate just as well in a sedentary person, as long as he keeps his body in motion to produce a sufficient amount of heat, the second factor is only incidental to physical training. It is possible to improve vasomotor reactions by the judicious use of cold showers without indulging in physical activities.

ALLERGIC REACTION TO EXERCISE

Some athletes develop rash and wheals (urticaria) after exercise. The areas usually involved are the skin of the neck just behind the ear lobes, and the upper parts of the thighs. Other portions of the body may also be affected. It seems that pressure exerted by the belt of the athlete's trunks aggravates this condition. The incidence of allergic skin reactions among athletes is, fortunately, rare. In some of them, such a rash is noticeable at the beginning of training and may disappear later. In others there seems to be no relation to training, and hives may appear after very light exercise, and sometimes even when a man is sitting relatively quietly studying.

A convenient explanation for urticaria after physical exertion has been based on the theory that histamine, which is produced during muscular work, is not readily destroyed in allergic persons. Such persons do not have sufficient amounts of histamine-esterase, an enzyme which destroys histamine. This theory has received some support because allergic patients have improved after systematic injections of small, but gradually increased doses of histamine. These injections supposedly have stimulated the production of histamine-esterase. This treatment, however, may not lead to a cure, since some patients have to rely on injections of epinephrine, which brings about relief within thirty to sixty minutes. Additional material on this topic may be found in Reference 273.

INDISPOSITION AND COLLAPSE AFTER STRENUOUS EXERTION

When physical exertion becomes excessive for an individual, fatigue serves as a safety device. It slows down the speed of movements or the intensity of contraction, and the organism again is able to function adequately.

Sometimes, however, this adjustment is achieved in a different manner. Instead of the normal symptoms and signs of fatigue, a person who has exerted himself excessively may experience an acute indisposition. He feels weak and complains of a headache and pain in the abdomen. Profuse sweating, nausea, and vomiting may be present. The acute state usually lasts a few minutes; the after-effects may last an hour or more.

Jokl[272] suggests, as a name for this condition, "athlete's sickness," because, in his opinion, this harmless disturbance occurs only in athletes, since non-athletes do not exert themselves sufficiently to get these reactions.

The present authors, however, have observed non-athletes who have developed these symptoms after running as fast as possible a distance of 300 yards. It is hoped that the name "athlete's sickness" will not be adopted. As it is, there are already two misleading names: "athlete's heart" and "athlete's foot." Why add one more?

The nature of this indisposition is not known. Various explanations have been offered, ranging from a fall in the sugar content of the blood to a vasomotor collapse. Jokl believes that a simple indisposition is due to a drop in the sugar content.

The authors have observed some students who showed marked indisposition after a short strenuous exertion on a bicycle ergometer. The blood sugar level in some of these men was normal, or elevated. Blood pressure also behaved normally. There is a convenient explanation that seems to be logical: cerebral disturbance caused by an excess of lactic acid in the blood.

Collapse due to physical exertion alone, as a rule, is not accompanied by nausea and vomiting. Conditions are aggravated by a vertical position and alleviated by a recumbent one. The patient feels dizzy and may become unconscious. This state is probably due to an extensive vasodilatation caused by muscular activity. Since collapse occurs only infrequently, it is justifiable to suspect that it is a pathological rather than physiological reaction, and that the person who has collapsed should have a careful medical examination for a possible cardiac lesion.

LONGEVITY OF ATHLETES

The death of an outstanding former college athlete from a cause other than old age immediately becomes news, and focuses public opinion on the question: Is participation in strenuous athletics harmful? Physical directors and coaches usually take it for granted that physical education and athletics help to prolong life. There are, however, many people who, admitting that moderate indulgence in exercise is hygienic, believe that participation in strenuous athletics is definitely harmful and shortens life. It is, therefore, of practical importance to review this question.

The study of the effect of participation in college athletics upon the life span is almost a futile task. First, the period of college life represents only a small fraction of a man's life; second, it is very difficult or impossible to isolate the influence of other factors such as heredity, occupational hazards, diet, and dissipation.

At the present time, some promoters of the physical fitness movement state that exercises delay the onset of old age and therefore prolong life. It might be so, except that convincing evidence is lacking. Women exercise less than men, yet they live longer. The "youthfulness" of a more active old man may be only apparent, representing a higher degree of motor fitness which this individual has maintained. The relation of this higher level to the unwinding of the Spring of Life is another matter.

Athletes are a selected group. A sickly student who may be a candidate for untimely death is less likely to join this group, and stays among the non-athletes, bringing their average span of life down. Dublin[142] says that perhaps an athlete is not predestined for a long life. Upon discontinuing his regular training, he has a tendency to put on weight, which may shorten his life. He is also a more physically adventurous man, and is likely at times to take undue chances, believing in his superior health, strength, and agility, and thus exposes himself to unnecessary risks. It is a common thing to see former athletes, who have passed the prime of life, engaging in strenuous games with much younger opponents. The adage that "one is as old as he feels" is a dangerous rule to follow.

Non-athletes, who may be small and physically less fit, may outlive more physically vigorous men by taking better care of themselves and avoiding risks.

In conclusion, it may be said that more is known about how to prevent shortening of life than about how to lengthen life. A report by

Montoye[382] and his associates not only gives an excellent review of the literature on the subject of longevity, but also furnishes results of a thorough investigation made on 628 athletes and 563 non-athletes from Michigan State University. The authors came to a definite conclusion that there is no difference in longevity between athletes and non-athletes. Seven years later, a follow-up study was published by Montoye and co-workers.[381] During this period 47 athletes and 30 controls had died. Again no difference in longevity between the groups was found.

ATHLETIC CONTESTS FOR CHILDREN AND ADOLESCENTS

There is a widespread tendency to lower the age limit for participants in interscholastic or interclub competitions. In some communities it has already reached the elementary school age level.

The official stand taken by the Society of State Physical Directors in the United States is that interscholastic competitions for elementary school children should not be practiced. The defenders of competition criticize this attitude as untenable and not even logical. They say that children compete anyway, so why not make competition safer by providing supervised team play. Moreover, there is usually a great deal of pressure from the parents and community which compels a school to have "varsity" teams and bring the school into the limelight.

Our knowledge regarding the relation between the age of the participant and the kind and the amount of exercise which he needs, or can withstand, is inadequate. Moreover, the situation is complicated by various factors.

1. Chronological age as compared with physiological growth. Chronological age cannot be taken as an adequate measure of physiological development.

2. The great degree of resistance on the part of the human organism. What appears to be an abuse may leave no detectable mark whatever.

3. The great weight given to the "voice of experience" on the part of physical educators and physicians. This experience may support or condemn one and the same thing.

During the Seventh International Congress on Sports Medicine held in Prague in 1948, the problem of age limit in competitive sports was discussed by a number of scientists pursuing research in physical education. The papers revealed the meagerness of objective data and, therefore, conclusions had to be based on the opinions of the investigators. Two years later, at the eighth congress of the same organization held in Florence, Kral,[326] a leading cardiologist and sports doctor, read a paper in which he outlined practical recommendations concerning training and competition among young children and adolescents. Although these recommendations will not be acceptable to everyone, they represent an attempt to supervise and control competitions among youngsters and, therefore, deserve our consideration. Kral's recommendations are presented here in part, and in a condensed form.

All pupils should have a careful medical examination before the beginning of training. Their participation in competitions depends on age, as shown in Table 19. When a pupil is obviously developed beyond his chronological age, an exception may be made to permit him to take part in competitions regardless of his age.

A gradual training in all these activities begins several years earlier than the age shown in Table 19.

Some American physical educators will find that the age limit is too high. In contrast to these recommendations is the extreme youth of the holders of many world swimming records for women. On the other hand, there are physiologists and medical men who believe that no child younger than sixteen should participate in competitive athletics.

Shuck[473] studied the effect of participation in intramural athletic

TABLE 19. *Age Requirements for Children and Adolescents for Participation in Athletic Contests (After Kral)*

I

Age	*Running (in Meters)* Sprint		Long-distance		Cross country		*Swimming (in Meters)*		*Diving* Height of diving board	
	BOYS	GIRLS	BOYS	GIRLS	BOYS	GIRLS	BOYS	GIRLS	BOYS	GIRLS
11–13	40						50*	50*		
13–16	60	60	1000		1800		100	100	1	
16–18	200	150	1200		3000	1000	200†	200‡	3	
20				800						

II

	BOYS	GIRLS		BOYS	GIRLS
Basketball	16 yrs.	16 yrs.	Rugby	16 yrs.	Not suitable
Boxing	16 yrs.	Not suitable	Soccer	16 yrs.	Not suitable
Fencing, foils	11 yrs.	13 yrs.	Volleyball	16 yrs.	16 yrs.
sabre	16 yrs.	Not suitable	Water polo	16 yrs.	Not suitable
Field hockey	16 yrs.	16 yrs.	Wrestling	14 yrs.	Not suitable
Golf	13 yrs.	13 yrs.			
Ping-pong	16 yrs.	13 yrs.			

*Without public admittance; †total distance in one day, 600 m.; ‡total distance, 200 m.

contests by boys in grades 7 to 9, and found that growth and development, judging by the Wetzel grid, had not been deleteriously affected.

An American reader, naturally, would be interested in the age level at which competition in baseball may begin. From the standpoint of physiology, training in this sport may start when the boy can hold the bat and hit the ball. This statement may be misconstrued as an indication that physiology is the sole basis for deciding when physical contests should begin. On the contrary, an educator should use at least three more types of criteria: psychological, social, and economic.

THE AGE OF MAXIMAL PROFICIENCY IN SPORTS AND ATHLETICS

One would think that a study of the records of the Olympic games would furnish the necessary data concerning the age of maximal proficiency in athletic performance. This is, however, not the case.[341] A comparison of the ages of Olympic champions with those of professionals has shown that the average age of the former is somewhat lower than that of the latter. It means that amateurs discontinue participation in contests at an early age. On the other hand, financial rewards compel a professional to continue as long as possible.

The most proficient age among Olympic athletes is between seventeen and thirty. The best age for activities requiring speed and agility is lower than for those requiring endurance. Short-distance swimmers become champions in their teens, while marathon runners do better when they have passed the middle twenties.

The capacity for moderate work does not decline with age, but the limit for hard work is considerably lowered.[128] However, people accustomed to physical activity are capable of a surprising level of physical performance.

Thirteen Bulgarians, 55 to 80 years of age (average 63), hiked across mountainous terrain at altitudes between 4900 and 8900 feet for

TABLE 20. *The Age of Best Performance*

Activity	*Age*	*Activity*	*Age*
Baseball		Golf, amateurs	25–29
Excluding pitchers	28	professionals	30–34
Pitchers only	27	Ice hockey, prof.	24–28
Bowling	30–34	Rifle and pistol	25–29
Boxing, heavyweight	26–30	Roller skating	14–18
Football, professional	23–27	Tennis, singles	22–26

(From Lehman: Am. J. Psychol. *64,* 1951.)

30 days, covering 435 miles. After the hike they were in good health. The number of red blood corpuscles increased by 0.3 million, the hemoglobin by 3 per cent. Eleven persons lost about 3 pounds of weight and two did not change.[334]

SOURCES OF FITNESS

The equipment that enables a man to combat adverse influences and meet the requirements of his labor is partly inborn and partly acquired. This equipment is divided into three main categories: morphological, physiological, and psychological. The physical form and structure of the body constitute the morphological aspects of the equipment. In large part these are determined by heredity, but no one today doubts the statement that "use makes the organ." Heredity determines the possible course and limits of development, but the use of an organ is absolutely necessary for its proper and full development. Graded and frequent use of organs is the instrument of physiological development by which the capacity for activity is enlarged and a nicety of adjustment obtained. The mind is the master of the bodily machine; it, too, acquires greater capacity and a better equilibrium and adjustment with graded and proper use.

Just how far one should strive to develop the inherited capacities of his body is a question that cannot be entirely satisfactorily answered. The ideal goal is to be sufficiently fit to accomplish each day's work with a minimum of fatigue, and to remain active to a good old age. This may mean that some individuals must train for heavy physical labor and others for light sedentary work. In either case life must be so ordered that the body maintains a normal physiological status. If it is not so ordered, the body becomes pathological—that is, unhealthy. A low degree of fitness seems inadvisable, for it leaves no margin of safety for the experiences of adversity which frequently descend upon mankind.

FITNESS OF THE AMERICAN CHILD

During the past decade, a number of reports[310, 322] have indicated that the American child has a lower level of physical fitness than the child in Europe or Asia. This statement was based on tests of motor performance which considered elements of muscular strength and endurance. These reports caused a great deal of concern among parents, educators and political leaders, and rightly so.

Do our children deteriorate physically? If they do, then the future of the nation is at stake. Yet, on the other hand, the life expectancy of the American youth is greater than for those abroad, and undoubtedly, this life will be energetic and productive and not merely prolonged vegetation of decrepit old age.

It is evident that motor performance tests, important as they are, do not tell the entire story. Several investigators have given a work capacity test to American, European, and Asian children and have found no significant difference.[1, 2] The work capacity test was that described by Wahland[523] (see Chapter 18, on testing).

Fowler and Gardner[175] made a dramatic study of 46 children with cardiac conditions and 14 with muscular dystrophy, 7 to 17 years of age. They gave them work capacity tests, the Kraus-Weber test and the American Association for Health, Physical Education and Recreation fitness test. The latter test consists of pull-ups, sit-ups, a shuttle run, a standing broad jump, a 50-yard dash, and a 600-yard walk and run. Children with muscular dystrophy showed a definite decrease in both the work capacity and motor performance tests, but "children with congenital heart disease or asthma had a marked decrease in physical work capacity but only slight changes from their predicted scores on most of the motor performance tests."

These findings demonstrate that one should be careful in generalizing results obtained with just one type of testing.

QUESTIONS

1. Define health in general and perfect health in particular.
2. Define disease.
3. Define physical fitness.
4. What is the relation between health and physical fitness?
5. How much physical fitness is indispensable for living? What can you say about Pépin of Paris? What does it prove?
6. What does: "Mens Sana in Corpore Sano" mean? Give the complete quotation.
7. Can one improve scholastic ability by improving physical fitness? How and why? What did Asmussen and Heebøll-Nielsen find regarding a relation between physical fitness and I.Q.?
8. Does a high degree of physical fitness mean an immunity to diseases?
9. Does exercise prevent or cure colds?
10. Discuss allergic reactions to exercise.
11. Discuss indisposition and collapse after physical exertion.
12. Do former varsity athletes live longer than non-athletes?
13. What is your attitude regarding athletic competitions at various levels: elementary; junior and senior high school; college?
14. At what age is the peak of athletic performance reached?

Chapter Eighteen

TESTS OF PHYSICAL FITNESS

The degree of physical fitness of an individual depends on the integration of innumerable functions of the tissues and the organs. Therefore, an appraisal of the fitness of an individual for an occupation cannot be made without a test that includes an amount of physical effort equal to that necessary for the occupation in question. Besides fitness for the *intensity* of the effort, fitness for the *quality* of the effort is of practical importance. This type of fitness may be ascertained by properly designed tests of skill. Without minimizing their importance, we shall omit a discussion of tests of skill from this text and discuss only physiological tests.

CLASSIFICATION OF TESTS

At present there are many different kinds of physical fitness tests. Although a rigid classification is almost impossible, they can be divided roughly into three groups: (1) muscular performance, (2) organic function, and (3) a combination of muscular performance and organic function. Examples of muscular performance tests will be found in chinning, sit-ups, running, and so forth. Examples of organic tests are: measure of vital capacity at rest, response of pulse rate to exercise, rate of oxygen consumption during exercise, and so on. The third type is well illustrated by the Harvard Step-Up Test, in which muscular endurance and pulse rate response to exercise are measured.

The weakness of the performance type of test is its great dependence on the cooperation of the subject, without which the test has no value. Some organic tests also suffer from this weakness; for instance, a measure of vital capacity cannot be obtained without the full cooperation of the subject. The other weakness of organic tests is that they may reflect emotional disturbances. For instance, a true measure of resting pulse rate or resting blood pressure cannot be obtained if the subject is emotionally disturbed. A combination of muscular performance and organic type of test may suffer from the weaknesses of both.

Although scientific testing of physical fitness is one of the newest adjuncts of physical education, it has developed to such dimensions that schools of physical education now have courses in the techniques of testing, usually called "Tests and Measurements," and books have been written on the subject. Therefore no attempt will be made in this chapter to cover this field, and only the broad physiological bases of various groups of tests will be discussed.

MUSCULAR STRENGTH TESTS

Muscular strength of certain muscle groups measured by dynamometers is employed as an index for estimating general condition. Physiology has shown that physical exertion overtaxes the circulatory mechanism long before it exhausts the skeletal musculature; and that, while it is not easy to overwork muscles, the heart may be overworked. The convalescent from infectious disease may be limited in his exercise, not by what his muscles can do, but by the strength of his heart. Hence, today the general opinion is that *strength tests do not permit us to draw satisfactory conclusions regarding the efficiency of the entire body.*

HEART TESTS

With advances in physiological knowledge, it was only natural that one seeking a physical fitness test should turn to the heart for evidence of physical condition. Among clinicians, the cardiologist has also felt the need of a test to determine the heart's working capacity or reserve power; he recognized that the capacity for physical exercise is undoubtedly a valid criterion of the condition of heart efficiency, because the capacity for exercise depends largely on the ability of the heart to increase its output.

While it is true that a man's ability to take exercise is, as a rule, determined by the functional capacity of his heart and that the stress of exercise usually falls more heavily on that than on any other part of the body, yet our knowledge of crest load and over load shows that the body often overdraws its oxygen account while the heart is still far from being overtaxed. Investigations on the effect of lack of oxygen indicate that the heart tolerates a high degree of oxygen want, but that the nervous system is more sensitive to a deficiency in oxygen supply; in fact, it is the most sensitive tissue of the body. A capacity test of the heart would be useful for determining the maximum capacity of the whole body to endure a heavy load for a short period of time, but it is questionable whether such a test can ever be standardized in terms of general fitness of the body and used to predict the approach of nervous exhaustion. It should be added that many of our best cardiologists believe that a test which would definitely determine the functional capacity of the heart is still an accomplishment for the future.

In addition, it should be mentioned that the behavior of the heart during exercise depends, not only on the degree of fitness of the heart itself, but also on the state of the other components of the circulatory system. For example, the excessive pulse response of an anemic person depends on the condition of the blood rather than on that of the heart.

PULSE RATE

If a test employs pulse rate at rest, pains should be taken to secure an actual resting pulse. Causes for excitement should be eliminated and a sufficient period of rest should be allowed. If only the post-exercise pulse rate is used, the exercise should be strenuous enough to eliminate the possible effects of emotional factors.

There seems to be an agreement that the pulse rate curve during the period of recovery after exercise is the most useful single measure of circulatory fitness.

Brouha[69] found pulse observations to be indispensable in industrial physiology for evaluation of the stress imposed by exertion and a high ambient temperature.

BLOOD PRESSURE

As in pulse rate studies, emotional disturbances should be eliminated in studies of blood pressure because they may cause an increase in the systolic pressure. During recent years blood pressure seems to have been used as a test less frequently than in the past. One of the assumed criteria of fitness—that in normal people systolic blood pressure on standing should be higher than in a recumbent position—has been found not to be true in many perfectly healthy athletes. A large drop may, however, indicate inefficient circulatory adjustment.

Bürger suggested a test of fitness based on the behavior of the systolic blood pressure while the subject blew against a 40-mm. column of mercury for twenty seconds. In this test, blood pressure is taken four times—before, at the beginning, immediately after, and twenty seconds after the "blow." In normal people, the blood pressure drops 20 to 30 mm. of mercury during the test. In physically unfit individuals, the blood pressure may drop below 40 mm. In trained athletes, the blood pressure rises; this has held true among Olympic athletes.

As was shown in an earlier chapter, there is normally a drop in the systolic blood pressure immediately after work, followed by a sharp rise that reaches its maximum within the first forty seconds after work. From then on the pressure gradually falls, returning to normal within approximately two minutes. Barringer[43] showed that physically deficient individuals experience a delay in the post-exercise blood pressure rise. Even when a healthy person continues to work until he has nearly reached the limit of capacity, the after-exercise maximum of the systolic pressure will not be obtained until sixty to eighty seconds after stopping work; and the pre-exercise level will not be reached until the end of three to six minutes. This is an example of what is now spoken of as the "delayed rise" and "prolonged fall." Barringer believed that a "delayed rise" indicates overtaxing of the heart's power.

RESPIRATORY TESTS

Some of the most common tests are those measuring vital capacity by means of a spirometer, by blowing into a flarimeter, or by breath holding. Although these tests may differentiate between weak and strong individuals, an interpretation of the data obtained on well subjects is difficult. As indicated in the chapter on respiration, vital capacity varies with the size of the body. Furthermore, some men with relatively small vital capacities have high degrees of fitness in running. It should also be mentioned here that the flarimeter, as ordinarily used, is not an accurate instrument. The readings obtained with this apparatus are affected by fluctuations of pressure of the expired air. Many subjects cannot maintain a steady pressure and tend to keep it *below* that required. As a result, vital capacity readings are larger than the actual. An experimenter has to be careful and consistent in order to obtain reliable data.

BREATH HOLDING

Breath holding was used to produce cardiovascular distress, in order to test fitness for flying in the Royal Air Forces during World War I. Although Hambley and his associates[206] in 1925 criticized this test as of no value for the selection of flyers, the test was not abolished until 1939. The element of determination or "will power" plays an important part in the test. A modification of breath holding, which involved blowing through a flarimeter, was used by McCurdy and Larson,[372] who found a significant correlation between the time of swimming 440 yards and breath-holding time. They also observed that breath-holding time decreased during confinement to the infirmary and increased during training. Karpovich,[288, 289] however, observed that breath-holding time could not be used for prediction of either running endurance time (treadmill) or Harvard Step Test score.

LUNG VENTILATION

Pulmonary reserve has been used as a test of fitness. The reserve may be expressed as the difference between the maximal ventilation during a voluntary hyperventilation and the maximum ventilation determined during muscular work. The pulmonary reserve may also be taken as the difference between the maximal ventilation during work and that during rest.[477]

OXYGEN USE

The ultimate function of the cardiorespiratory chain running from the lungs to the tissues is the delivery of oxygen. Any weak link (such as a deficiency in ventilation, heart pumping action, blood transport capacity, or gas diffusion) will reduce the ability to do so.[249] For this reason, the maximum oxygen intake test is perhaps the most valid single indicator we have of overall cardiorespiratory function. This test does *not* show us the site of such a deficiency, only that a deficiency exists.

Procedures for measuring the maximum oxygen intake are described in Chapter 6. Unfortunately, the test is difficult to administer, requires expensive equipment, and sometimes may be dangerous to the subject. Because of these limitations, several attempts have been made to devise a testing method whereby the maximum can be estimated from response to submaximal exercise using minimal equipment.

Perhaps the most commonly used of these tests is the one devised by Åstrand and Ryhming.[30] (Fig. 96). Submaximal exercise, either in the form of a step-test or bicycle ergometer ride, is performed until a steady state is reached. This phase of the test is usually six minutes in duration. The mean heart rate is taken during the final two minutes. The oxygen consumption is measured or estimated on the basis of a subject's body weight, for a step-test, or his work load, for a bicycle test. A fine string or straightedge is placed between the appropriate oxygen intake, and the attained pulse rate and the maximum oxygen intake are read from the scale.*

---

*This test is described in detail for use with the Monark bicycle ergometer. See reference 30.

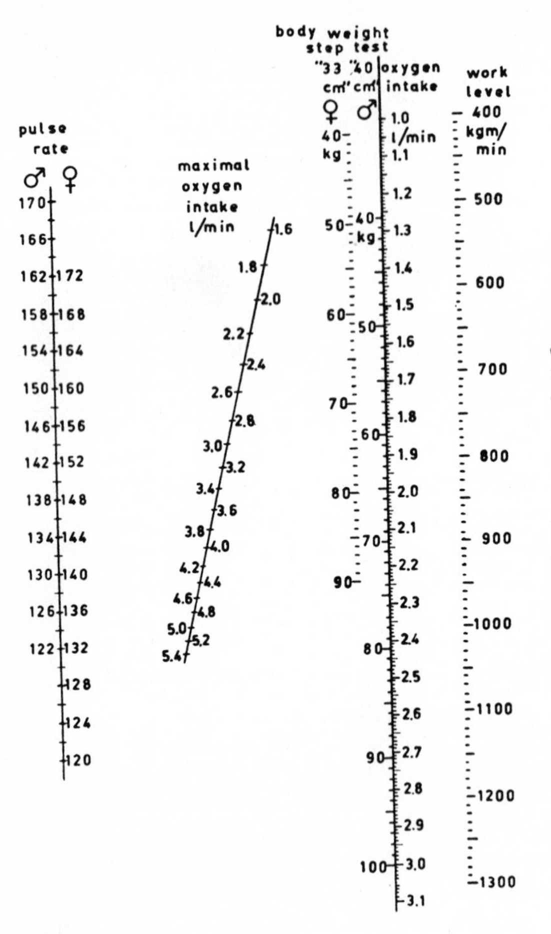

Figure 96. Nomogram for calculation of aerobic capacity from values of pulse rate and oxygen intake during submaximal work. Work level in kgm./min. indicates measured work on a bicycle ergometer. Step tests are performed at 22.5 steps per minute on a 40-centimeter bench for men and a 33-centimeter bench for women. (Åstrand and Ryhming: J. Appl. Physiol. 7:218, 1954.)

Certain steps may be taken to improve the accuracy of the prediction. Estimates of the aerobic capacity should not be made at heart rates below 120 beats per minute. Also, better accuracy will be attained with bicycle ergometer tests if the average of two work loads is used. It is suggested that women start at a work load of 600 kgm./min. and men at 900, and perform additional exercise bouts if necessary. Correction factors should be used to allow for normal decrement in maximum heart rate with age. Factors given by Åstrand[23] for use in making the correction are: 15 years, 1.10; 25 years, 1.00; 35 years, 0.87; 40 years, 0.83; 45 years, 0.78; 50 years, 0.75; 55 years, 0.71; 60 years, 0.68; and 65 years, 0.65. (Multiply estimated aerobic capacity by correction factor.)

This nomogram was constructed on an empirical basis and is, therefore, based on certain assumptions. For this nomogram, it was assumed that there is a constant mechanical efficiency when exercising on a bicycle ergometer, or when bench stepping at a constant rate and height. In the latter case, the variable determining oxygen consumption is body weight. It was also assumed that there is a linear relationship between heart rate and oxygen intake and that both reach their maximum almost simultaneously, the former being age-dependent.

Various factors can contribute to error in this technique. For best results all assumptions must be met. For example, if a subject has an unusually high maximum heart rate, the maximum intake will be underestimated; if the rate is low, the converse. Also, factors which affect normal heart rate response, such as heat or emotional tension, will adversely affect the estimate.

The validity and reliability of this test have been studied by several investigators. Teräslinna. Ismail, and MacLeod[505] found a correlation coefficient of 0.69 between estimated and measured values. The coefficient increased to 0.92 when the estimated maximum intakes were corrected for age. Rowell, Taylor, and Wang[444] found that true values were underestimated by 27 and 14 per cent in a sedentary group, and by 5.6 per cent in athletes. Davies[119] showed that the Åstrand-Ryhming technique, like similar nomograms constructed by other investigators, underestimated true values consistently; and he concluded that, when an accuracy greater than ± 15 per cent is required, the oxygen intake must be measured directly.

These procedures obviously have value for screening purposes. However, when research precision is desired or subjects obviously fail to meet test assumptions, there is no alternative to making the actual measurement.

Physical Work Capacity (PWC)

The physical work capacity at a given heart rate is also frequently used as a test of cardiovascular fitness. Heart rates of 170 and 150 beats per minute are frequently used, the latter especially with older subjects. A typical procedure is described.[1]

The subject is asked to perform two or three 6-minute bouts on a bicycle ergometer, with a pedaling rate of 60 to 70 revolutions per minute. The bicycle is braked so that the load varies from 100 to 800 kgm./min.; the loads selected so that the highest one will cause a pulse rate approximating 170.

The pulse rates obtained at the end of each trial are plotted against work in kilogram meters per minute. A best-fit line is then drawn through the three points, and the work capacity corresponding to the pulse rate of 170 beats per minute is found by extrapolation.

Cummings and Danzinger[117] questioned the validity of the 170 beats per minute pulse as an indication of work capacity; they compared the amounts of energy spent by their subjects when the pulse reached the 170 limit and when the subject performed an all-out test. During both tests, the expired air was collected and analyzed and the rate of oxygen intake was calculated. The amount of energy used during the all-out test exceeded by 27 per cent the amount used when the pulse rate reached 170 beats per minute. In similar tests reported fifteen years before that study, Wahlund[523] showed that, during all-out tests on adults, oxygen consumption was 20 per cent higher than when the pulse rate was 170.

THE TUTTLE PULSE-RATIO TEST

In 1931, Tuttle[517] introduced a modified Hambley step-up test. This test is based on an observation that, for the same number of steps, a less fit person will have a relatively higher pulse rate during the two-minute period of recovery immediately after exercise. The total number of pulse beats after exercise, divided by the resting pulse rate, is called the "pulse-ratio," and cardiac fitness is evaluated by determining the amount of exercise required to obtain a 2.5 pulse-ratio.

The method of the test is as follows:

1. The resting pulse in sitting position is taken.
2. The subject makes twenty complete steps in one minute on a bench 13 inches high (twenty steps for males; fifteen for females).
3. Immediately after exercise, the subject sits down, and the pulse is counted for two minutes.
4. The total pulse for two minutes is divided by the resting rate. This is called the first "pulse-ratio."
5. The subject rests until the pulse returns to normal.
6. The subject again steps up and down for one minute, making thirty-five to forty complete steps on the bench. (The number of steps is recorded.)
7. Immediately after exercise, the subject sits down, and the pulse is counted again for two minutes.
8. The pulse obtained in two minutes is divided by the resting pulse. The dividend is called the second "pulse-ratio."
9. The number of steps required to obtain a 2.5 pulse-ratio is calculated. The formula used for the calculation, and an illustrative example are as follows:

$$S_0 = S_1 \frac{(S_2 - S_1)\,(2.5 - r_1)}{r_2 - r_1}$$

S_1 = The number of steps used in the first test
S_2 = The number of steps used for the second test

S_0 = The number of steps required to obtain a 2.5 ratio
r_1 = The pulse-ratio for S_1
r_2 = The pulse-ratio for S_2

Let us assume that the sitting normal pulse = 70; the two-minute pulse after twenty steps = 154; $r_1 = 154 \div 70 = 2.2$; the two-minute pulse after forty steps in the second test = 189; and $r_2 = 189 \div 70 = 2.7$. Then

$$S_0 = 20 + \frac{(40 - 20)\,(2.5 - 2.2)}{2.7 - 2.2}$$

$$S_0 = 32 \text{ steps.}$$

The norms for this test are: boys, ages ten to twelve, thirty-three steps; boys, ages thirteen to eighteen, thirty steps; adult *men*, twenty-nine steps; adult *women*, twenty-five steps.

Subsequently, Tuttle and Dickinson[517] suggested the use of just one part: namely, stepping up and down thirty times per minute. A pulse-ratio obtained in the usual manner will serve as an index of fitness.

The simplified form of this test deserves further investigation because of certain features which may make the test practical for the purpose of medical diagnosis. The height of the bench is convenient for most patients. The technique is simple. It may, however, be necessary to consider extension of the duration of the stepping-up to two minutes and reduction in the number to twenty-four per minute (see the following discussion).

THE HARVARD STEP-UP TEST

This test consists of measuring the endurance in stepping up and down on a bench 20 inches high, and the pulse reaction to this exercise.

1. A subject steps up and down on a 20-inch bench at the rate of thirty complete steps per minute as long as he can, but not in excess of five minutes. Stepping up and down is done so that the lead foot may be alternated. The cadence of 120 counts per minute may be maintained by watching the swinging of a 39-inch pendulum.

2. Immediately after the test, the subject is seated and his pulse is taken. The pulse may be taken in two different manners: the "slow" form[70] and the "rapid" form.[271] (1) In the "slow" form the pulse rate is taken for three periods, each one thirty seconds in duration. The first period is from one minute to one minute and thirty seconds after the exercise; the second period from two minutes to two minutes and thirty

seconds after the exercise; and the third period from three minutes to three minutes and thirty seconds after the exercise.

$$\text{Index of Fitness} = \frac{\text{Time of stepping in seconds} \times 100}{2\ (\text{Sum of 3 counts of the pulse})}$$

(2) The "rapid" form consists of taking the pulse count only once—from one minute to one minute and thirty seconds after the exercise. The score is obtained from the formula:

$$\text{Index of Fitness} = \frac{\text{Time of stepping in seconds} \times 100}{5.5\ \text{pulse count}}$$

The interpretation of scores is as follows:

(1) For the "Slow" form:
- Below 55 = Poor physical condition
- 55–64 = Low average
- 65–79 = High average
- 80–90 = Good
- Above 90 = Excellent.

(2) For the "rapid form:
- Below 50 = Poor
- 50–80 = Average
- Above 80 = Good

Computations for the "rapid" form test may be avoided by the use of Table 21.

The pulse counts in both the "slow" and "rapid" forms of the Harvard Test developed as a simplification of the original method of counting pulse beats continuously for ten minutes after exercise.[274] It has been found that the three-pulse counts and even the one-pulse count may be used to estimate the total pulse count for the ten minutes. This was verified by Ronkin,[437] who found, on 132 subjects, that the coefficient of correlation between the ten-minute pulse count and the three-pulse count was 0.98, and between the ten-minute pulse count and the one-pulse count 0.92. Under laboratory conditions (test-retest), the reliability for the three-pulse counts was 0.83, and for the one-pulse count 0.89. Under conditions outside the laboratory Karpovich[287] found that on the test-retest of the "rapid" form the coefficient was 0.73 on 187 men.

A slight difference in the scales for the "slow" and the "rapid" forms depends on a slight variation in the judgments of the investigators involved in the development of these two variations of the test. On the whole, they indicate a close degree of agreement.

One of us (P.V.K.) and his associates has tested several hundred well and convalescing young men and considers the scores by the rapid form

TABLE 21. *Scoring Table for Harvard Step-Up Test (Rapid Form)*

INSTRUCTIONS: (1) Find the appropriate line for duration of effort; (2) find the appropriate column for the pulse count; (3) read off the score where the line and column intersect; (4) interpret according to the scale given below.

Duration of Effort	*Heart Beats from 1 Minute to 1½ Minutes in Recovery*										
	40–44	45–49	50–54	55–59	60–64	65–69	70–74	75–79	80–84	85–89	90–over
0 - 29″	5	5	5	5	5	5	5	5	5	5	5
0′30″-0′59″	20	15	15	15	15	10	10	10	10	10	10
1′ 0″-1′29″	30	30	25	25	20	20	20	20	15	15	15
1′30″-1′59″	45	40	40	35	30	30	25	25	25	20	20
2′ 0″-2′29″	60	50	45	45	40	35	35	30	30	30	25
2′30″-2′59″	70	65	60	55	50	45	40	40	35	35	35
3′ 0″-3′29″	85	75	70	60	55	55	50	45	45	40	40
3′30″-3′59″	100	85	80	70	65	60	55	55	50	45	45
4′ 0″-4′29″	110	100	90	80	75	70	65	60	55	55	50
4′30″-4′59″	125	110	100	90	85	75	70	65	60	60	55
5′	130	115	105	95	90	80	75	70	65	65	60

Below 50 = Poor general physical fitness.
50–80 = Average general physical fitness.
Above 80 = Good general physical fitness.

to be preferable to those by the slow form. Moreover, a score of 75 is accepted as the minimum for "good" condition.

Several modifications of the Harvard Test have been suggested. Clarke[100] used a bench 16 inches high for college girls and found that the scoring formula could be applied without a change.

Karpovich and his associates[305] reduced the stepping rate from 30 to 24 per minute. This was done for two reasons: this rate is easier for the subject to maintain; and it is easier for the tester to keep the count without either pendulum or metronome. The original Harvard Scoring Table (Table 21) was slightly modified, and the modification is shown in Table 22.

Some of the drawbacks of the Harvard Test are that it may produce acute local muscular fatigue and that the bench is too high.

TABLE 22. *Scoring Table for Step-Up Test (24 Complete Steps per minute on a 20-Inch Bench)*

Duration of Exercise	*Pulse Rate One Minute After Exercise*	*Score*
Below 2′	Any Rate	Low
2′ to 2′29″	100 and above Below 100	Low Fair
2′30″ to 2′59″	130 and above Below 130	Low Fair
3′ to 3′29″	Above 140 100 to 140 Below 100	Low Fair Good
3′30″ to 3′59″	Above 170 110 to 170 Below 110	Low Fair Good
4′ to 4′29″	130 and above Below 130	Fair Good
4′30″ to 4′59″	140 and above Below 140	Fair Good
5′	150 and above Below 150	Fair Good

THE McCURDY-LARSON TEST

The formulation of this test was preceded by an extensive statistical study of the reliability of functional tests and the significance of twenty-six items, as determined from tests on 409 students. The correlations having been determined, a selection of five items was made from those ranking highest. The test, announced in 1935,[372] requires the following determinations: (1) sitting diastolic pressure, (2) breath holding twenty seconds after a stair-climbing exercise, (3) difference between standing normal pulse rate and pulse rate two minutes after exercise, (4) standing pulse pressure, and (5) vital capacity measured by a flarimeter.[372] The description of the details and the scoring tables occupy several pages, and for this reason are not given here.

The construction of this test is a good illustration of the use of statistical methods in physiological problems. A would-be designer of a new test could profit by studying the process of development of this test.

THE KRAUS-WEBER TEST

The Kraus-Weber Strength Test[328] consists of six items: two sit-ups (one with straight knees and one with bent knees); two leg lifts (one lying on the back and one prone); one trunk lift in prone position; and one forward bend in standing position aimed at touching the floor without bending the knees.

While this test may be of some value in medical examinations of abnormal children, it is of doubtful value for examining normal children. Moreover, the name "Strength Test" is a misnomer, because most failures occur in a test of flexibility, the forward bend. More boys than girls fail in this test. The Kraus-Weber test has been applied even to adults. This fact indicates that fashion affects even physical education.

The Kraus-Weber test has a certain historical value because it focused interest of the nation on physical fitness. For this Dr. Kraus deserves sincere and lasting gratitude.

WHAT TYPE OF TESTS SHOULD BE USED?

It is beyond the scope of this book to go into detailed discussion on this topic. Therefore, only a brief suggestion will be made. Most physical educators agree in principle that testing is desirable; but, when it comes to selecting test items, there is no agreement.

From the standpoint of physiology, it is important to select tests that measure the strength and endurance of some important large muscle groups and also the functioning of circulatory and respiratory systems under physical stress.

The important muscle groups are those of legs, trunk, and arms. No disagreement is expected on this score, because these three groups will cover most of the muscles of the body.

For mass use, these tests should be few and simple and should require only minimal equipment that can be easily procured.

The tests which may be strongly recommended are

1) Either standing broad jump or jump and reach;
2) Sit-ups;
3) Push-ups;
4) A fast step-up-and-down test to test cardiorespiratory function under stress.

All these tests require minimal space and equipment, and are not affected by weather. It is advisable to use national standard scores, but the most important scores are those of the individuals themselves. Improvement in these scores indicates progress.

In closing this chapter, the authors would like to quote a statement made by Schneider in 1923: "In conclusion let us remind ourselves of the fact that there is a disposition to demand more of a fitness measure than is demanded of a mechanic who provides the engine for the automobile or airplane. The mechanic can measure the maximum load that a machine may carry, but how long this can be carried he does not undertake to state; a minimum endurance may be predicted but never the maximum. So must it be with the human machine. We have methods of measuring the maximum effort that may be tolerated for a short time, we can determine what constitutes an overload, we can even determine the presence of poor adjustments in the machine; but we are unable to predict how long the human machine will be able to carry a normal load; we cannot even say how long it can carry a moderate over-load. The best we can hope for is a measure for actual accomplishment and present perfection of adjustment."

QUESTIONS

1. How can tests of physical fitness be physiologically classified?
2. Discuss the values of the following tests: muscular strength, pulse rate, blood pressure, breath holding, lung ventilation, oxygen consumption.
3. What is the physiological basis of the Tuttle pulse-ratio test? The Harvard step-up test?
4. What are the strong and weak points of the Kraus-Weber test?
5. What minimum battery of tests can you recommend, and why? (You don't have to agree with the minimum set in the book.)

Chapter Nineteen

BODY CONSTITUTION AND COMPOSITION

That some people have physical characteristics especially well adapted for successful performance in certain sports can be readily verified by observation. Because of this, considerable study of the exact significance of physical characteristics relative to performance potential, exercise needs, and exercise effects has been pursued. Two related areas of study that have dealt with these problems are those of body constitution and body composition. The former deals with the structure of a supposedly unchangeable frame, while the latter deals with the analysis of the existing composition or changes in composition within that frame.

BODY CONSTITUTION

The history of the classification of body types is a long one. Before the year 400 B.C., Hippocrates distinguished two extreme types: *habitus apoplecticus* (short, thick) and *habitus phthisicus* (long, thin). Halle, in 1797, recognized four types: abdominal, muscular, thoracic (long, slender), and nervous (cephalic). Rostan, in 1828, also recognized four types: digestive, muscular, respiratory, and cerebral. Kretschmer's classification, which was suggested in 1925, has received considerable attention. He suggested three types: pyknic (round, compact), athletic, and leptosome (asthenic).[471]

The method of body-build classification used most frequently in studies reported in the English-language literature is the one developed by Sheldon and his collaborators.[471] Sheldon developed a new technique because of the disappointing results he obtained when he attempted to classify 400 undergraduate students according to Kretschmer's system. He had to classify 72 per cent of his sample as being of mixed body type, a category of undifferentiated subjects, which provided no meaningful information.

Sheldon, like Kretschmer, included three components of body build in his system but identified them as endomorph, mesomorph, and ectomorph. The extremes of the three types are shown in Figure 97. An endomorph displays soft roundness throughout the body; while a mesomorph is rectangular in build, with a heavy, hard musculature. The ectomorph is linear and fragile, without clear muscular definition.

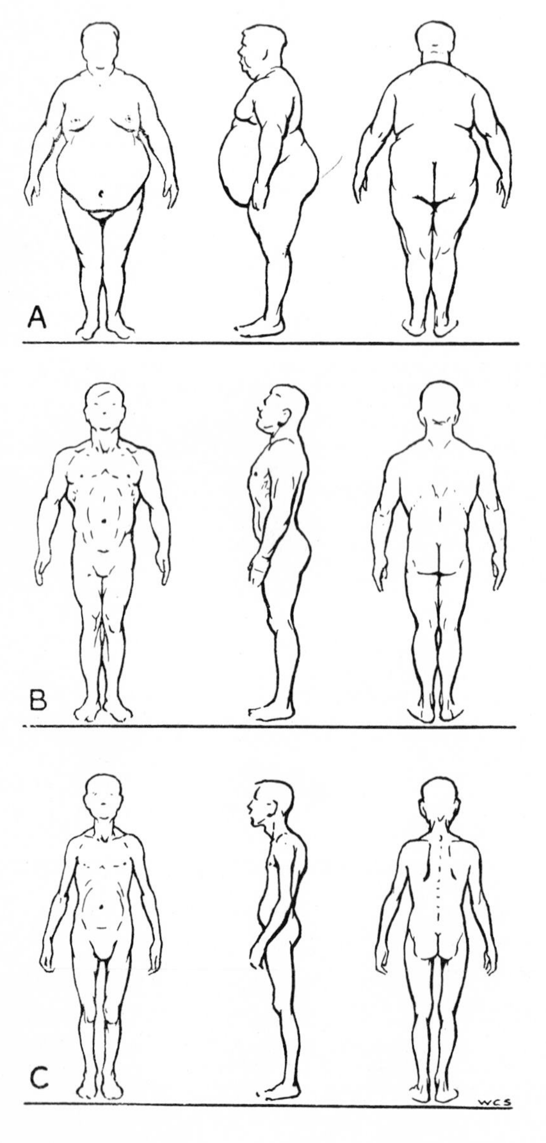

Figure 97. The extreme varieties of human physique. *A*, Endomorph (711); *B*, mesomorph (171); *C*, ectomorph (117). (Modified from Sheldon.)

Names of the three descriptive components relate to the three germinal layers from which each component is supposedly derived.

In his early work, Sheldon rated each component on a 7-point scale, the highest rating being 7 and the lowest 1. The extreme types illustrated in Figure 97 would be 711, 171, and 117 for endomorphy, mesomorphy, and ectomorphy, respectively. The middle of the scale for each component is 4. It is now common to use half-units, so that ratings such as 4½ 5½ 1 are frequently found.

Extreme somatotypes are uncommon, and everyone is a combination of the three types. With a 7-point-scale system of three items, there are 343 different types possible. Even though the three components are supposedly independent of one another, some numerical combinations are impossible. As an example, a somatotype of 777 would suggest a person who is maximally endomorphic, mesomorphic, *and* ectomorphic. Sheldon has described seventy-six somatotypes, of which fifty are fairly common. Compound terminology is used to describe mixed somatotypes; the dominant type being used as a second term, so that a person who is highly mesomorphic, with a strong endomorphic component, for example, is identified as an endomorphic-mesomorph. Those who are relatively equal in all components (such as 444) are referred to as mid-types.

Sheldon's technique, which is described in detail in *Atlas of Men,*[470] requires each subject to be photographed in rigidly standardized positions on a single negative. The ponderal index (Height/$\sqrt[3]{\text{weight}}$) is first computed, which gives a general category of somatotype for the subject's build and reflects especially the ectomorphic component. After the general category is determined, the photographs of the subject are compared to specified reference pictures in *Atlas of Men* until the correct somatotype is matched.

Sheldon refers to his technique as being "thoroughly objective," but it is somewhat esoteric and only a few trained people are considered qualified to use it. Heath,[215] who is a qualified somatotyper, has pointed out some of the limitations and presents a case for needed modifications. In addition to technical changes, Heath recognizes the need for rating scales that have no end point so that components may be higher than 7, as well as the changing nature of the somatotype rating during the growing years.

Sheldon's method of somatotyping has not come into common use because of the difficulty in becoming a trained somatotyper and because of the need for specialized photographic equipment. Consequently, simplified techniques using live anthropometric measurements have been developed.

Parnell[403] used measures of height, weight, skinfolds, bone width, and muscle circumferences to determine "Fat," "Muscularity," and

"Linearity," three components which are very similar to Sheldon's. A special chart, which Parnell named the "M.4 Deviation Chart," is used to make the somatotype rating from selected anthropometric measures.

More recently, Heath and Carter[215A] described a modification of Parnell's method that we have found to be more applicable to an athletic population. The rating is made by use of the Heath-Carter Somatotype Rating Form which is shown in Figure 98. The example is one given by Heath and Carter. The ratings derived by this method may be as low as ½ in any one component; the upper ends of the scales are not limited. The form may be used for either males or females at all ages.

To use the Heath-Carter rating form, the measurements shown to the left of the chart (Fig. 98) are first taken. Skinfold sites are measured to 1/10 millimeter on the right side of the body as follows: triceps, a vertical fold midway between the olecranon and acromion processes over the triceps muscle; subscapular, an oblique fold at the inferior angle of the scapula; suprailiac, a horizontal fold one to two inches above the anterior-superior iliac spine; and calf, a vertical fold on the medial side at the greatest circumference.

Bone and muscle measures are taken in centimeters on both sides of the body, and the larger of the two measurements is used. Humerus width is the most lateral distance between the epicondyles just above the elbow joint, whereas femur width is the most lateral distance between the femoral epicondyles just above the knee joint. Both measurements are taken with the joints flexed to 90 degrees. The greatest circumference of the biceps is measured with the muscle fully contracted. The greatest calf circumference is measured while the subject stands with the weight equally distributed on both feet.

Determination of the first and third components (endomorphy and ectomorphy) is quite simple. As indicated in Figure 98, to find endomorphy merely sum the three skinfolds and encircle the appropriate value; the rating is read directly below on the line marked "first component." To find ectomorphy (third component), merely divide the height in inches by the cube root of the weight in pounds. Encircle the appropriate tabled value and read the third component directly below.

The determination of the second component is more complex. The middle section of the rating form is used. The height, in inches, is first marked on the form. This scale is continuous so the mark (arrow in the example) should be placed to represent its exact value. Next, the other skeletal and muscular measurements are encircled. Muscle circumferences are first corrected for subcutaneous fat by subtracting the triceps and calf skinfolds, in centimeters, from the measured values for biceps and calf circumferences. If the value is exactly midway or less between the given values, encircle the lower number; if more, encircle the higher number.

HEATH-CARTER SOMATOTYPE RATING FORM

NAME J. B. AGE 45-2 SEX: (M) F NO: 135
OCCUPATION Teacher ETHNIC GROUP Caucasian DATE 11 Nov 1966
PROJECT: A. T. P. MEASURED BY: L. C.

Skinfolds (mm):
Triceps = 13.0
Subcapular = 15.3
Supraliac = 9.9
TOTAL SKINFOLDS = 38.2
Calf = 9.8

TOTAL SKINFOLDS (mm)

Upper Limit	10.9	14.9	18.9	22.9	26.9	31.2	35.8	40.7	46.2	52.2	58.7	65.7	73.2	81.2	89.7	98.9	108.9	119.7	131.2	143.7	157.2	171.9	187.9	204.0
Mid-point	9.0	13.0	17.0	21.0	25.0	29.0	33.5	38.0	43.5	49.0	55.5	62.0	69.5	77.0	85.5	94.0	104.0	114.0	125.5	137.0	150.5	164.0	180.0	196.0
Lower Limit	7.0	11.0	15.0	19.0	23.0	27.0	31.3	35.9	40.8	46.3	52.3	58.8	65.8	73.3	81.3	89.8	99.0	109.0	119.8	131.3	143.8	157.3	172.0	188.0
FIRST COMPONENT	½	1	1½	2	2½	3	3½	4	4½	5	5½	6	6½	7	7½	8	8½	9	9½	10	10½	11	11½	12

Height (in.) = 67.1
Bone: Humerus (cm) = 6.84
Femur = 9.27
Muscle: Biceps 33.0 (cm) −1.3 − (triceps skinfold) = 31.7
Calf 35.1 − (calf skinfold) .95 = 34.1

Height	55.0	56.5	58.0	59.5	61.0	62.5	64.0	65.5	67.0	68.5	70.0	71.5	73.0	74.5	76.0	77.5	79.0	80.5	82.0	83.5	85.0	86.5	88.0	89.5
Humerus	5.19	5.34	5.49	5.64	5.78	5.93	6.07	6.22	6.37	6.51	6.65	6.80	6.95	7.09	7.24	7.38	7.53	7.67	7.82	7.97	8.11	8.25	8.40	8.55
Femur	7.41	7.62	7.83	8.04	8.24	8.45	8.66	8.87	9.08	9.28	9.49	9.70	9.91	10.12	10.33	10.53	10.74	10.95	11.16	11.37	11.58	11.79	12.00	12.21
Biceps	23.7	24.4	25.0	25.7	26.3	27.0	27.7	28.3	29.0	29.7	30.3	31.0	31.6	32.2	33.0	33.6	34.3	35.0	35.6	36.3	37.1	37.8	38.5	39.3
Calf	27.7	28.5	29.3	30.1	30.8	31.6	32.4	33.2	33.9	34.7	35.5	36.3	37.1	37.8	38.6	39.4	40.2	41.0	41.8	42.6	43.4	44.2	45.0	45.8

SECOND COMPONENT	½	1	1½	2	2½	3	3½	4	4½	5	5½	6	6½	7	7½	8	8½	9

Weight (lb.) = 147.0
Ht. / $\sqrt[3]{\text{Wt.}}$ = 12.76

Upper limit	11.99	12.32	12.53	12.74	12.95	13.15	13.36	13.56	13.77	13.98	14.19	14.39	14.59	14.80	15.01	15.22	15.42	15.63
Mid-point	and	12.16	12.43	12.64	12.85	13.05	13.26	13.46	13.67	13.88	14.01	14.29	14.50	14.70	14.91	15.12	15.33	15.53
Lower limit	below	12.00	12.33	12.54	12.75	12.96	13.16	13.37	13.56	13.78	13.99	14.20	14.40	14.60	14.81	15.02	15.23	15.43
THIRD COMPONENT	½	1	1½	2	2½	3	3½	4	4½	5	5½	6	6½	7	7½	8	8½	9

	FIRST COMPONENT	SECOND COMPONENT	THIRD COMPONENT	
Anthropometric Somatotype	4	5	2½	BY: L. C.
Anthropometric plus Photoscopic Somatotype				RATER:

Figure 98. The Heath–Carter Somatotype Rating Form for the assessment of somatotype from live, anthropometrical measures. See test for explanation. (Heath and Carter: Am. J. Phys. Anthrop. 27:57, 1968.)

The determination of mesomorphy is based on the average deviation of the muscle and skeletal measures from the recorded height. First, find the encircled number that is to the left (33.9 in the example) and then find the total number of columns the other numbers are displaced to the right of it. In the example, 6.80 is three columns, 9.28 is one column, and 31.6 is four columns displaced to the right, for a total of eight. Divide this total by four (quotient in the example is two). Mark, with an asterisk, the distance this quotient is to the right of the number originally selected. Next determine the number of columns between the height (arrow in Fig. 98) and the asterisk. In the example this is 2, and it represents the average deviation of the skeletal and muscle measures from the height.

Once the average deviation has been determined, go to the second component rating scale and, using "4" as the starting point, count the number of columns equal to the computed average. If the asterisk is to the right of the arrow, count to the right from 4; if it is to the left, count to the left from 4. In the example, 5 is two columns to the right of 4, hence, it is encircled. The final rating is recorded in the space provided on the chart.

This method is adaptable to class work and research. In practice, it is desirable to have the conventional somatotype pictures which are of value when making marginal assessments, but such an examination apparently plays a role only in determining half-units.[215A] The fledgling investigator should note that taking anthropometric measurements requires skill that can be gained only through instruction and careful, intensive practice. It is not possible to discuss all of the nuances of such procedures here, and the research should be thoroughly reviewed before an attempt is made to use them in new investigations.

The Somatotype and Exercise Physiology. The observation that people with relatively similar body types tend to participate in selected sports is substantiated by somatotype studies. Carter,[87] using Parnell's method, found somatotype and size differences between playing positions in college football players, but the dominant physique was the extreme endomorphic-mesomorph. Backfield players were lower than linemen in endomorphy, but higher in ectomorphy. Parnell[403] showed that sprinters tended to be endomorphic-mesomorphs, while distance runners, long jumpers, and high jumpers were more ectomorphic in build. Weight men also tended to be endomorphic-mesomorphs.

Tanner[496] studied the physiques of 137 Olympic track and field athletes. Of the runners, sprinters tended to be endomorphic- or ectomorphic-mesomorphs; 400-meter runners ectomorphic-mesomorphs; while runners at distances of 800 meters and above tended to be either mesomorphic-ectomorphs or ectomorphic-mesomorphs. Weight men tended to be high in mesomorphy and fell into the general classification of endomorphic-mesomorphs (Fig. 99).

Pugh and co-workers[417] noted that channel swimmers tend to have high ratings in endomorphy and mesomorphy. In a sample of eleven swimmers who participated in the 1955 Cross-Channel Race, all fell into the upper and middle portion of the somatogram. These swimmers also had a relatively thick layer of subcutaneous fat, a condition that helped them maintain heat balance during the long period of exposure to the cold water.

Body type is also apparently related to the endowment of certain motor-performance characteristics essential for sports participation. Clarke and Broms[99] studied strength in 722 boys at the ages of 10, 13, and 16 years. They divided their groups into high, average, and low strength groups on the basis of Strength Index scores, and found that high strength groups had the highest mesomorphy means. The low strength group at 16 years and the average strength groups at 10 and 13 years were high in ectomorphy. Similar results were found for cable-tension strength tests. Those low in the Physical Fitness Index tended to be high in the endomorphic component. Laubach and McConville[337] found that mesomorphy was the only body-type component that correlated significantly with muscle strength, although the amount of correlation was inadequate to use the somatotype for the precise prediction of strength. They also found that flexibility was not related to somatotype.[338] Jones[275] found that static strength was dependent on both mesomorphy and body size when he studied extreme ectomorphs and mesomorphs.

Performance success on physical fitness tests has also been an item of study. Perbix[408] correlated results of tests of flexibility, agility, strength, and power with somatotype components in college women. All correlations were low. She found, however, that physical education majors tended to have more dominant mesomorphic traits than nonmajors, although endomorphy was the dominant component of all the women studied. The mesomorphic component was apparently directly related to success in the strength and power items of the physical fitness tests. When endomorphy was high and mesomorphy low, there was an inverse relationship between endomorphy and the strength and agility items. No relationship was shown between somatotype components and flexibility. Garrity[180] found that women who were mesomorphic-ectomorphs did better on physical fitness test items than women of other body types, while ectomorphic-endomorphs performed consistently low on all items. Sills and Mitchem[474] proposed a predictive formula for fitness scores for men on the basis of somatotype ratings.

Several authors have indicated that body build is also related to a predisposition for coronary heart disease, a relationship which has special significance for physical educators involved in adult physical fitness programs. In 1959, Parnell,[402] on the basis of data from several studies, concluded that men who are muscular have a greater tendency toward coronary heart disease than those who are less muscular. Men who were of a "fat-muscular" build have the greatest risk. Parnell

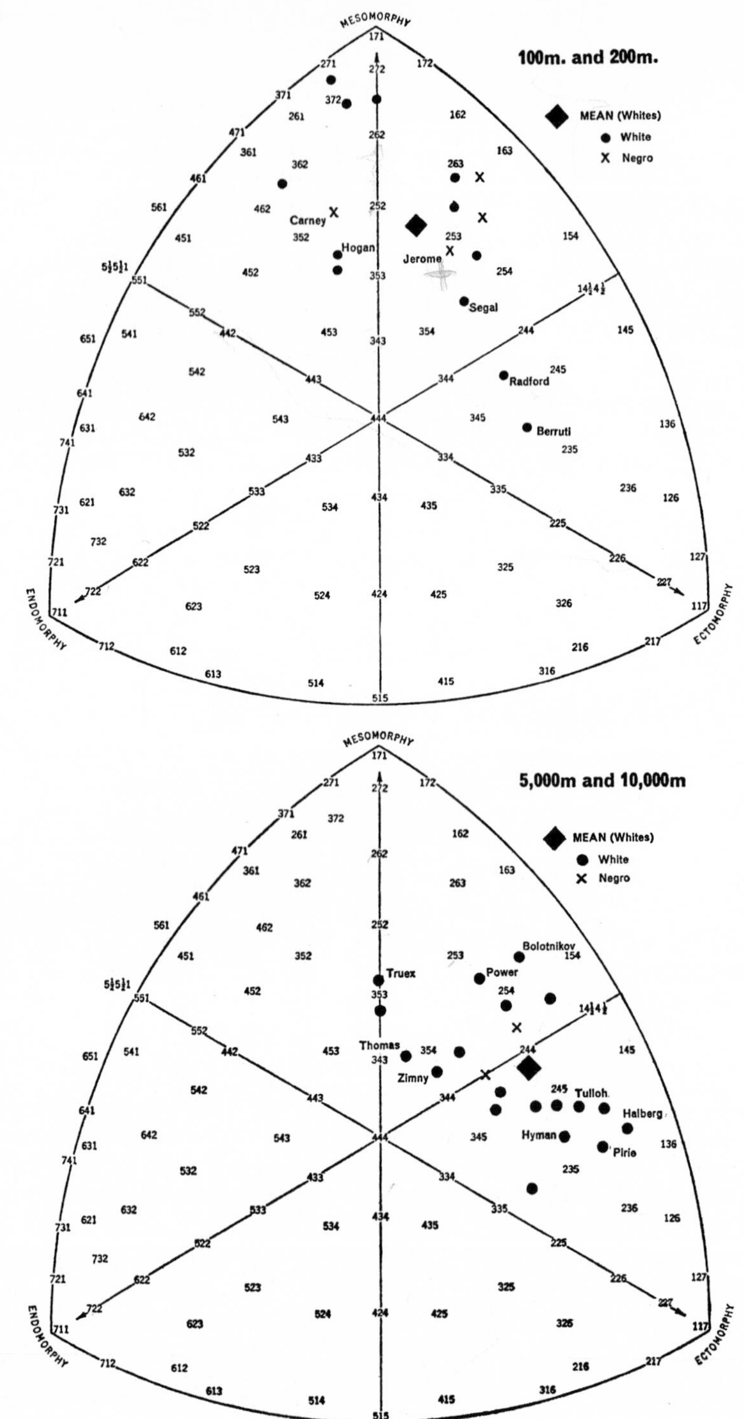

Figure 99. Somatotypes of groups selected from the data of Tanner to show clustering of somatotypes according to events. (Tanner, J. M.: The Physique of the Olympic Athlete. George Allen and Unwin Ltd., 1964.)

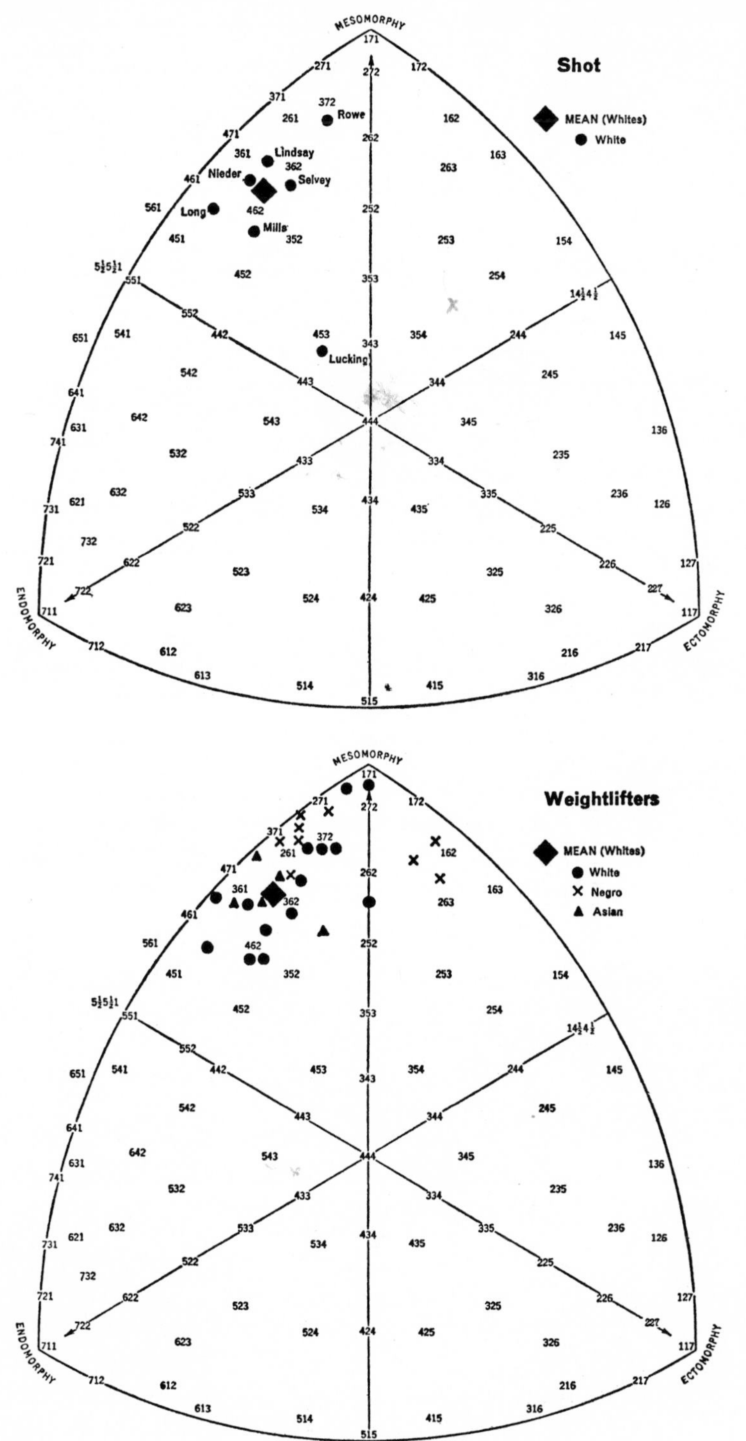

Figure 99. (Continued)

proposed that such men needed more exercise and attention to diet than others in order to prevent coronary attacks. Since then, Spain[489] and his co-workers have studied intensively a group of 5000 men and found that endomorphic-mesomorphs have a higher prevalence for coronary atherosclerotic heart disease than ectomorphs. There is danger in over-generalizing relative to implications from body-type findings, however, since it was shown by Spain's group that overweight, hypertension and diabetes mellitus were also significant factors in conjunction with body type in the incidence of heart disease. However, the endomorphic-mesomorphs showed almost as high a disease incidence even when hypertension was excluded as a causative factor.

Gertler[182] reported that mesomorphy is associated with abnormal cholesterol metabolism and ischemic heart disease and concluded that mesomorphy, abnormal serum lipids and the tendency toward ischemic heart disease are transmitted as a genetic unit. He suggested that mesomorphs require constant exercise to maintain physical fitness and favorable control of body weight and serum lipids.

At least some mesomorphic adult males apparently sense this need and gravitate toward exercise programs. Carter et al.[88] studied phenotypes (Parnell's Method) of two samples of men in the age range of 35 to 68 years who participated in two adult fitness programs, one in New Zealand (19 subjects) and the other in the United States (California, 55 subjects). Both samples contained a number of potentially "coronary-prone" physiques. Most of the subjects fell into the upper-left three sectors of the somatochart.

BODY COMPOSITION

The study of body composition can be approached in a number of ways: organ systems, fluid compartments, kinds of tissue, and so forth. An important consideration for teachers and coaches is the assessment of desirable body weight. The inadequacies of commonly used age-height-weight tables have been discussed by Keys and Brožek[312] as well as by Kandel.[278] Such tables, for example, generally make no provision for skeletal dimensions other than length. Because of these inadequacies, numerous studies have been completed to find better ways of evaluating body weight.

The body can be divided into metabolically active and inactive parts.[312] The inactive component consists of extracellular fluids, bone minerals, and "depot" fat. The depot fat is an extremely variable quantity, accounting for most long-term body weight changes. Some storage fat is normal and desirable. Men normally have about 15 per cent[545] of their total body weight in depot fat, while women have a larger quantity, comprising approximately 23 per cent.[307]

The sum of the body constituents other than depot fat may be referred to conceptually as the fat-free mass, quantitatively as the fat-free weight.[312] Behnke[48] uses the term "lean body mass" as the conceptual expression and "lean body weight" as the quantitative expression. The relative amounts of the fat-free and depot-fat masses is of concern here, since both may be affected by exercise.

Methods of Assessment. Direct analysis of body composition is impossible except in cadavers, so several indirect techniques have been developed.[73] These techniques differ; some require the injection of chemicals that are selectively absorbed by different body compartments; others are based on measuring the normal irradiation of radioactive potassium from the body. The most practical approach for the teacher or coach is the use of physical anthropometry.

Body densitometry is commonly used in research and is the method by which other anthropometric techniques are frequently validated. The procedure is based on knowledge of the densities of different body tissues. "Density" is the mass of a substance per unit of volume. The densities of tissues such as bone and muscle, for example, are obviously different. The densities of the lean and fat body components have been established experimentally.[425, 75] Fat, which is the main variable in body weight, is of low density which is affirmed by its tendency to float in water. As the amount of depot fat increases body density decreases.

If the body volume and weight are known, it is easy to compute density by dividing the former into the latter. However, it is difficult to measure body volume. Siri[482] has developed a dilution technique in which helium replaces air surrounding the subject while he sits in an air-tight chamber of known volume. This technique is not commonly used, but the underwater weighing technique is.

Underwater weighing is based on Archimedes' principle: an immersed or floating body is buoyed up by a force equal to the weight of the water it displaces. A subject is weighed on land and in water. Since water has a density of one gram per cubic centimeter, a body immersed in water will be supported by a force of one gram for every cubic centimeter of water displaced. In application, if a person weighs 60 kg. in air and 3 kg. in water, he must be buoyed up by a force of 57 kg., which is equivalent to a volume of 57,000 cc. His density would be 1.052 gm./cc. By use of the following or a similar standardized equation,[75] the per cent of the total body weight that is fat can be computed:

$$\text{per cent fat} = \frac{4.570}{\text{body density}} - 4.142$$

The foregoing is only a simplified overview of the application of Archimedes' Principle. Underwater weighing requires two sensitive scales and a suitable water tank or swimming pool. Corrections must be made for variations in water density due to temperature. The volume of air in the lungs at the time of weighing must be measured or

TABLE 23. *Computation of Lean Body Weight According to the Method of Behnke—Male Subject*

Diameter	k(LBW) Values[a]	
	Men	*Women*
Biacromial	21.6	20.4
Bi-iliac	15.6	16.7
Chest width	15.9	14.8
Bitrochanteric	17.4	18.6
Wrists	5.9	5.6
Ankles	7.4	7.4
Knees	9.8	10.3
Elbows	7.4	6.9

Diameter	*diameter* [*cm.*]	k^b	*d(LBW)* [*c/k*]
Biacromial	39.3	21.6	1.82
Bi-iliac	27.9	15.6	1.79
Chest width	30.0	15.9	1.89
Bitrochanteric	32.5	17.4	1.87
Knee[c]	18.2	9.8	1.86
Ankle[c]	14.3	7.4	1.93
Elbow[c]	13.4	7.4	1.81
Wrist[c]	10.4	5.9	1.76

$D = \Sigma\, d(LBW)/8 = 14.73/8 = [1.84]$ $\Sigma\, d(LBW) = 14.73$

Note: Height 170 cm. (17.0 decimeters); Weight 70 kg.

$LBW = D^2 \times h = [1.84]^2 \times 17.0 = 3.38 \times 17.0 = 57.46$ kg.

% Fat = (Body Weight − LBW)/Body Weight = (70 − 57.5)/70 = 12.5/70 = 17.8%

[a]From Behnke, A. R.: J. Appl. Physiol. *16*:960, 1961.
[b]From upper part of table.
[c]Use the sum of the right and left sides.
(Form of table and data adapted from Wilmore, J. H., and Behnke, A. R.: J. Appl. Physiol. *25*:349, 1968.)

estimated, since the lungs act as a built-in flotation device that reduces the weight of the body in water. For utmost accuracy, measuring the air volume requires the use of special respiratory apparatus. Also, some people are too apprehensive about complete body immersion to subject them to this procedure. For these reasons, the underwater weighing technique is not readily adaptable to the teaching situation.

Several methods of estimating body fat from anthropometric measures have been developed. These techniques can be used by the fitness-program leader, classroom teacher, or coach. Two general approaches have been used: skeletal diameters and skinfold measures.

Behnke's[48, 546] technique for estimating lean body weight is based on the concept that a given skeletal structure reflects a relatively constant lean body mass. He uses eight measures of skeletal diameter which are converted into "*d* quotients" by dividing by a constant "*k*", which has been derived from a "reference man." The reference man's dimensions and lean body mass were computed from measures of the skeletal diameters and body density of many subjects. The *d* quotients and the height are used to estimate the lean body weight, according to the equation:

$$LBW = D^2 \times h$$

where LBW = lean body weight in kilograms
h = height in decimeters
D = average of d values obtained when $d = c/k$ and
c = specific diameter in centimeters
k = conversion constant.

The computational steps for the Behnke technique are shown in Table 23 with the *k* values for men and women.[48, 546]

The measurement of skinfolds is based on the knowledge that approximately 50 per cent of the depot fat is stored in specialized cells within the subcutaneous areas.[312] A fold consisting of two layers of skin and subcutaneous structures can be picked up with the thumb and index finger. The thickness of the fold will depend upon the amount of stored fat and can be measured with a special instrument called a skinfold caliper. The number of sites at which skinfolds can be measured is practically limitless, but only a few have regularly been found to be of value in estimating body density. Sites frequently used are shown in Figure 100.

Several equations have been developed for the estimation of body density from skinfold measures. Only two are presented here. We have found both of these to give fair estimates. Sloan[485] developed the following equation for men (skinfolds measured in millimeters):

$$\text{Body Density} = 1.1043 - 0.001327 \text{ (thigh skinfold)} - 0.001310 \text{ (scapular skinfold)}.$$

The per cent of the body weight that is fat is then computed from the estimated density according to the equation given previously. The procedure is the same for the following equation for women, which was developed by Sloan, Burt, and Blyth[486] (skinfolds measured in millimeters):

$$\text{Body Density} = 1.0764 - 0.00081 \text{ (iliac skinfold)} - 0.00088 \text{ (triceps skinfold)}.$$

Hall et al.[205] developed a technique for estimating "desirable" total body weight for both sexes in the age range of ten to eighteen years.

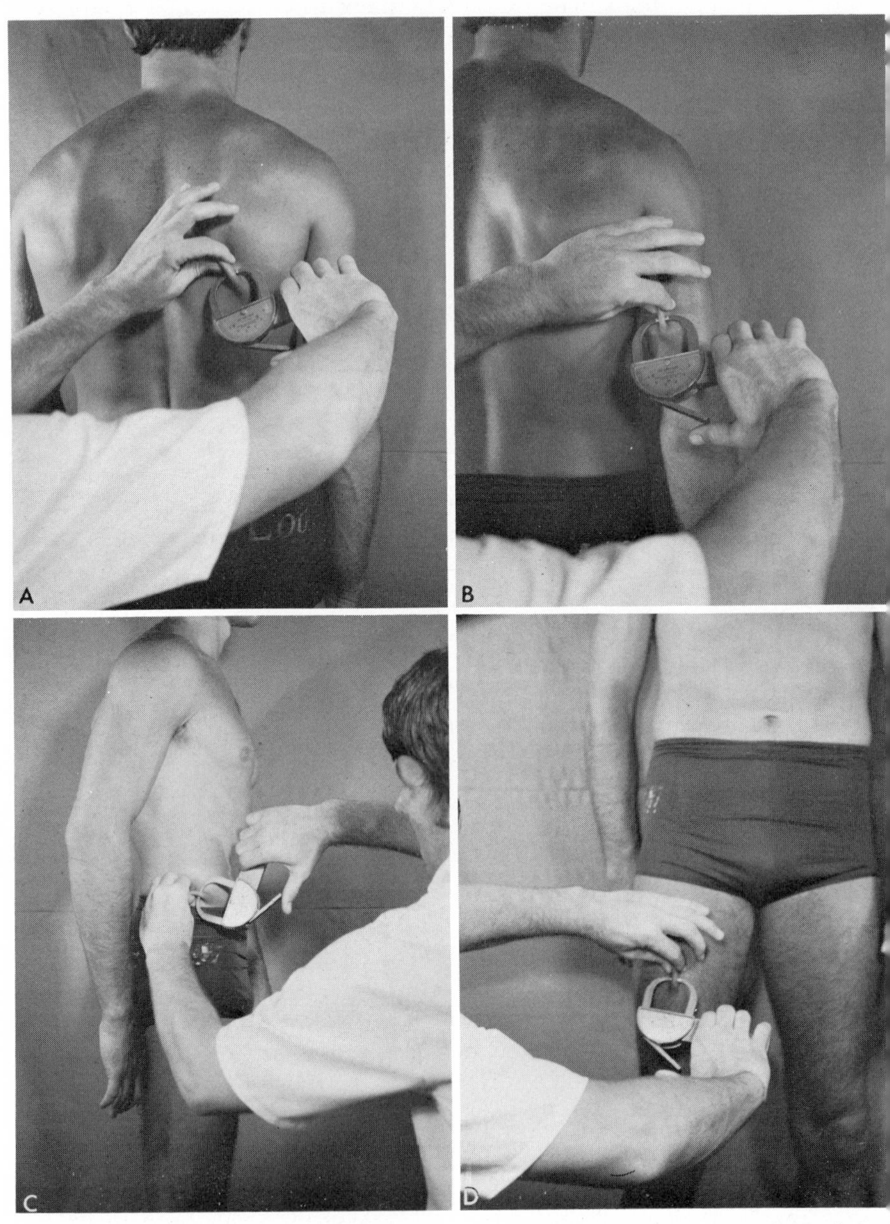

Figure 100. Sites for measuring skinfolds described in the text: *a*, subscapular; *b*, triceps; *c*, iliac crest; *d*, thigh. Skinfolds are grasped between the tester's thumb and index finger so that a double fold of skin and subcutaneous fat is lifted. The calipers are placed approximately 1 cm. from the fingers, at a point on the fold where the surfaces are parallel, avoiding the expanded base or rounded crest of the fold.

The equations are based on data gathered from over 30,000 Illinois 4-H Club members over a period of more than twenty years. Only those members adjudged to be physically fit by physical performance tests were used to develop the equations. The equations have been shown to have good accuracy. To use this method, measurements are taken in inches and are height, thigh circumference, hip width, chest width, and chest depth (Fig. 101). The equations are as follows:

Boy's Weight = 2.4919 (height) + 2.0825 (thigh circumference) + 3.3925 (hip width) + 5.3074 (chest width) + 5.8129 (chest depth) − 205.621.

Girl's Weight = 1.9814 (height) + 3.1144 (thigh circumference) + 3.2198 (hip width) + 4.1373 (chest width) + 3.4066 (chest depth) − 165.138.

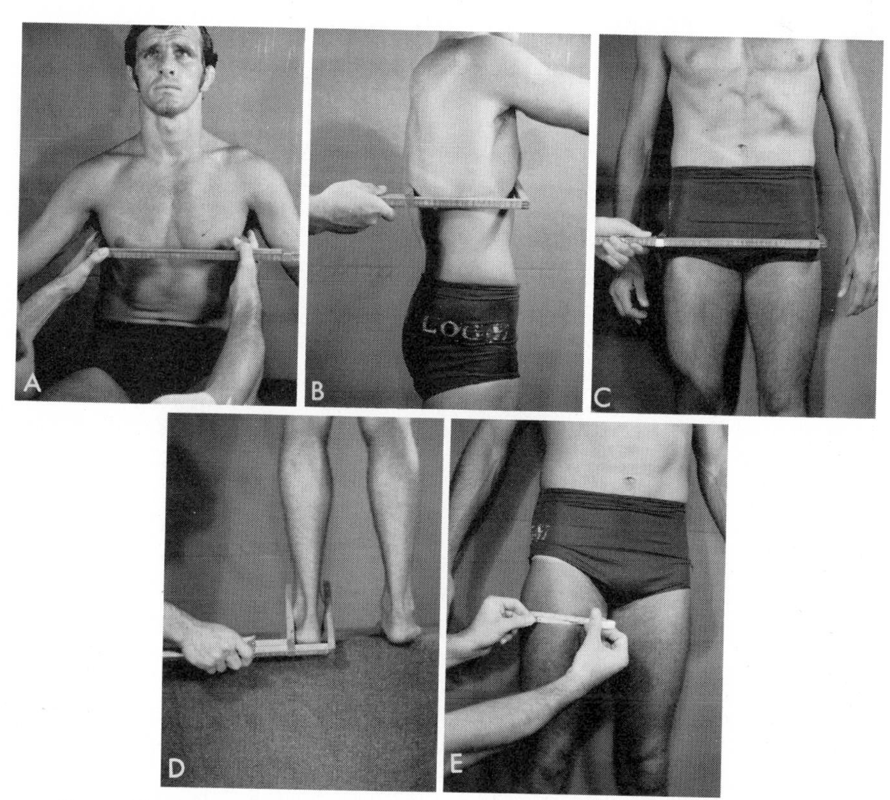

Figure 101. Selected anthropometric measures described in the text. The calipers are placed only to show position; all measurements are taken with the hands (as in *a*), which push the caliper blades into position after palpating the point of application to assure proper positioning. *a*, chest width; *b*, chest depth (level of the xiphoid); *c*, bitrochanteric width (palpate the greater trochanter of the femur on each side); *d*, ankle width; *e*, thigh circumference.

Space limitations do not permit a complete review of all of the equations that have been developed for the estimation of the lean and fatty compartments of the body. The foregoing procedures, which have been shown to be reasonably accurate, do give an overview of the approaches used in anthropometric assessment appropriate for both sexes at different age levels. It must be remembered, however, that maximum accuracy in estimating body components is attained only when the population to which the equation is being applied is similar to the one from which it was derived.[545]

Developmental Considerations. A knowledge of the physical development of students is important to the teacher or coach in evaluating performance capacity or in anticipating the physical changes that result from participation. Changes in body composition during the growth period have been shown by different analytical techniques. Anderson and Langham[8] found that boys and girls show similar increases in the lean protoplasmic mass up to age nine. From ages nine to twelve, both sexes show a rapid decline, which is explained by an acceleration of the skeletal growth resulting in a smaller proportional contribution of the lean mass to the total body weight. After age twelve, girls continue to decline, while boys increase in density up to age 16 as a result of an increase in muscle mass. From ages sixteen to twenty, there is a second decline in the relative lean mass in males because of a filling out with connective tissue and fat. Beyond twenty years, both sexes show a life-long, continuous decline in the lean mass.

The amount of lean body mass is apparently affected by physical exercise: it is the tendency of the body to increase its musculature with activity. Pařízková[399] showed that from the ages of eleven to fifteen years, boys who were very active had a higher absolute amount of lean body mass, less body fat, and a higher aerobic capacity than boys who were much less active. Brožek and Keys[74] have shown similar findings for middle-aged men.

Lack of physical activity as a person ages leads to a reduction of the portion of the total body weight that is lean mass and an increase in the portion that is fat, even though total body weight might show no marked changes. This is one of the reasons for the inadequacy of age-height-weight tables when used with older people. Participation in exercise, however, can reverse the trend.[410] As people age, there may be a time when the body no longer has the ability to respond to exercise in such a manner. Pařízková and Eiselt[400] found that, in men older than sixty-five years, significant differences in lean body mass and limb circumference no longer existed between physically active and inactive groups. They later found similar results in a three-year longitudinal study of men who were in the seventh and eighth decades of life.[401]

Body Composition of Athletes. The well-trained athlete is typified as being lean and hard, and research has tended to support this generalization. One of the classic studies on the body composition of athletes was carried out by Welham and Behnke[528] when they examined twenty-five professional football players. Seventeen of the twenty-five athletes were not considered as being physically qualified for military duty or first-class insurance risk, since all of their body weights were more than 15 per cent above height-weight table values. However, high body density values indicated that the subjects' high weights were due to large amounts of muscular rather than fatty tissue.

More recently, Behnke and Royce[49] studied four weight lifters, three basketball players, and three distance runners with anthropometric, potassium-40, and body densitometry methods. Weight lifters tended to have excessive quantities of muscle which would be underestimated by anthropometric means of assessing lean body mass. One of the weight lifters had developed the characteristic body-builder's physique after having normal body conformation at 27 years of age. The track men were typified by leanness, small body size, and small arm girth compared to chest and leg development, while the basketball players exhibited no extreme deviations. For the track men, fat values ranged from 8.4 to 9.6 per cent; for the basketball players 8.9 to 14.2 per cent.

In our laboratory (unpublished data) average per cent fat values for athletes participating in spring sports were as follows: fifteen track team members, 9.4 per cent; seven middle-distance and long-distance runners from the track team, 8.0 per cent; seventeen baseball players, 12.1 per cent; six tennis players, 14.6 per cent; and twelve spring football participants, 14.5 per cent. All groups tended to carry less fat than the male population average, although some individuals had high fat values. The latter observation was especially characteristic of some of the larger football players.

Skinfolds have been shown to change with participation in sports training and tend to support body densitometry findings about the leanness of athletes. Thompson, Buskirk, and Goldman[507] found that basketball and hockey players maintained a relatively constant body weight during the season, but lost subcutaneous fat. Women basketball players and field hockey players studied by Lundegren[350] showed reduction in sites specific for the respective sports. Basketball players decreased in fat at the arm, iliac, and thigh sites; field hockey players at the arm, umbilicus, and thigh. Basketball players decreased in thigh girth while hockey players decreased in girth at the level of the umbilicus.

Changes in Composition with Changes in Weight. The gain and loss of body weight is not entirely dependent upon the gain and loss of fat. The increase in weight due to an increase in muscle mass with training has been demonstrated in the studies previously reviewed.

Increases and decreases in obesity tissue are accompanied by increases and decreases in body water and protein as well.[193] Consequently, gains in fatty tissue are accompanied by gains in the lean body mass. The relative amounts of lean and fatty tissue do, however, change; the fatty tissue making up a greater proportion of the total body mass when obesity exists. The converse is true with weight loss. On starvation diets consisting of carbohydrates, early weight losses are largely due to losses of water, and the caloric equivalent is less than 4500 kilocalories per kilogram of weight lost, whereas a kilogram of fat has a caloric equivalent of over 9000 kilocalories.[193] After about three weeks, the loss is about 85 per cent fat and 15 per cent protein with a relatively constant body water content.

The latter observation may explain the success some "exercise parlors" have when they promise miracle weight losses and, for the ladies especially, a reduction of at least one clothing size within a short period of time. At the beginning of a weight-loss program, the energy expenditure per pound of weight lost is relatively low. Good clothing fit is also dependent upon good muscle tone, a condition that can be restored rather quickly with carefully planned exercises if the person has been inactive previously. As the program continues, however, the challenge becomes greater.

MAKING WEIGHT

Setting realistic weight goals in sports such as wrestling, where participants must compete at set weight limits, is a continual problem and frequently a source of criticism of such sports. Some states have set rather arbitrary, empirically derived procedures for setting weight limits for high school wrestlers. There are no such limits for college and amateur wrestlers, and weight goals are usually mutually set by the coach and athlete, neither of whom may have any real objective basis for arriving at a decision.

The development of simplified methods for measuring body composition can provide an objective means of establishing individual weight goals. Tipton, Tcheng, and Paul[513] have been working with the Iowa State High School Athletic Association and the Iowa Medical Society on the use of the Hall[205] and Behnke[48] methods for determining ideal body weight, as previously discussed. They used these techniques to measure more than 2000 wrestlers during a 3-year period, including more than 500 state finalists, and found that the original equation used by Hall was unsatisfactory for predicting the weight of Iowa wrestlers. In addition, they reported that the use of a thigh circumference measure was more of a hindrance than an asset in predicting minimal weights. Emerging from this study was a multiple regression equation that used measures of bitrochanteric width (Behnke), chest width (Hall and Behnke), chest depth (Hall), ankle width (Behnke), and height (Hall and Behnke). The intent of the equation was to predict a minimal wrestling weight six to ten weeks before official certification, in order that an individual could "make weight" in a systematic manner. In the event that an individual for any reason wished to compete at a weight lower than predicted, the wrestler had to secure medical authorization and be supervised during this time interval. All measurements were taken in centimeters. The equation follows[504] (if the weight is to be expressed in kilograms, do not multiply by the constant 2.2):

$$\begin{aligned}\text{Weight in pounds} = [&1.443\ (\text{bitrochanteric width}) + 1.793\ (\text{chest width}) \\ &+ 1.545\ (\text{chest depth}) + 1.723\ (\text{ankle width}) \\ &+ 0.253\ (\text{height}) - 126.560] \times 2.2\end{aligned}$$

Use of body composition measuring techniques and medical advisement is a rational approach to setting weight limits. Because these techniques are based upon relatively constant measures, participants cannot cheat by cutting drastically to meet certain weight limits at a specified date. Such a procedure also makes it possible for the boy who is markedly overweight to cut to a weight limit that would be beneficial to his health; and it protects, at the same time, the boy whose weight is so close to his minimum allowable weight that to lose more would require him to experience true starvation and its consequences.

QUESTIONS

1. Distinguish between body composition and body constitution.
2. What somatotype component would you expect to be high in a high jumper? in a heavyweight wrestler?
3. What is the ponderal index?
4. How is somatotype related to strength? flexibility?
5. What are "mid-types" in reference to somatotyping?
6. Define somatotype.
7. What is the lean body mass? lean body weight?
8. Define density.
9. Why are age-height-weight tables inadequate, in many instances, for determining whether or not a person is at his best weight?

10. Why can we get a reasonable estimate of body fat from skinfolds?
11. How does the Hall method differ from the Behnke and skinfold techniques in evaluating body composition?
12. How could you use the techniques described in this chapter in a school physical education program? in an adult fitness program?
13. How does body composition change during growth? Would you expect to see similar changes in somatotype? Why?
14. How would weight training affect body composition?
15. How could you use body composition procedures in a school athletic program?

Chapter Twenty

PHYSICAL ACTIVITY FOR CONVALESCENTS

World War II focused national attention on the fact that the average patient remains physically inactive too long. It was realized that, as far as the armed forces were concerned, it was a waste of manpower to keep convalescing patients inactive in hospitals and to discharge them for limited service while the "natural processes" of rehabilitation took their course.

On the initiative of Howard A. Rusk, an active rehabilitation program was introduced in 1942 in the Army Air Force hospitals.[448] This program spread rapidly throughout the other branches of the armed forces.

There was one great difficulty, however, in operating these "rehabilitation" programs: the vagueness in the definition of "convalescent." Who can definitely state in every case when patients begin to be convalescent? The period of convalescence traditionally has started either at that moment when the doctor allows his patient to sit up in bed or, more frequently, when the patient is allowed to leave his bed for the shortest period of time.

But what have been the traditional criteria determining the length of time during which a patient has had to remain bedridden? Tillet[508] has pointedly said that, if an attending physician were to explain why each of his patients was kept at complete bed rest, his reasons would be vague in a surprising number of instances. Even enthusiasts of early rehabilitation would not deny that bed rest is imperative for certain conditions. On the other hand, there is enough evidence to show that the abuse of bed confinement not only may lead to "deconditioning," but may actually prolong the need of therapeutic procedures, and cause complications which sometimes may be fatal.

An impressive symposium, published in the Journal of the American Medical Association,[495] gives sufficient evidence that early physical activity, judiciously applied, is indispensable for successful treatment. This symposium covers cardiovascular diseases, obstetrics, surgery, orthopedics, and psychiatry. It also shows that medical men started questioning the values of enforced physical inactivity long before the beginning of World War II.

According to this symposium, more harm than good may result from prolonged rest, even during such serious diseases as angina pectoris and myocardial infarction. As to the rest imposed on pregnant women, it is considered to be "an unjust and unnecessary penalty for motherhood."

However, Miasnikov[374] reported that, in animals with an experimentally induced hypercholesteremia and lipid deposits in aorta and coronary arteries, a three-minute run caused extensive myocardial damage. Repeated runs caused changes similar to infarction. Thus, the relation between cardiac infarct and exercise should be re-investigated.

One of the contributors to the symposium calls attention to bone atrophy, muscular wasting, and vasomotor instability as not infrequent sequelae of bed confinement. In addition to this, he states that constipation with subsequent cathartic habituation may develop. He also calls attention to the fact that a recumbent position after surgical operations, during which numerous clots have formed, may lead to thrombosis, because in the horizontal position veins of the legs are compressed, while on standing they are dilated.

He also calls attention to the danger experienced by a patient in the horizontal position who tries to use a bed pan. This uncomfortable position requires so much strain that blood pressure may be considerably increased. It is much safer if the patient is allowed to leave the bed and use a commode, or allowed to walk slowly to a toilet.

In psychiatric and neuropsychiatric cases, restriction of physical activities shuts off one of the most natural and important outlets for available energy. Menninger[495] makes the rather blunt statement that "the death of some hypertensive patients has been hastened by physicians who removed from them the only available or acceptable form of aggression to which they had access." A person who has had a chance to observe the sedative effect of physical effort involving large muscle groups upon men in states of anxiety will never need any additional proof. Whether or not an anxiety state is essentially a fear reaction—and the most natural primary reaction to fear is to run, an activity involving large muscles—is still to be decided. The truth of the matter is that strenuous muscular activity gives relief.

Probably the most important single beneficial factor derived from physical activities by patients is the effect on the circulation and distribution of body fluids. The best illustration of this is the relief of congestion and edema in the lungs which have resulted from a prolonged supine position. On assuming a vertical position, patients begin to cough, thus clearing the respiratory passages, and edema, if it develops, will be found around the ankles and the lower legs. Sometimes just a change from a supine to a sitting position may prevent complications in the lungs.

Probably the most striking results of early ambulation are described by Leithauser.[342] His patients are encouraged to get out of bed and to stand for a few moments on the floor beside the bed, and to cough as early as three to four hours after an appendectomy operation. Ambulation for his patients also begins some time on the day of the operation, or on the day immediately after. While in bed, the patient is instructed to exercise his legs by bending his knees, and flexing and extending his feet four times every hour from the moment he is conscious after the anesthetic wears off.

It is beyond the scope of this text to go much further into a discussion of various pathological conditions and the corresponding application of physical activities. The final decision, and rightly so, will be the responsibility of the attending physician. The physical instructor or the physical therapist will have to follow his prescription.

Students interested in the physiology of muscular exercise as applied to pathological conditions are referred to an excellent review by Simonson and Enzer.[477]

Beyond any doubt, physical reconditioning is here to stay, although it may take a long time to overcome the inertia of tradition. The welfare of the patient demands this. The patient and the community will also benefit economically from this.

Another important benefit derived from properly conducted physical rehabilitation is overcoming or preventing of psychological damage caused by overemphasizing pathological conditions. Patients who know that they *can* do things feel better than those who are convinced that they *cannot* do anything. One of the best illustrations of this may be observed on so-called cardiac patients. Cabot[78] said once, "Most 'heart disease' is imaginary . . . Myocarditis was recognized six times as often as it was present, valvular disease twice as often."

Even at the present time, with all the advancements in technical aids for diagnosis, what cardiologist can be always sure that heart disease is present or absent? Therefore, there are many people who may become invalids because of difficulties encountered in diagnosis.

One can hardly blame an earnest physician for "playing safe" in a case of doubtful diagnosis. It is his duty. It is always advisable, however, to weigh all the components of safety in a less routine manner. A good example may be found in a study of the reconditioning of adult patients convalescing from rheumatic fever.[305] Traditionally, such men were condemned to a life of physical inactivity, which made many of them hypochondriacs and psychoneurotics. A carefully graded system of physical training used on adult patients in an Army Air Force hospital brought about undeniably good results. The physical and mental well-being of the patients markedly improved. The patients took part in physical activities, and from their own experiences they knew that they "*could do it.*" The most striking cases were those patients who had spent several months in bed and were so convinced that they had to be physically inactive that they broke into tears when the first attempts were made to test their physical abilities by having them take a few steps on a bench 12 inches high. When they realized that physical exertion was possible for them, their attitudes changed completely. As a matter of fact, the investigators had a rather hard time to persuade one of these crying patients not to take part in competitive basketball. Just to be safe!

A person engaged in rehabilitation work will soon realize that his experience in directing physical activities among the well may lead him astray. He may tend to give more strenuous exercise than is advisable. He will have to relearn the concept of gradualness. It is true that muscular overexertion will not hurt the muscles themselves, but an intensive muscle soreness may discourage patients from participation in physical activities.

In patients weakened by bed confinement, it is wise to start with exercises of low intensity and gradually increase the dosage. A convenient method of evaluating the intensity of exercise in terms of resting metabolism has been presented in Chapter 7. Detailed information may be obtained from the original reference.[527]

Roughly speaking, there are two main types of reconditioning: kinesiological and physiological. They are not always well defined, and may often overlap each other. The kinesiological type strives to restore normality in the function of muscles and joints. A typical illustration is the effort to restore the normal (or at least the best possible) range of movement in an affected limb. They physiological type has as its chief aim physical endurance of the whole body.

Whereas in kinesiological reconditioning one may use passive movements, massage and heat, as well as active exercises, physiological reconditioning is based on graded physical activities, whether in the form of calisthenics, modified sport exercises or occupational pursuits.

During World War II many men had experience in the physical reconditioning of convalescing patients. Some of them may be expected to follow the same type of work in civilian hospitals and rehabilitation centers. The systematic and energetic continuation of this reconditioning program by Army and Navy hospitals and by the Veterans Administration is giving encouragement to such programs.

Under usual conditions the average physician or surgeon in charge of rehabilitation does not do the original research necessary to support this developing field. He may accumulate valuable clinical material, but there is always the possibility that there may not be sufficient objective appraisal of this material, resulting in the following of narrow routines. The ordinary physical reconditioning instructor, in most instances, will follow suit unless he is stimulated to do otherwise.

It is the sincere hope of the authors of this book that there will be a sufficient number of large hospitals and centers where special research into the physiological basis of reconditioning will be conducted. Progress in knowledge is impossible without research.

Undeniably, training in any profession makes observers keener and results in that uncanny tool called experience. An experienced medical man can tell whether a patient is getting better or worse, and can detect even shades of change. This ability has been recognized by physical educators only too well, so that when one of them attempts to propose a test of physical fitness for groups, he often uses the opinions of one or more physicians as a standard for the appraisal of the validity of his own test. In a way, it is a curious situation, because the physicians themselves are in need of objective tests which will give a solid foundation to their verdicts regarding patients' physical fitness.

As it is, no single test will meet all situations. For instance, there is a place for special tests of muscles and joints which can be used by orthopedic surgeons. There is a place for a test of physiological responses after all forms of surgery. The general surgeon may still rely on merely watching his patient make a certain number of steps and then counting the pulse rate and ascertaining the degree of acceleration caused by the exertion.

At the present time, it seems that the most practical test of "fitness" will consist of some stepping up on a rather low bench, 12 to 13 inches in height. One can determine the endurance of his subject by timing the duration of the exercise. The pulse rate response to this exercise will show two things: (*a*) the degree of cardiac adjustment; and (*b*), provided the subject could not perform a standard amount of work, whether it was caused by local fatigue of the leg muscles or by a circulatory inadequacy.

A simple test of this type was successfully used on Army patients convalescing from rheumatic fever.[305] Patients were asked to make twelve complete step-ups (with both feet) on a bench 12 inches high in thirty seconds. If the pulse rate, taken six to ninety seconds after exercise, was 100 per minute or less, the patient was considered fit for physical activity. The second part of the test, given after a lapse of several days, consisted of stepping up and down (with both feet) on a bench 20 inches high at the rate of twenty-four steps per minute, as long as the patient could, but not in excess of five minutes. It is felt, however, that this height is too great for civilian use and that a lower bench should be substituted. The test will remain the same in principle and will serve as an objective guide to the physician in charge.

Figure 102 illustrates how testing helped to evaluate the physical

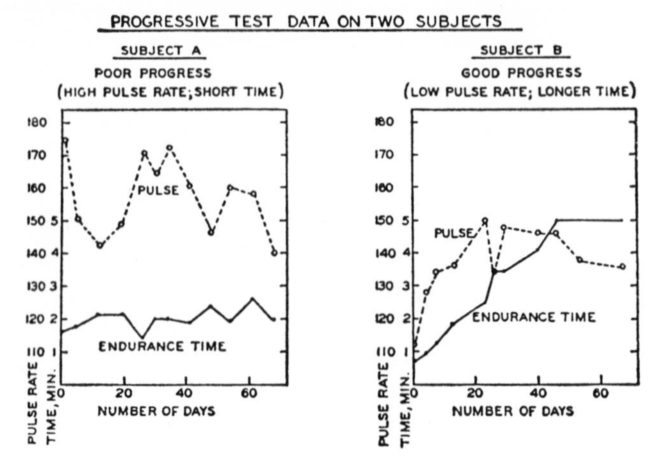

Figure 102. Records of improvement in physical fitness of two convalescing patients. Subject A showed slow and incomplete improvement and was discharged for limited military duty. Subject B showed rapid and complete improvement and was discharged for full military duty. (Karpovich et al.: J.A.M.A. *130,* 1946.)

condition of two patients before they were discharged from the hospital. Patient A hardly improved in endurance, although, judged by his high pulse rate, he tried hard. On the other hand, patient B reached the required maximum time of five minutes, and his post-exercise pulse showed a decline. These records were used as partial evidence that patient A should be discharged to limited military duty, whereas patient B could be discharged to full duty. Incidentally, a five-minute step-up test is equivalent to walking up to the twenty-first floor of a skyscraper (allowing 10 feet per floor) and coming down, in five minutes. A person who can perform this amount of work with ease is surely physiologically able to perform full military duty.

QUESTIONS

1. What is the aim of rehabilitation?
2. When does convalescence start?
3. Is there any relation between physical activities and mental health?
4. What are the advantages of early ambulation for surgical patients?
5. What is the psychological effect of physical rehabilitation on patients in general?
6. Are there imaginary heart diseases? How are they produced?
7. Should a pregnant woman avoid physical activities?
8. Is there any place for physical fitness tests in hospitals, civilian and military?
9. Would it help hospital work if a scientific guide book for the prescription of physical activities were available? What are the *two* basic parts necessary for the preparation of such a book?

Chapter Twenty-One

ERGOGENIC AIDS IN WORK AND SPORTS

In a situation in which excellence in physical performance is of great importance, several questions immediately present themselves: Are there any special foodstuffs, drugs, or other means which will increase work capacity? What are they? Are they dangerous?

Newspapers, and even scientific periodicals, from time to time carry articles describing the remarkable effects of various "aids" which increase muscular strength, speed and endurance, and hasten recovery from fatigue.

In most cases, waves of enthusiasm affect the investigators of these "aids" and result in poorly controlled experiments which unfortunately might lend support to the original questionable communications. Critical and contradictory articles soon appear, but, as usual, the negative findings are slow to affect the practical field, and the impetus gained by the "positive" observations may prevail for a long time, especially if supported by commercial interests.

On the surface, merely testing the subjects before and after the administration of the substance in question may seem enough to determine the effect of the substance upon muscular performance. This may be sufficient in cases in which the effect of a big dose of a powerful drug is tested, but in most cases the doses are relatively small, and their effects are not obvious. The common error in many investigations is the absence or inadequacy of control in the experiments.

Practical men—athletes and their coaches—are especially guilty whenever they ascribe success in games to the use of some substance. The weakness of such assumptions is evident. There are so many factors operating in sports involving skill and team coordination, as well as changes in team personnel, that it is practically impossible to discern the effect of any substance upon team performance. Even in experiments with table salt, which has reduced the incidence of cramps among football players, it is possible to make but one deduction: it helps only those who would lose by their sweat more salt than is taken with their regular diet. In the men who either consume enough salt with their food or lose little in sweat, it is unnecessary, if not harmful.

In measurable events, such as swimming or track and field sports, it still is not easy to discover the effect of some supposed aids upon performance. One should make allowance for the influences of training, excitement, and unpredictable and inexplicable changes in the athlete which make him excel on one day and fail on another. The results obtained during contests should be compared with experiments made during time trials. Sham tests should be employed in which, instead of the "real stuff," an inert substitute, or placebo, is given, and the psychological factors should be controlled as much as possible.

Attention may be called here to the early reviews of the literature on this subject by Baur[45] and Bøje.[62, 63] Baur speaks of all the possible aids to muscular performance as "drugs." Bøje calls them "dopes." It seems improper to refer to table salt, vitamins, and ultraviolet rays as either drugs or dopes. Although the word "doping" is frequently used in connection with the administration of various substances to athletes, and may eventually become a proper term, at present it is objectionable since it connotes an administration of drugs akin to opium. In order to avoid this unfortunate association, it may be advisable to refer to these aids as *ergogenic* aids, or work-producing aids. As will be seen, many of these aids have either doubtful or no effects at all upon muscular work. Therefore, the terms "ergogenic aid" or simply "aid" should be used advisedly.

The question has often been raised concerning the ethics of the use of so-called ergogenic aids. It may be stated here that the use of a substance or device which improves a man's physical performance without being injurious to his health, can hardly be called unethical. As for taking advantage of other contestants who do not use these aids, this should be regarded in the same light as the use of special diets, massage, special exercises, and so forth. All these means are available to everyone, and may be used if desired. On the other hand, no one would consider the drinking of coffee or tea as unethical, yet the amount of caffeine consumed in these beverages may be considerable, and the effects harmful.

Some difficulties of classification have arisen in organizing the material for discussion. In this chapter all material will be arranged in alphabetical order, according to the dominant chemical component. Also, for convenience, some of the aids will be considered together. Thus, all alkalies will be discussed under one heading, all fruit juices under another, and so on for all large groups.

The following will be discussed: alcohol; alkalies—bicarbonate of soda, and sodium and potassium citrates; amphetamine (Benzedrine);

caffeine; cocaine; fruit juices; gelatin and glycine; hormones; lecithin; oxygen; phosphates; sodium chloride; sugar; tobacco smoking; ultraviolet rays; vitamins. The effect of tobacco smoking has been included in this chapter because many smokers insist that smoking "quiets their nerves," thus improving performance.

The purpose of ergogenic aids is to improve performance or hasten recovery, or both. The nature of their action is not always well known, and may involve one or several of the following possibilities: (1) direct action upon the muscle fibers; (2) counteraction of fatigue products; (3) supply of the fuel needed for muscular contraction; (4) effect upon the heart and circulatory system, increasing their efficiency and thus facilitating the transport of oxygen, fuel, and wastes; (5) effect upon the respiratory center; (6) delay of the onset of the feeling of fatigue by action on the nervous system; and (7) counteraction of the inhibitory effect of the central nervous system upon maximal muscle contraction, thus allowing a muscle to develop greater force.

A search through the literature reveals a great complexity of problems involved in the evaluation of the effects of any type of aid. Data obtained under seemingly identical conditions vary a great deal, and the interpretations are often contradictory. It is true that one cannot measure all possible changes, and many so-called subjective effects have an objective basis which eventually may be discovered and measured. Nevertheless, one is forced to accept with reservation any references to subjective sensations, because their reliability is often questionable.

ALCOHOL

Alcohol has been used since time immemorial to bolster courage, to counteract fatigue, to help one forget worries, and "to warm up."

Old experiments showed that small amounts of alcohol increased muscular endurance. Hellsten, using Johansson's ergograph, which involves the pulling of weights with both hands, found that small doses of alcohol taken five to ten minutes before the exercise did increase work for the first twelve to forty minutes, after which there was a definite drop in performance lasting for two hours. Up to 80 cc. of 38 per cent brandy taken half an hour before Hellsten's test caused a decrease in work output from the beginning of the test. On the other hand, Atzler and Meyer[33] found that even 240 cc. of alcohol in the form of beer or brandy given immediately, or as much as four hours before their test, would increase the work output, provided the men were habitual drinkers. The same amount of alcohol taken the night before caused a marked drop in work capacity.

Herxheimer[233] observed a deleterious effect of alcohol upon speed in swimming or running short distances. Herring observed the same effect in 100-, 400-, and 1500-meter races. The amount given was rather large: 100 cc. of 52 per cent alcohol, or the equivalent of about half a tumbler of whiskey. Simonson[476] reported a decrease in oxygen debt after work done following the intake of a small amount of alcohol, whereas Meyer[373] found no change. As to the question whether alcohol may be utilized as a source of energy for muscular contraction, opinions differ. Investigations by Carpenter and co-workers[86] and Canzanelli and others[82] indicate that such utilization does not take place.

The common use of alcohol for warming-up seems to draw supporting evidence from observations that small doses of alcohol increase the endurance of chilled muscles. This coincides with a feeling of superficial warmth, caused by a larger amount of blood coming to the skin blood vessels, which have been dilated by alcohol. However, the loss of heat is also increased, and the danger of greater chilling is enhanced.[9]

There seems to be agreement that large amounts of alcohol are detrimental to muscular performance, but differences of opinion do exist as to the influence of small amounts. Also, the size of the dose is relative, since a "small" dose may be large for a total abstainer, and a "large" dose may be relatively small for the habitual drinker. There is no question but that alcohol is definitely detrimental in skill exercises; numerous tests on drivers may be used as evidence.

In conclusion it may be said that alcohol cannot be recommended for use by athletes. However, additional experiments are necessary to determine how deleterious alcohol is to athletes accustomed to weak wine or beer with their meals, as is customary in some countries.

ALKALIES

During intensive muscular exertion, acids accumulate in the blood. To take care of these, more buffer alkalies are needed.

It is logical to assume that an artificial increase in the amount of alkalies in the body would raise the level of muscular capacity. Such an assumption has been responsible for the alkali feeding of athletes.

Dill and his co-workers[131] found that an intake of sodium bicarbonate allowed a greater oxygen debt, but they could not notice any increase in muscular performance. Dawson[120] cited the case of an athlete who could not complete a long race after taking 10 gm. of sodium bicarbonate.

Dennig and co-workers[122] on the other hand, found an increase in endurance after alkali intake. They recommended this prescription:

Sodium citrate	5 gm.
Sodium bicarbonate	3.5 gm.
Potassium citrate	1.5 gm.

This represents a daily dose to be taken after a meal for two days before a test and two days after the test to avoid an acidotic reaction. A longer preliminary period of intake may lower the performance. Dennig's experiments were based on treadmill and stationary bicycle tests. One of the authors, in cooperation with Mr. Charles Silvia, swimming coach at Springfield College, tried Dennig's formula on varsity swimmers. No definitely beneficial effect was observed.

AMPHETAMINE (BENZEDRINE)

The chemical composition and physiological action of amphetamine are closely related to those of epinephrine. In doses of from 5 to 20 mg. it is capable of abolishing the sense of fatigue, especially when this has been caused by lack of rest and sleep. It should be remembered, however, that it is a powerful and dangerous drug and that excessive use may lead to insomnia, hypertonia, and circulatory collapse.

Foltz et al.[173] gave 10 to 15 mg. of amphetamine intravenously to two trained subjects, 30 seconds or 30 minutes before rides on a stationary bicycle. They rode to exhaustion twice in a row, with a 10-minute rest between. Altogether 10 double rides were made. No ergogenic effect was observed after either the first or the second ride. Four men received injections immediately after the first ride; this, however, did not improve the second ride in 17 trials.

Smith and Beecher[487] gave 14 mg. of amphetamine per 70 kg. (154 lb.) of body weight to swimmers, runners, and weight throwers two to three hours before tests and found an improvement in 75 per cent of the cases.

Karpovich[294] gave 10 to 20 mg. of amphetamine per person to swimmers, track runners, and all-out runners on a treadmill, 10 to 30 minutes before the tests (mostly 30 minutes), and found no beneficial effect on performance. All experimental tests were double, with a 10-minute rest between.

Golding and Barnard[186] used 15 mg. of amphetamine on 20 men, two to three hours before two bouts of all-out runs on a treadmill, and found no beneficial effect.

One may wonder why the findings of Smith and Beecher are at variance with those of other investigators. Was it because of a difference in dosage or a difference in time interval between medication and testing?

The difference in dosage can hardly be called critical in those cases in which 15 to 20 mg. of amphetamine per man was given. For men weighing 170 to 180 lb. the amount of the drug per 70 kg. of body weight was 14 to 18 mg. and 13 to 13.6 mg., respectively.

As to the time, Foltz and associates injected the drug into a vein, so that there was no waiting time. Golding and Barnard used the same time interval as Smith and Beecher. Only Karpovich used a much shorter time; however, most of the subjects were able to feel the effect of the drug. Therefore, the difference in results may be dependent on the manner in which the tests were conducted.

Incidentally, Rasch and co-workers[423] studied the effect of 20 mg. of amphetamine on reaction time and speed of movement in 26 subjects, when the drug was administered 2 to 3 hours before the test, and found no effect. Thus, it must be concluded that the ergogenic effect of amphetamine has not been proved. It is true that many subjects feel "pepped-up" after amphetamine; but this sensation, as has been shown by Karpovich and also by Foltz, does not necessarily lead to a better or a best performance.

CAFFEINE

Caffeine acts upon the blood vessels, heart, and nervous system. It causes general vasoconstriction, with simultaneous dilatation of the coronary artery, and increases the contractile power of the heart. It stimulates the central nervous system, accelerating the respiratory rate and shortening the reaction time. In small doses it acts beneficially upon psychic processes.

Early experiments showed that caffeine increased muscular performance in ergograph tests. Schirlitz[456] found that 0.3 gm. of caffeine-sodium-salicylate caused a slight increase in the work output of subjects riding bicycle ergometers. It has been observed that tea, because of its caffeine content, is beneficial in prolonged exertion.

Caffeine and cola-nuts (which also contain caffeine) were used in a well-controlled experiment by Graf.[192] Subjects riding ordinary or stationary bicycles were given chocolate, either plain or with the addition of caffeine or cola. He found that chocolate with cola had a more noticeable effect than caffeine, raising the work output 20 to 30 per cent. Foltz and co-workers[173] gave 500 mg. of caffeine intravenously to four subjects riding a stationary bicycle, and found an increase in endurance and a faster recovery from fatigue. Sprint running is not affected by caffeine: Herxheimer[233] gave 250 gm. of caffeine-sodium-benzoate to forty-six subjects running a 100-meter race, and could observe no effect on performance.

Incidentally, a cup of coffee may contain 97 to 195 mg. of caffeine.

COCAINE

Cocaine has a powerful stimulating effect upon the central nervous system, increasing the activity of the cerebrum. It accelerates the respiratory and circulatory rates, has a direct sympathomimetic effect, and increases muscular tension.

In the form of the coca-leaf, cocaine is used extensively by the South American Indians. This enables them to perform prodigious feats of endurance. They can march for days with little food or rest if they have coca leaves to chew.

Mosso[45] showed that 0.1 gm. of cocaine postpones the onset of fatigue. Thiel and Essig[506] found that the endurance of men and women riding bicycle ergometers was increased when they were given 0.1 gm. of cocaine hydrochloride by mouth. The maximum effect of the drug was noticed thirty minutes after intake. Herbst and Schellenberg,[228] using the same amount of cocaine, noted that the speed of recovery after riding stationary bicycles was increased.

Since cocaine is a dangerous, habit-forming drug, its use in athletics cannot be recommended.

FRUIT JUICES

Dietitians rightfully extol all kinds of fruit juices for their vitamin and mineral content. They are also supposed to "alkalize the blood," thus increasing capacity for work. Hewitt and Callaway[236] reported an improvement in swimming speed after drinking either orange or tomato juice. They attributed this to an increase in blood alkalinity.

Numerous tests made by one of the authors (P.V.K.), in collaboration with Pestrecov and LeMaistre, showed that the liberal use of various juices, grapefruit, orange, and tomato, had no effect upon the muscular performance of twenty-eight bicycle riders. No significant change in buffer alkalies in the blood of eighteen men could be observed after administration of these juices to the extent of 3 quarts in four hours.

Fruit juices should be considered an important part of the diet. They may help in building strength and endurance as long as they supply the needed amount of minerals and vitamins, but the experimental evidence shows that they are not directly connected with an increase in muscular performance. A psychological effect may be suspected.

GELATIN AND GLYCINE

Gelatin is an incomplete protein, rich in amino-acetic acid (glycine), which constitutes about 25 per cent of its weight. Glycine is chemically related to creatine, a substance indispensable to muscular contraction. For this reason glycine and, later, gelatin have been used in pathological and normal cases in efforts to improve muscular action.

A number of reports on the administration of glycine in various muscular diseases stated that it was beneficial, that muscular strength increased and that fatigability decreased. A defatiguing effect of glycine on normal persons was reported by Wilder.[541] This report stimulated further investigations.

Ray and his co-workers,[426] using gelatin or pure glycine, reported an increase of up to 240 per cent work output in men and no appreciable increase in women. No control experiments were conducted, and the effect of training was not excluded. Hellebrandt and her associates[220] experimented on women, arranging the gelatin administration in such a manner that its effect would be distinguished from that of training; and, although some of the subjects improved as much as 200 per cent, the effect was clearly one of training. They also found that gelatin was of no value in the prevention of staleness. All these investigators exercised their subjects on bicycle ergometers.

Karpovich and Pestrecov[303] carried out a series of experiments on bicycle ergometer riders and on swimmers, heavyweight lifters, and wall-weight pullers. Diet was controlled in some of the groups. No effect of gelatin upon the working capacity of muscle could be observed. A group of county jail inmates, after twelve weeks of bicycle exercises, improved up to 4420 per cent, regardless of gelatin. The maximum time of uninterrupted riding was six hours and twelve minutes a day, the rate of work being 0.217 horsepower. College students, in experiments similar to those of Ray and his co-workers, improved up to 334 per cent; the subject showing the greatest improvement received no gelatin. The psychological effect of the administration of gelatin was also noticed. When farina was given under the guise of "concentrated gelatin," a noticeable "stimulating" effect resulted.

It may be stated positively that the addition of gelatin to a normal diet does not act as a special source of extra power, nor does it increase endurance. In animal experiments, gelatin has again failed to show any beneficial effect upon the strength or fatigability of skeletal muscles. In spite of this evidence, some weight lifters still use gelatin to get "extra power."

Since gelatin has been used for its glycine content, it indirectly proves the inefficacy of glycine.

Hilsendager and Karpovich[243] tested the effect of 1.5 gm. of glycine and 150 gm. of niacin, separately and in combination, on 66 subjects. Twenty subjects rode a stationary bicycle, and the other 46 worked on the elbow flexors ergometer. Each test consisted of two all-out bouts, with a five-minute rest between. No ergogenic effect was found after ingestion of either of these chemicals taken separately or together.

HORMONES

The profound effects of various hormones upon the vital bodily functions are well understood. Thyroxin increases metabolism; epinephrine causes a rise in blood pressure and a greater contractility in the muscles. Lack of active agents in secretion of the adrenal cortex is associated with great muscular weakness. All this has led to a hope that hormones may increase muscular strength and endurance in normal people.

Simonson[476] gave 3 tablets of thyroidin daily for two to four days. All subjects had an increase in basal metabolism averaging 20.1 per cent. Recovery after exercise was accelerated.

Press[411] experimented with Sympatol (oxyphenylethanolmethylamine), a substance closely related to epinephrine, on normal and on hypotonic persons, who were given 0.15 gm. orally before a 2000-meter race. The data showed that Sympatol did not increase the work capacity of either normal or hypotonic people.

Campos and his co-workers[81] observed that an injection of epinephrine in dogs before a run had no favorable effect upon performance. However, the injection, when made after the animals had reached a stage of exhaustion, quickened the recovery and enabled them to continue running. A large dose of epinephrine (0.174 mg. per 1 kg. of body weight) caused great excitement and a decrease in the capacity to work.

Dill's group[133] found that an injection of epinephrine was beneficial because it stepped up the utilization of sugar in work and made the subject feel more energetic. Eagle, Britton, and Kline[147] found that an injection of adrenal cortex extract greatly increased the work capacity in dogs. Little improvement could be noticed in men after injections of 0.5 to 1 cc. of adrenal cortex extract (Missiuro, Dill, and Edwards[375]).

There is sufficient evidence that the administration of hormones may raise the level of physical fitness. Further research in this field will undoubtedly be fruitful.

LECITHIN

Lecithin belongs to the so-called phosphatides, containing fatty and phosphoric acids. It seems to play an important part in the oxidation of neutral fats. It also is a fine source of phosphorus, which may be utilized in the chemical changes involved in muscular contraction.

Atzler and Lehmann[32] studied the effect of lecithin of soya bean on five persons, giving them 44 gm. daily for several days; an increase in the strength and the skill of the hands was reported. Dennig[121] also used this type of lecithin and claimed that the effect was favorable. On the other hand, no favorable effect from lecithin was observed by Staton.[491] As with many other substances, an extra supply does not mean an extra utilization.

OXYGEN

One of the main limiting factors in physical activities is the amount of oxygen which the organism can take up. Therefore, it seems logical to suppose that breathing pure oxygen will increase one's capacity for exertion and recovery.

Hill and Flack[242] reported that when oxygen was given for three minutes immediately before exercise and also for four to five minutes during recovery, the athletic performance was improved and the recovery from fatigue was quickened. They found that a man holding his breath was able to run 470 yards after oxygen inhalation. Moreover, Feldman and Hill[164] noticed that preliminary oxygen inhalation resulted in lower lactic acid accumulation, and also stated that the effects of preliminary oxygen inhalation may last as long as fifteen minutes. Karpovich[285] found that, two minutes after oxygen breathing, there was only 20.3 per cent of that gas in the expired air. A longer effect will be observed only if subjects *do not breathe*, but remain still. An impressive illustration of the effect of preliminary inhalations of oxygen on subsequent breath holding was obtained in Schneider's laboratory. After three deep inhalations of oxygen, one subject was able to hold his breath for *twenty minutes and forty-five seconds.*

Miyama[379] reported that preliminary oxygen breathing was beneficial before a 120-meter run, and also in recovery after that run. Unfortunately, in his test the effect of "getting used" to peculiarities of the run in a long corridor had not been eliminated. His decisive experiment consisted of testing two men. Both of them ran faster after several trials, whether oxygen was inhaled or not, and the degree of improvement was about the same.

Simonson and collaborators[476] noticed that a preliminary inhalation of oxygen, in spite of some favorable sensation, had no effect upon speed in a 100-yard dash.

Karpovich[281] experimented with preliminary oxygen breathing immediately before the start of a 100-yard swim. There was definite improvement in speed. Obviously this was due to the ability to hold the breath longer while swimming.

The suggestion has been made that athletes should be able to exert themselves to a greater extent in an atmosphere rich in oxygen. Nielsen and Hansen,[394] experimenting with subjects riding bicycle ergometers, found this to be true only when the rate of work became strenuous.

In conclusion: Oxygen breathing immediately before a short swimming race and during strenuous work is beneficial. Breathing air containing 66 per cent oxygen during strenuous work is more beneficial than breathing 100 per cent oxygen.[40] Since there is no storing up of oxygen, the preliminary three deep inhalations of oxygen are just as effective as prolonged breathing; and since the effect of preliminary oxygen breathing wears out in three minutes, there is no basis for the assertion that Japanese swimming victories during the 1936 Olympics were made possible by oxygen breathing. Forced breathing of ordinary air at the start enriches the lungs with oxygen, and therefore should be helpful, especially before sprinting events. The present fad of breathing oxygen to hasten recovery after a physical exertion in football or other activities is based on salesmanship rather than physiology.

PHOSPHATES

Phosphates are indispensable to physical activities, because the breakup of phosphorus compounds furnishes the energy for muscular contraction, and because phosphates also function as buffers in the blood.

Embden and his associates[157] used chiefly sodium phosphate in the form of Recresal, and reported a 20 per cent increase in working capacity in ergometric tests. Sham feeding was used to eliminate any psychological effect.

Phosphates also appeared to be beneficial to soldiers and coal miners. In addition to a greater work capacity, the subjects experience a sense of well-being and elation. These results can be obtained with daily doses of 3 gm.; larger doses may cause insomnia and other disturbances.

Flinn[172] repeated Embden's experiments on industrial workers. Although he found no increase in work output, he noticed that many felt better, especially those who, before the experiment, had been subject to chronic constipation. Since phosphates are mild laxatives, they help to produce "regularity" and a feeling of well-being.

Puni[418] confirmed most of the work of Embden, and found that a small dose of phosphate taken one to three hours before a psychomotor or motor test increased the endurance of the subject. Atzler,[32] using the phosphate preparation Sanatogen, also noticed an improvement in muscular work. Loewy and Eysern[347] experimented with Evianis, a drink containing phosphorus, and noticed a quicker recovery after exercise and sometimes less fatigue.

Observations made by Riabuschinsky[428] on the effect of ingestion of phosphates on the capacity of normal young adults for work showed that the amount of work accomplished was considerably increased by taking sodium phosphate in amounts sufficient to approximately double the daily phosphorus intake. Freeman,[179] on perfusing a frog's heart with solutions of phosphate, or glucose and phosphate, obtained well-defined increases in muscular efficiency.

In addition to the negative findings similar to those reported by Flinn, Schorn[465] also noticed no effect of phosphates on either muscular or mental work, and explained the results of Poppelreuter on the basis of autosuggestion. Marbe[360] also pointed out the possibility of a psychological effect, for in his experiments a drop of Congo red in distilled water had the same effect as 3 to 5 gm. of Recresal. Krestovnikoff and his co-workers[331] could not discover any definite effect of phosphates after a three-month intake, except a questionable shortening of the recovery period.

The preponderance of evidence seems to favor the idea that phosphate preparations are beneficial. Yet there is no definite proof that the administration of phosphates per se was responsible for improved performance in sports and athletics.

SODIUM CHLORIDE

It has been known for a long time that profuse sweating, causing a large loss of sodium chloride, may lead to muscular cramps. Sweat contains from 0.05 to 0.5 per cent sodium chloride, and it is possible to lose from 3 to 30 gm. of salt per day. Since the average intake with food is 10 to 20 gm. daily, it is clear that excessive sweating may cause a serious depletion of body sodium chloride.

Dill[127] has contributed much to our knowledge of the importance of sodium chloride, and has been instrumental in the promotion of the prophylactic use of 0.1 per cent salted water in industries where sweating is profuse. This has resulted in a practical prevention of "heat" cramps. On Dill's initiative, football teams began using well-salted bouillon and salt tablets, with favorable results. Of interest is the experience of a southern football team, whose fall practice cramp epidemic disappeared when more salt was taken by members of the team.

Some coaches erroneously believe that salt tablets can be responsible for extra energy and endurance in athletes in general. Unfortunately this is too much to expect. Addition of salt to the diet is only a precautionary measure to insure a normal output of energy, which may be lowered if there is excessive loss of sodium chloride. An extra amount of salt is advisable for athletes, especially at the beginning of the season, except for those who usually like and take salty food.

SUGAR

Sugar furnishes the fuel for muscular contractions, and this fact has been responsible for its use by athletes. Quantities taken per day vary a great deal; some coaches have attempted to give up to one-half pound of sugar, thereby causing severe gastric disturbances. The usual intake is a few lumps of cane sugar, a few tablets of dextrose or two spoonfuls of honey.

An improvement in the condition of long-distance runners has been observed after sugar administration. Voegler and Ferguson[521] claimed that, when 4.8 gm. of dextrose, equivalent to one cube of sugar, was given for two days, the speed in a 40-yard dash and muscular strength as tested by dynamometer increased. Pampe,[398] however, could not observe any favorable effect of the ingestion of 50 to 100 gm. of sugar upon muscular work. Numerous tests by Karpovich showed no effect of sugar in tests of short duration, such as swimming 100 yards or riding a stationary bicycle for one to three minutes at 0.5 horsepower. The tests were negative for glucose, cane sugar, and honey.

The use of sugar is most indiscriminate, and has no physiological support, except when given during prolonged exercise. The beneficial effect of sugar given before a test of short duration may be ascribed to a psychological influence. Although Voegler and Ferguson attempted to eliminate any psychological effect, it is inconceivable that 4.8 gm. of sugar, which constitutes about 1 per cent of the total daily carbohydrate intake, can be physiologically responsible for an improvement in muscular work. In experiments of this type it may be advisable to eliminate psychological factors affecting the investigators themselves. The tested substance and the substitute should be given in such a manner that the person who actually tests the subjects does not know what is being administered in each case.

TOBACCO SMOKING AND ATHLETIC PERFORMANCE

As a rule coaches are against smoking during training. While there is no general agreement regarding the harmfulness of smoking, *no one as yet has seriously recommended smoking as an adjunct to training.*

In spite of a great deal of interest in the effect of smoking on athletes, our knowledge regarding this topic is still inadequate. The chief reason for this inadequacy is the scarcity of objective experiments. The most frequently quoted reference is a report,[311A] made of students in a military school, stating that non-smokers were faster in cross-country running than smokers.

On the other hand, after a marathon race at Pittsburgh, it was reported that the first five winners were smokers.[120] To this we may add that many athletes who continue smoking during training do it without any apparent detrimental effect on performance.

Three investigators have shown that smoking had no apparent

effect on strength of the hand grip,[308, 428] speed of tapping a telegraph key, the Sargent jump test, the Harvard Step Test score,[428] oxygen intake, oxygen debt, and the net oxygen cost of an exercise performed on a bicycle ergometer.[225]

A priori one would not expect much, if any, effect of smoking on physical activities of the strength type and of short duration. On the other hand, because evidence has been presented that smoking may reduce the oxygen-carrying capacity of the blood, one might suppose that activities requiring endurance would be affected by smoking. The negative reports regarding the Harvard step-up test and bicycle ergometer rides seem to make this supposition incorrect. Such, however, is not the case. One can rightfully question conclusions drawn from a few tests made on a few men. The effect of smoking evidently may not be too well pronounced and not always the same. Moreover, man's physical performance has a range of fluctuation which may mask the effect of tobacco.

Although, in theory, experiments dealing with the effects of smoking upon physical performance would seem to be very simple, in practice they offer great difficulties. One has to find absolutely dependable men: habitual smokers who would completely abstain from smoking and non-smokers who would smoke when required to do so by experimental schedules. Even after a careful selection, several men may fail to adhere to regulations imposed by an experiment and drop out; therefore, at the end, only a small group may be left. For this reason, the number of tests given to each subject should be sufficiently large.

One such study was conducted by Karpovich and Hale.[297] It continued for two years, but only thirteen subjects were tested—eight men who were habitual smokers and five men who were non-smokers. All of the men practiced riding a bicycle ergometer for one to two months, two or three times a week, until they became proficient, as judged by their riding time with prescribed load and number of pedal revolutions. After the training period was over, the experiment itself started. The subjects were asked to complete a prescribed number of pedal revolutions in the shortest possible time, while the load remained the same during each test. (For most subjects it was 425 pedal revolutions with an 8-pound load, an activity similar to a 1-mile run on the track.) Each man served as his own control, smoking during certain periods of the test and abstaining from smoking during other periods. The number of tests made on individual men varied from 24 to 31.

On the strength of this experiment it is possible to draw the following conclusions:

1. Although the average performance of the group was worse after smoking than without smoking, the difference was not statistically significant.

2. Three habitual smokers and two non-smokers did better when they did not smoke. This difference was statistically significant.

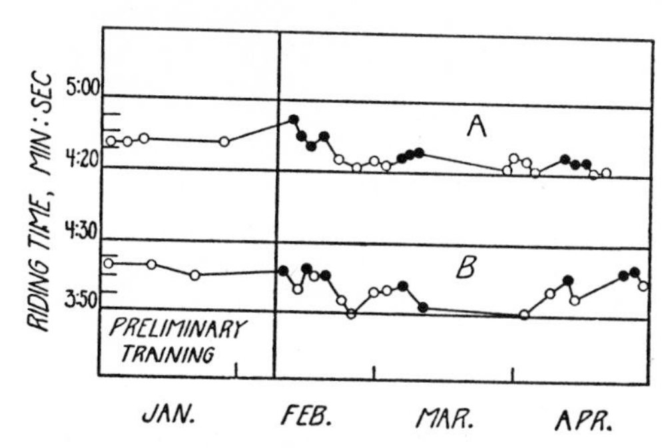

Figure 103. Performance curves of two subjects. *A*, Habitual smoker; *B*, non-smoker; ●, performance after smoking; ○, performance without smoking. Exercise consisted of doing 48,928 foot pounds of work on a bicycle ergometer in the shortest possible time. Therefore, when curve goes down, performance is better. (Karpovich and Hale: J. Applied Physiol. *3*, 1951.)

Thus it appears that not all people are noticeably affected by smoking. Some are tobacco-sensitive and some are not. This study indicated that as high as 37.5 per cent of young men may be tobacco-sensitive, and their speed will be slowed by smoking. True enough, this slowing may not be observed each time, as is shown by Figure 103. One may see that, occasionally, performance with smoking (dark disks) was better than without smoking, but that these occasions were exceptions. As a rule, speed was lower after smoking.

This means that by smoking, a tobacco-sensitive athlete may jeopardize his own and his team's chances for victory. Since the percentage of tobacco-sensitive men is relatively high, a non-smoking rule for athletic teams is a wise precaution which should be firmly supported.

ULTRAVIOLET RAYS

The effect of ultraviolet rays has been tried on runners and on swimmers, on oarsmen and on stationary bicycle riders. All reports indicate a beneficial effect of irradiation upon muscular performance and general well-being. The explanations for this are not clear, and vary from a mere "psychological" effect to an action through the central nervous system leading to a definite increase in vagotonus.

Hettinger[234] reported that, if vitamin D is administered, ultraviolet rays cease to have any effect upon the subjects.

VITAMINS

Most of the known vitamins are indispensable to normal existence, and the disastrous results of their lack are evident in pellagra, rickets, beriberi, scurvy, and other avitaminosic diseases. The rapid improvement from such conditions upon administration of a small amount of the needed vitamin dramatically illustrates how little of the substance is needed and how powerful is its action.

There are also many cases of borderline vitamin deficiencies. Although the symptoms are not so definite as in the prolonged complete absence of vitamins, nevertheless many bodily functions become abnormal and physical fitness is lowered. Administration of needed vitamins is followed by marked improvement. It is possible that improvement in these borderline cases is responsible for the popular notion that excess vitamin intake by normal people will increase their well-being.

Vitamin B Complex. Bickel and Collazzo[55] found an increase in muscle and liver glycogen of rabbits fed on concentrated yeast. Sugar storage may be increased if yeast is given with small quantities of sugar, but if large quantities of sugar are given, no relationship can be observed. Csik and Benesik[116] experimented with vitamin B extract on two subjects, using the ergograph, dynamometer, weight lifting, and treadmill tests. Although an increase in work capacity was noted, the results are nevertheless doubtful, since the effect of training was not completely isolated.

Vitamin B_1. Hard-working men and athletes should have more than 300 International Units of vitamin B_1 daily, although a large number of people in the United States live on a diet poor in this essential.

McCormick[371] claimed that the daily administration of 5 mg. of vitamin B_1 (1665 International Units) increased speed and endurance in swimming. Administration of this daily dose of vitamin B_1 for one week slightly increased the breath-holding capacity and greatly increased endurance in static work (holding the arms steadily outstretched). Karpovich and Millman[301] repeated McCormick's experiments, using vitamin B_1 tablets and also placebos for control, but could not notice any beneficial effect. Of special interest was a subject on a diet definitely poor in vitamin B_1. He was able to hold his arms outstretched for four hours and twenty minutes, without any preliminary vitamin B_1 feeding. It is possible that neither arm holding nor breath holding is an adequate test for the effect of vitamin B_1.

Vitamin B_{12}. Montoye and co-workers[380] experimented on 51 subjects between 12 and 17 years of age. They found no effect on the half-mile running time, or the Harvard step-up test score, from taking vitamin B_{12}.

Vitamin C. When symptoms of scurvy developed among students of a school of physical education in Russia in 1920, Karpovich observed a sharp decline in athletic performance. This is in agreement with the findings of Schroll[466] that a lack of ascorbic acid in the diet of guinea pigs produced a more rapid onset of fatigue and a greater accumulation of lactic acid in the blood.

Dutch investigators reported that saturation of the blood with vitamin C caused an increase in work capacity and mechanical efficiency.[250] They administered 5 gm. of vitamin C a day for five days and found a beneficial effect on the sixth day. According to these investigators, the amount of oxygen needed for the same intensity of work was less with vitamin C than without it. From this, they concluded that work capacity increased under the influence of vitamin C. Reference was made to a book describing ergogenic values of vitamin C, published by Hoitink in 1946. There does not seem to be substantiation of these claims by other laboratories. No other experimental evidence can be found, however, showing that excess of vitamin C in any way affects muscular performance.

It is clear that a lack of vitamin B_1 or vitamin C causes a drop in physical fitness and in athletic performance. Since the possibility of a lack of vitamin B_1 is always present, it would seem to be a good idea to administer it in all doubtful cases. A marked improvement in physical condition may be the result. There is no adequate evidence that excess vitamin supply would increase athletic performance.

Vitamin E. Recently vitamin E, usually in the form of wheat germ oil, has become popular as an ergogenic aid. When experiments failed to substantiate the ergogenic properties of vitamin E, promoters started praising the unknown factors present in the wheat germ oil. As proof of the beneficial action they cite a *fact* that some champion athletes use this *oil*. It is the same argument that is exploited in promoting hair tonics, tooth pastes, and even cigarettes.

Niacin did not show any ergogenic effect on normal people. (For details, see the preceding section on gelatin and glycine.)

CONCLUSIONS

Few of the substances just discussed have an ergogenic action. The most powerful are hormones, caffeine, and cocaine. With the exception of caffeine, when consumed in moderation in tea and coffee, these substances are dangerous and their use should not be encouraged. According to our present knowledge of these drugs, they have no practical application except for therapeutic reasons. Alkalies have been found effective in quiet laboratory experiments, but so far have failed to show their influence upon muscular exertion involving emotional stress.

Many substances are helpful only when they are used to replenish a previous lack, or as a precaution for a possible depletion during work. To this group belong sugars, sodium chloride, and vitamins.

Oxygen is helpful during work at high altitude. If inhaled before a sprint, it enables one to hold the breath longer and to move faster.

Some substances should be considered as questionable aids for normal people. Among them is Benzedrine. Others, such as gelatin and glycine, definitely have no ergogenic effect in normal persons.

Ultraviolet light exerts a beneficial effect upon a muscular performance, but further studies are needed to determine "why" and "how" this is brought about.

QUESTIONS

1. What does the word "ergogenic" mean?
2. How can you define a drug?
3. Is the use of ergogenic aids in sports ethical or not?
4. When does it become definitely unethical?
5. What will be your judgment regarding the ergogenic properties of a substance X in the following instances:
   (a) You watched an athlete who, in your presence, took substance X and won a contest or broke a record.
   (b) You have been told about an athlete using X and winning a contest or breaking a record.
   (c) You know of an athlete who always takes X and is always victorious.
6. Discuss alcohol. Incidentally, is it a stimulant or a depressant?
7. Is there any evidence that alkalies may have an ergogenic action?
8. Discuss caffeine.
9. Why do some South American Indians chew coca leaves?
10. Is gelatin an ergogenic aid? Does it help in weight lifting?
11. Is oxygen inhalation before or during athletic performance beneficial? Why do some football teams use it?
12. Discuss sodium chloride. (Incidentally, is it wise to take salt on an empty stomach?)
13. When may sugar or sugar products be beneficial for an athletic contest?
14. Should athletes smoke? If not, why not? Does smoking impair physical performance?
15. Is there any vitamin that definitely improves the physical performance of normal athletes? Is there any evidence of an ergogenic action of vitamins A, B, or E?
16. Is there any proof that Benzedrine has been responsible for setting athletic records? Does Benzedrine produce an improvement in speed or in endurance in running and swimming?
17. Does niacin have an ergogenic effect on normal people?

BIBLIOGRAPHY

1. Adams, F. H., Bengtsson, E., Bervern, H., and Wegelius, C.: The Physical Working Capacity of Normal School Children. II. Swedish City and Country. Pediatrics *28*:243, 1961.
2. Adams, F. H., Linde, F. M., and Miyoka, H.: The Physical Working Capacity of Normal School Children. Pediatrics *28*:55, 1961.
3. Adrian, M., Singh, M., and Karpovich. P. V.: Energy Cost of Leg Kick, Arm Stroke, and Whole Crawl Stroke. J. Appl. Physiol. *21*:1763, 1966.
4. Agostino, E., and Fenn, W. O.: Velocity of Muscle Shortening as a Limiting Factor in Respiratory Air Flow. J. Appl. Physiol. *15*:349, 1960.
5. Alam, M., and Smirk, F. H.: Observations in Man on Pulse Accelerating Reflex from the Voluntary Muscles of the Legs. J. Physiol. *92*:167, 1938.
6. Alteveer, R.: A Natographic Study of Some Swimming Strokes. Unpublished Master's Thesis, Springfield College, Springfield, Massachusetts, 1958.
7. Alvarez, W. C., and Stanley, L. L.: Blood Pressure in Six Thousand Prisoners and Four Hundred Prison Guards. Arch. Intern. Med. *46*:17, 1930.
8. Anderson, E. C., and Langham, W. H.: Average Potassium Concentration of the Human Body as a Function of Age. Science *130*:713, 1959.
9. Andersen, K. L., Hellstrom, B., and Lorentzen, F. V.: Combined Effect of Cold and Alcohol on Heat Balance in Man. J. Appl. Physiol. *18*:975, 1963.
10. Asa, M.: Effect of Isotonic and Isometric Exercises Upon the Strength of Muscle. Unpublished Doctoral Dissertation, Springfield College, Springfield, Massachusetts, 1958.
11. Asmussen, E.: Muscular Performance. *In* Muscle as a Tissue, Edited by Rodahl, K., and Horvath, S. M. New York, McGraw-Hill Book Company, 1962.
12. Asmussen, E.: Observations on Experimental Muscular Soreness. Acta Rheum. Scandinav. *2*:109, 1956.
13. Asmussen, E., and Bøje, O.: Body Temperature and Capacity for Work. Acta Physiol. Scandinav. *10*:1, 1945.
14. Asmussen, E., Christensen, E. H., and Nielsen, M.: Pulsfrequenz und Korperstellung. Skand. Arch. f. Physiol. *81*:190, 1939.
15. Asmussen, E., and Heebøll-Nielsen, K.: Phys cal Performance and Growth in Children. Influence of Sex, Age and Intelligence. J. Appl. Physiol. *8*:371, 1956.
16. Asmussen, E., and Nielsen, M.: The Cardiac Output in Rest and Work at Low and High Oxygen Pressures. Acta Physiol. Scandinav. *35*:73, 1955.
17. Asmussen, E., and Nielsen, M.: Experiments on Nervous Factors Controlling Respiration and Circulation during Exercise Employing Blocking of the Blood Flow. Acta Physiol. Scandinav. *60*:103, 1964.
18. Asmussen, E., Nielsen, M., and Wieth-Pedersen, G.: Cortical or Reflex Control of Respiration during Muscular Work? Acta Physiol. Scandinav. *6*:168, 1943.
19. Asmussen, E., Nielsen, M., and Wieth-Pedersen, G.: On the Regulation of Circulation during Muscular Work. Acta Physiol. Scandinav. *6*:353, 1943.
20. Asprey, G. M., Alley, L. E., and Tuttle, W. W.: Effect of Eating at Various Times

on Subsequent Performances in the One-Mile Run. Res. Quart. Amer. Ass. Health Phys. Educ. *35*:227, 1964.
21. Asprey, G. M., Alley, L. E., and Tuttle, W. W.: Effect of Eating at Various Times on Subsequent Performances in the Two-Mile Run. Res. Quart. Amer. Ass. Health Phys. Educ. *36*:233, 1965.
22. Asprey, G. M., Alley, L. E., and Tuttle, W. W.: Effect of Eating at Various Times on Subsequent Performances in the 440-Yard Dash and Half-Mile Run. Res. Quart. Amer. Ass. Health Phys. Educ. *34*:267, 1963.
23. Åstrand, P.-O.: Ergometry—Test of "Physical Fitness." Varborg (Sweden), A. B. Cykelfabriken Monark.
24. Åstrand, P.-O.: The Respiratory Activity in Man Exposed to Prolonged Hypoxia. Acta Physiol. Scandinav. *30*:343, 1954.
25. Åstrand, P.-O., and Åstrand, I.: Heart Rate during Muscular Work in Man Exposed to Prolonged Hypoxia. J. Appl. Physiol. *13*:75, 1958.
26. Åstrand, I., Åstrand, P.-O., Christensen, E. H., and Hedman, R.: Intermittent Muscular Work. Acta Physiol. Scandinav. *48*:488, 1960.
27. Åstrand, I., Åstrand, P.-O., Christensen, E. H., and Hedman, R.: Myohemoglobin as an Oxygen-Store in Man. Acta Physiol. Scandinav. *48*:454, 1960.
28. Åstrand, P.-O., Cuddy, T. E., Saltin, B., and Stenberg, J.: Cardiac Output during Submaximal and Maximal Work. J. Appl. Physiol. *19*:268, 1964.
29. Åstrand, P.-O., Ekblom, B., Messin, R., Saltin, B., and Stenberg, J.: Intra-arterial Blood Pressure during Exercise with Different Muscle Groups. J. Appl. Physiol. *20*:253, 1965.
30. Åstrand, P.-O., and Ryhming, S.: A Nomogram for Calculation of Aerobic Capacity (Physical Fitness) from Pulse Rate during Submaximal Work. J. Appl. Physiol. *7*:218, 1954.
31. Atwell, W. O., and Elbel, E. R.: Reaction Time of Male High School Students in 14-17 Year Age Group. Res. Quart. Amer. Ass. Health Phys. Educ. *19*:22, 1948.
32. Atzler, E., and Lehmann, G.: Die Wirkung von Lecithin auf Arbeitsstoffwechsel und Leistungsfähigkeit. Arbeitsphysiol. *9*:76, 1935.
33. Atzler, E., and Meyer, F.: Schwerarbeit des Alkoholgewohnten unter dem Einfluss des Alkohols. Arbeitsphysiol. *4*:410, 1931.
34. Bailie, M. D., Robinson, S., Rostorfer, H. H., and Newton, J. L.: Effects of Exercise on Heart Output of the Dog. J. Appl. Physiol. *16*:107, 1961.
35. Bainbridge, F. A.: The Physiology of Muscular Exercise, 3rd ed. Rewritten by Bock, A. V., and Dill, B. D. London, Longmans Green and Company, 1931.
36. Balke, B.: Correlation of Static and Physical Endurance. Report No. 1, USAF School of Aviation Medicine. Randolph Field, Texas, April, 1952.
37. Balke, B., Daniels, J. T., and Faulkner, J. A.: Training for Maximum Performance at Altitude. *In* Exercise at Altitude. Edited by Margaria, R. Amsterdam, Excerpta Medica Foundation, 1967.
38. Balke, B., Grillo, G., Korecci, E., and Luft, U.: Work Capacity after Blood Donation. J. Appl. Physiol. *7*:231, 1954.
39. Balke, B., Nagle, F. J., and Daniels, J.: Altitude and Maximum Performance in Work and Sports Activity, J.A.M.A. *194*:176, 1965.
40. Banister, R. C., and Cunningham, D. J. C.: The Effect on the Respiration and Performance during Exercise of Adding Oxygen to the Inspired Air. J. Physiol. *125*:118, 1954.
41. Barcroft, J., and Margaria, R.: Some Effects of Carbonic Acid on the Character of Human Respiration. J. Physiol. *72*:174, 1931.
42. Barcroft, J., and Stephens, J. G.: Observations upon the Size of the Spleen. J. Physiol. *64*:1, 1927.
43. Barringer, T. R.: Studies of the Heart's Functional Capacity. Arch. Intern. Med. *20*:829, 1917.
44. Basmajian, J. V.: Muscles Alive, Their Functions Revealed by Electromyography, 2nd ed. Baltimore, Williams and Wilkins Company, 1967.
45. Baur, M.: Pharmakologische Beeinflussung der Korperleistung im Sport. Arch. f. Exper. Path. u. Pharmacol. *184*:51, 1936.
46. Bazett, H. C.: Physiological Responses to Heat. Physiol. Rev. *7*:531, 1927.
47. Becklake, M. R., Varvis, C. J., Pengelly, L. D., Kenning, S., McGregor, M., and

Bates, D. K.: Measurement of Pulmonary Blood Flow during Exercise Using Nitrous Oxide. J. Appl. Physiol. *17*:579, 1962.

48. Behnke, A. R.: Quantitative Assessment of Body Build. J. Appl. Physiol. *16*:960, 1961.
49. Behnke, A. R., and Royce, J.: Body Size, Shape, and Composition of Several Types of Athlete. J. Sports Med. *6*:75, 1966.
50. Bell, G. H., Davidson, J. N., and Scarborough, H.: Textbook of Physiology and Biochemistry, 6th ed. Baltimore, Williams and Wilkins Company, 1968.
51. Berger, R.: Effect of Varied Weight Training Programs on Strength. Res. Quart. Amer. Ass. Health Phys. Educ. *33*:168, 1962.
52. Bergström, J., Hermansen, L., Hultman, E., and Saltin, B.: Diet, Muscle Glycogen and Physical Performance. Acta Physiol. Scandinav. *17*:140, 1967.
53. Best, C. H., and Partridge, R. C.: Observations on Olympic Athletes. Proc. Roy. Soc. [Biol.] *105*:323, 1930.
54. Bevegård, S., Holmgren, A., and Jonsson, B.: Circulatory Studies in Well-trained Athletes at Rest and during Heavy Exercise, with Special Reference to Stroke Volume and the Influence of Body Position. Acta Physiol. Scandinav. *57*:26, 1963.
55. Bickel, A., and Collazzo, J. A.: Wirkungen eines Hefekonzentrationsproduktes nach parenteraler und enteraler Gabe auf dem Kohlenhydratstoffwechsel. Biochem. Ztschr. *221*:195, 1930.
56. Bigland, B., and Lippold, O. C. J.: The Relation between Force, Velocity and Integrated Electrical Activity in Human Muscles. J. Physiol. *123*:214, 1954.
57. Black, W. A., and Karpovich, P. V.: Effect of Exercise upon the Erythrocyte Sedimentation Rate. Amer. J. Physiol. *144*:224, 1945.
58. Blyth, C. S., and Burt, J. J.: Effect of Water Balance on Ability to Perform in High Ambient Temperature. Res. Quart. Amer. Ass. Health Phys. Educ. *32*:301, 1961.
59. Bohm, W. H. S.: Opinions of Experienced Coaches and Athletes in Training Track and Field Athletes. Unpublished Master's Thesis, Springfield College, Springfield, Massachusetts, 1938.
60. Böhmer, D.: Enzymatic Activity in Normal, Trained and Inactivated Muscle. *In* Biochemistry of Exercise, Medicine and Sport, Volume 3. Baltimore, Maryland and Manchester, England, University Park Press, 1968.
61. Bouisset, S., Goubel, F., and Lestienne, F.: Tension-length Curve at Variable Velocity in Normal Human Muscle. J. Physiol. *197*:46P, 1968.
62. Bøje, O.: Doping: Use of Various Stimulating Drugs for Improvement of Performance in Sports. Nord. Med. Tidskr. *2*:1963, 1939.
63. Bøje, O.: Doping. Bull. Health Organ. League of Nations *8*:439, 1939.
64. Booyens, J., and Keatinge, W. R.: Energy Expenditure during Walking. J. Physiol. *138*:165, 1957.
65. Bowen, W. P.: Changes in Heart-Rate, Blood Pressure, and Duration of Systole Resulting from Bicycling. Amer. J. Physiol. *11*:59, 1904.
66. Bowles, C. J., and Sigerseth, P. O.: Telemetered Heart Rate Responses to Pace Patterns in the One-Mile Run. Res. Quart. Amer. Ass. Health Phys. Educ. *39*:36, 1968.
67. Bramwell, C., and Ellis, R.: Clinical Observations on Olympic Athletes. Arbeitsphysiol. *2*:51, 1929.
68. Brezina, E., and Kolmer, W.: Ueber den Energieverbrauch der Geharbeit unter dem Einfluss verschiedener Geschwindigkeiten und verschiedener Belastungen. Biochem. Ztschr., Berlin *38*:129, 1912.
69. Brouha, L.: Physiology in Industry. New York, Pergamon Press, 1960.
70. Brouha, L., Heath, C. W., and Graybiel, A.: Step Test: Simple Method of Measuring Physical Fitness for Hard Muscular Work in Adult Man. Rev. Can. Biol. *2*:86, 1943.
71. Brouha, L., Smith, P. E., Jr., De Lanne, R., and Maxfield, M. E.: Physiological Reactions of Men and Women during Muscular Activity and Recovery in Various Environments. J. Appl. Physiol. *16*:133, 1961.
72. Broun, G. O.: Blood Destruction during Exercise. I. Blood Changes Occurring in the Course of a Single Day of Exercise. II. Demonstration of Blood Destruction

in Animals Exercised after Prolonged Confinement. III. Exercise as a Bone Marrow Stimulus. IV. The Development of Equilibrium between Blood Destruction and Regeneration after a Period of Training. J. Exp. Med. *36*:481; *37*:113; *37*:187; *37*:207, 1922-23.

73. Brožek, J., and Henschel, A. (editors): Techniques for Measuring Body Composition. Proceedings of a Conference. Washington, D.C., National Academy of Sciences—National Research Council, 1961.
74. Brožek, J., and Keys, A.: Relative Body Weight, Age and Fatness. Geriatrics *8*:70, 1953.
75. Brožek, J. et al.: Densitometric Analysis of Body Composition: Revision of Some Quantitative Assumptions. Ann. N.Y. Acad. Sci. *110*:131, 1963.
76. Bruusgaard, C.: The Effects of Physical Exertion on the Blood Sugar Level. Norsk Mag. f. Laegevidensk. *90*:778, 1929.
77. Burch, G. E., and Sodeman, W. A.: Effect of Cooling Isolated Parts upon Comfort of Man Resting in Hot Humid Environment. Proc. Soc. Exp. Biol. Med. *55*:190, 1944.
78. Cabot, R. C.: Facts on the Heart. Philadelphia, W. B. Saunders Company, 1926.
79. Campbell, J. A., and Angus, T. C.: Some Physiologic Reactions to Cooling Power during Work, with Special Reference to Evaporation of Water. J. Indust. Hyg. *11*:315, 1929.
80. Campbell, J. M. H., Mitchell, G. O., and Powell, A. T.: Influence of Exercise on Digestion. Guy's Hosp. Rep., London *78*:279, 1928.
81. Campos, F. A. deM., Cannon, W. B., Lundin, H., and Walker, T. T.: Some Conditions Affecting the Capacity for Prolonged Muscular Work. Amer. J. Physiol. *87*:680, 1928.
82. Canzanelli, A., Guild, R., and Rapport, D.: Use of Ethyl Alcohol as Fuel in Muscular Exercise. Amer. J. Physiol. *110*:416, 1934.
83. Carlile, F.: Effect of Preliminary Passive Warming on Swimming Performance. Res. Quart. Amer. Ass. Health Phys. Educ. *27*:143, 1956.
84. Carlson, L. A., and Pernow, B.: Studies of Blood Lipids during Exercise. J. Lab. Clin. Med. *53*:33, 1959.
85. Carlsten, A., and Grimby, G.: The Circulatory Response to Muscular Exercise in Man. Springfield, Illinois, Charles C Thomas, 1966.
86. Carpenter, T., Burdett, M., and Lee, R.: The Effect of Muscular Exercise on the Disappearance of Ethyl Alcohol in Man. Amer. J. Physiol. *105*:17, 1933.
87. Carter, J. E. L.: Somatotypes of College Football Players. Res. Quart. Amer. Ass. Health Phys. Educ. *39*:476, 1968.
88. Carter, J. E. L., Ross, W. D., Kasch, F. W., and Phillips, W. H.: Body Types of Middle-aged Males in Training. J. Ass. Phys. Ment. Rehab. *19*:148, 1965.
89. Cathcart, E. P., Richardson, D. T., and Campbell, W.: On the Maximum Load to be Carried by the Soldier. Army Hygiene Advisory Committee, Report No. 3. J. Roy. Army M. Corps *40*:435; *41*:12, 87 and 161, 1923.
90. Cavagna, G. A., Saibene, F. P., and Margaria, R.: Effect of Negative Work on the Amount of Positive Work Performed by an Isolated Muscle. J. Appl. Physiol. *20*:157, 1965.
91. Cavagna, G. A., Saibene, F. P., and Margaria, R.: Mechanical Work in Running. J. Appl. Physiol. *19*:249, 1964.
92. Cerretelli, P.: Lactacid O_2 Debt in Acute and Chronic Hypoxia. *In* Exercise at Altitude. Edited by Margaria, R. Amsterdam, Excerpta Medica Foundation, 1967.
93. Cerretelli, P., and Margaria, R.: Maximum Oxygen Consumption at Altitude. Int. Z. Angew. Physiol. *18*:460, 1961.
94. Christensen, E. H., Hedman, R., and Holmdahl, I.: The Influence of Rest Pauses on Mechanical Efficiency. Acta Physiol. Scandinav. *48*:443, 1960.
95. Christensen, E. H., and Högberg, P.: Physiology of Skiing. Arbeitsphysiol. *14*:292, 1950.
96. Christensen, E. H., and Högberg, P.: Steady-State, O_2-Deficit and O_2-Debt at Severe Work. Arbeitsphysiol. *14*:251, 1950.
97. Chui, E.: The Effect of Systematic Weight Training on Athletic Power. Res. Quart. Amer. Ass. Health Phys. Educ. *21*:188, 1950.

98. Chusid, J. G., and McDonald, J. J.: Correlative Neuroanatomy and Functional Neurology, 12th ed. Los Altos, California, Lange Medical Publications, 1964.
99. Clarke, H. H., and Broms, J.: Differences in Maturity, Physical and Motor Traits of High, Average, and Low Gross and Relative Strength. J. Sports Med. *8*:143, 1968.
100. Clarke, H. L.: A Functional Physical Fitness Test for College Women. J. Health and Phys. Educ. *14*:358, 1943.
101. Clark-Kennedy, A. E., and Owen, T.: The Effect of High and Low Oxygen Pressures on the Respiratory Exchange during Exercise. J. Physiol. *62*:24, 1926.
102. Cogswell, R. C., Henderson, C. R., and Berryman, G. H.: Some Observations of the Effects of Training on Pulse Rate, Blood Pressure and Endurance in Humans, Using the Step Test (Harvard), Treadmill and Electrodynamic Brake Bicycle Ergometer. Amer. J. Physiol. *146*:422, 1946.
103. Comroe, J. H., and Schmidt, C. F.: Reflexes from the Limbs as a Factor in the Hyperpnea of Muscular Exercise. Amer. J. Physiol. *138*:536, 1943.
104. Costill, D. L.: Metabolic Responses during Distance Running. J. Appl. Physiol. *28*:251, 1970.
105. Costill, D. L., and Fox, E. L.: Energetics of Marathon Running. Medicine and Science in Sports *1*:81, 1969.
106. Cotton, F. S., and Dill, D. B.: On the Relation between the Heart-Rate during Exercise and That of Immediate Post-Exercise Period. Amer. J. Physiol. *111*:554, 1935.
107. Cotton, T. F., Lewis, T., and Rapport, D. L.: After-Effects of Exercise on Pulse-Rate and Systolic Blood-Pressure in Cases of "Irritable Heart." Heart *6*:269, 1917.
108. Courtice, F. C., and Douglas, C. G.: The Effect of Prolonged Muscular Exercise on the Metabolism. Proc. Roy. Soc. [Biol.] *119*:381, 1935-36.
109. Craig, A. B., Jr.: Causes of Loss of Consciousness during Underwater Swimming. J. Appl. Physiol. *16*:583, 1961.
109A. Craig, A. B., Jr.: Evaluation and Prediction of World Running and Swimming Records. J. Sports Med. *3*:14, 1963.
110. Craig, A. B., Jr.: Heart Rate Responses to Apneic Underwater Diving and to Breath Holding in Man. J. Appl. Physiol. *18*:854, 1963.
111. Craig, A. B., Jr.: Olympics 1968: A Post-Mortem. Medicine and Science in Sports *1*:177, 1969.
112. Craig, F. N., Cummings, E. G., and Blevins, W. V.: Regulation of Breathing at Beginning of Exercise. J. Appl. Physiol. *18*:1183, 1963.
113. Crescitelli, F., and Taylor, C.: The Lactate Response to Exercise and Its Relationship to Physical Fitness. Amer. J. Physiol. *141*:630, 1944.
114. Cruchet, R., and Moulinier, R.: Air Sickness. New York, William Wood and Company, 1920.
115. Cruickshank, E. W. H.: On the Output of Hemoglobin and Blood by the Spleen. J. Physiol. *61*:455, 1926.
116. Csik, L., and Bencsik, J.: Versuche, die Wirkung von B-Vitamin auf die Arbeitsleistung des Menschen Festzustellen. Klin. Wchnschr. *6*:2275, 1927.
117. Cumming, G. R., and Danzinger, R.: Bicycle Ergometer Studies in Children; Correlation of Pulse Rate with Oxygen Consumption. Pediatrics *32*:202, 1963.
118. Daniels, F., Jr., Vanderbie, J. H., and Bommarito, C. L.: Energy Cost of Load Carrying on a Treadmill. Fed. Proc. *11*:30, 1952.
119. Davies, C. T. M.: Limitations to the Prediction of Maximum Oxygen Intake from Cardiac Frequency Measurements. J. Appl. Physiol. *24*:700, 1968.
120. Dawson, P. M.: The Physiology of Physical Education. Baltimore, Williams and Wilkins Company, 1935.
121. Dennig, H.: Über Steigerung der körperlichen Leistungsfähigkeit durch Eingriffe in den Säurebasenhaushalt. Deutsche med. Wchnschr. *63*:733, 1937.
122. Dennig, H., Talbot, J. H., Edwards, H. T., and Dill, D. B.: Effect of Acidosis and Alkalosis upon Capacity for Work. J. Clin. Invest. *9*:601, 1931.
123. Dill, D. B.: Assessment of Work Performance. J. Sports Med. *6*:3, 1966.
124. Dill, D. B., Seed, J. C., and Marzulli, F. N.: Energy Expenditure in Bicycling. J. Appl. Physiol. 7:320, 1954.

125. Dill, D. B.: The Economy of Muscular Exercise. Physiol. Rev. *16*:263, 1936.
126. Dill, D. B.: Life, Heat and Altitude. Cambridge, Harvard University Press, 1938.
127. Dill, D. B., Bock, A. V., Edwards, H. T., and Kennedy, P. H.: Industrial Fatigue. J. Indust. Hyg. Toxicol. *18*:417, 1936.
128. Dill, D. B., and Consolazio, C. F.: Responses to Exercise as Related to Age and Environmental Temperature. J. Appl. Physiol. *17*:645, 1962.
129. Dill, D. B., Edwards, H. T., Bauer, P. S., and Levenson, E. J.: Physical Performance in Relation to External Temperature. Arbeitsphysiol. *4*:508, 1931.
130. Dill, D. B., Edwards, H. T., and de Meio, R. H.: Effects of Adrenalin Injection in Moderate Work. Amer. J. Physiol. *111*:9, 1935.
131. Dill, D. B., Edwards, H. T., and Talbott, J. H.: Alkalosis and the Capacity for Work. J. Biol. Chem. *97*:lviii, 1932.
132. Dill, D. B. Edwards, H. T., and Talbott, J. H.: Studies in Muscular Activity. J. Physiol. *69*:267, 1930.
133. Dill, D. B., Edwards, H. T., and Talbott, J. H.: Studies in Muscular Activity. VII. Factors Limiting the Capacity for Work. J. Physiol. 77:49, 1932.
134. Dill, D. B., Robinson, S., Balke, B., and Newton, J. L.: Work Tolerance: Age and Altitude. J. Appl. Physiol. *19*:483, 1964.
135. Dill, D. B., and Sactor, B.: Exercise and the Oxygen Debt. J. Sports Med. *2*:66, 1962.
136. Dixon, M. E., Stewart, P. B., Mills, F. C., Varvis, C. J., and Bates, D. V.: Respiratory Consequences of Passive Body Movement. J. Appl. Physiol. *16*:30, 1961.
137. Doss, W. S., and Karpovich, P. V.: A Comparison of Concentric, Eccentric, and Isometric Strength of Elbow Flexors. J. Appl. Physiol. *20*:351, 1965.
138. Douglas, C. G., and Haldane, J. S.: The Capacity of the Air Passages under Varying Physiological Conditions. J. Physiol. *45*:235, 1912.
139. Draper, J., Edwards, R., and Hardy, R.: Method of Estimating the Respiratory Cost of a Task by Use of Minute-Volume Determinations. J. Appl. Physiol. *6*:297, 1953.
140. Dreyer, G.: The Assessment of Physical Fitness. New York, Paul B. Hoeber, 1920.
141. Droese, W., Kofranyi, E., Kraut, H., and Wildemann, L.: Energetische Untersuchung der Hausfrauen Arbeit. Arbeitsphysiol. *14*:63, 1949.
142. Dublin, L. I.: Longevity of College Athletes. Harper's Magazine *157*:22, 1928.
143. Dublin, L. I.: Death Rate of College-Bred Men. New York Times Special Feature Section, July 20, 1930.
144. Duffner, G.: Medical Problems Involved in Underwater Compression and Decompression. CIBA Clinical Symposia *10*:99, 1958.
145. Düntzer, E.: Leibesubungen und Menstruation. Ztblt. Gyn. *54*:29, 1930.
146. Durand, J., Pannier, Cl., DeLattre, J., Martineaud, J. P., and Verpillat, J. M.: The Cost of the Oxygen Debt at High Altitude. *In* Exercise at Altitude. Edited by Margaria, R. Amsterdam, Excerpta Medica Foundation, 1967.
147. Eagle, E., Britton, S. W., and Kline, R.: Influence of Cortico-Adrenal Extraction on Energy Output. Amer. J. Physiol. *102*:707, 1932.
148. Eason, R. J.: Electromyographic Study of Local and General Muscle Impairment. J. Appl. Physiol. *15*:479, 1960.
149. Eccles, J.: The Synapse. Sci. Amer. *212*:56, 1965.
150. Edwards, H. T., Richards, T. K., and Dill, D. B.: Blood Sugar, Urine Sugar, and Urine Protein in Exercise. Amer. J. Physiol. *98*:352, 1931.
151. Edwards, H. T., Thorndike, A., and Dill, D. B.: The Energy Requirement in Strenuous Exercise. New Eng. J. Med. *213*:532, 1935.
152. Edwards, H. T., and Woods, W. B.: A Study of Leukocytosis in Exercise. Arbeitsphysiol. *6*:73, 1932.
153. Egoroff, A.: Die Myogene Leukocytose. Ztschr. f. Klin. Med. *100*:485, 1924.
154. Eimer, K.: Exercise and the Heart. Deutsche Med. Wchnschr. *54*:174, 1928.
155. Ekblom, B., Åstrand, P.-O., Saltin, B., Stenberg, J., and Wallstrom, B.: Effect of Training on Circulatory Response to Exercise. J. Appl. Physiol. *24*:518, 1968.
156. Elbel, E. R., and Green, E. L.: Pulse Reaction to Performing Step-Up Exercise on Benches of Different Heights. Amer. J. Physiol. *145*:521, 1946.
157. Embden, G., Grafe, E., and Schmitz, E.: Increase of Working Capacity through Administration of Phosphate. Ztschr. f. Physiol. Chem. *113*:67, 1921.

158. Erickson, L., Simonson, E., Taylor, H. L., Alexander, H., and Keys, A.: The Energy Cost of Horizontal and Grade Walking on the Motor-Driven Treadmill. Amer. J. Physiol. *145*:391, 1946.
159. Evdokimova, M. M.: Painful Liver Syndrome in Athletes. *In* Problems of Medical Control. Edited by Letounov, S. Fizkultura i Sport, Moscow, 1960.
160. Eysenck, H. J.: Experimental Study of Improvement of Mental and Physical Functions in Hypnotic State. Brit. J. Med. Psychol. *18*:304, 1941.
161. Eyster, J. A. E., and Hicks, E. V.: The Effect of Respiration on Cardiac Output. Amer. J. Physiol. *101*:33, 1932.
162. Faulkner, J. A., Daniels, J. T., and Balke, B.: Effects of Training at Moderate Altitude on Physical Performance Capacity. J. Appl. Physiol. *23*:85, 1967.
163. Faulkner, J. A., Kollias, J., Favour, C. B., Buskirk, E. B., and Balke, B.: Maximum Aerobic Capacity and Running Performance at Altitude. J. Appl. Physiol. *24*:685, 1968.
164. Feldman, I., and Hill, L.: The Influence of Oxygen Inhalation on the Lactic Acid Produced during Hard Work. J. Physiol. *142*:439, 1911.
165. Fenn, W. O.: Work against Gravity and Work due to Velocity Changes in Running: Movement of the Center of Gravity within the Body and Foot Pressure on the Ground. Amer. J. Physiol. *93*:433, 1930.
166. Fenn, W. O., Brody, H., and Petrilli, A.: The Tension Developed by Human Muscles at Different Velocities of Shortening. Amer. J. Physiol. *97*:1, 1931.
167. Ferguson, R. J., Faulkner, J. A., Julius, S., and Conway, J.: Comparison of Cardiac Output Determined by CO_2 Rebreathing and Dye-dilution Methods. J. Appl. Physiol. *25*:450, 1968.
168. Ferguson, R. J., Marcotte, G. G., and Montpetit, R. R.: A Maximal Oxygen Uptake Test during Ice Skating. Medicine and Science in Sports *1*:207, 1969.
169. Fiodorov, V. L.: A Study of Muscle Relaxation in Athletes. *In* Problemy Fiziologii Sporta. Edited by Gippenreiter, B. S. Fizkultura i Sport, Moscow, 1958.
170. Fischer, A., and Merhautova, J.: Electromyographic Manifestations of Individual Stages of Adapted Sports Technique. *In* Health and Fitness in the Modern World. Institute of Normal Human Anatomy. Chicago, Illinois, The Athletic Institute, 1961.
171. Fleigl, J., Knock, A. V., and Koopmann, E.: The Blood of Participators in an Army March. I. The Changes and Excretion of Blood-Pigment. Biochem. Ztschr. *76*:88, 1916.
172. Flinn, F. B.: The So-Called Action of Sodium Phosphate in Delaying Onset of Fatigue. Pub. Health Rep. *41*:1463, 1926.
173. Foltz, E. E., Ivy, A. C., and Barborka, C. J.: The Influence of Amphetamine Sulfate, D-Desoxyephedrine Hydrochloride and Caffeine upon Work Output and Recovery when Rapidly Exhausting Work is Done by Trained Subjects. J. Lab. Clin. Med. *28*:603, 1943.
174. Forbes, A.: The Place of Incidence of Reflex Fatigue. Amer. J. Physiol. *31*:102, 1912.
175. Fowler, W. M., Jr., and Gardner, G.: The Relation of Cardiovascular Tests to Measurements of Motor Performance and Skills. Pediatrics *32*:778, 1963.
176. Fox, I. J., Crowley, W. P., Jr., Grace, J. B., and Wood, E. H.: Effects of the Valsalva Maneuver on Blood Flow in the Thoracic Aorta in Man. J. Appl. Physiol. *21*:1553, 1966.
177. Fox, S. M., and Haskell, W. L.: Physical Activity and the Prevention of Coronary Heart Disease. Bull. N.Y. Acad. Med. *44*:8, 1968.
178. Francis, P. R., and Tipton, C. M.: Influence of a Weight Training Program on Quadriceps Reflex Time. Medicine and Science in Sports *1*:91, 1969.
179. Freeman, N. E.: The Role of Hexose Diphosphate in Muscle Activity. Amer. J. Physiol. *92*:107, 1930.
180. Garrity, H. M.: Relationship of Somatotypes of College Women to Physical Fitness Performance. Res. Quart. Amer. Ass. Health Phys. Educ. *37*:340, 1966.
181. Gasser, H. S., and Meek, W. J.: A Study of the Mechanism by Which Muscular Exercise Produces Acceleration of the Heart. Amer. J. Physiol. *34*:48, 1914.
182. Gertler, M. M.: Ischemic Heart Disease, Heredity and Body Build as Affected by Exercise. Canad. Méd. Ass. J. *96*:728, 1967.

183. Goddard, R. F. (editor): The Effects of Altitude on Physical Performance. Chicago, Illinois, The Athletic Institute, 1967.
184. Goff, L. G., Brubach, H. F., and Specht, H.: Measurements of Respiratory Responses and Work Efficiency of Underwater Swimmers Utilizing Improved Instrumentation. J. Appl. Physiol. *10*:197, 1957.
185. Golding, L. A.: Effect of Physical Training upon Total Serum Cholesterol Levels. Res. Quart. Amer. Ass. Health Phys. Educ. *32*:499, 1961.
186. Golding, L., and Barnard, J. R.: The Effect of D-Amphetamine Sulfate on Physical Performance. J. Sports Med. *3*:221, 1963.
187. Gollnick, P. D., and Karpovich, P. V.: Electrogoniometric Study of Locomotion and of Some Athletic Movements. Res. Quart. Amer. Ass. Health Phys. Educ. *35*(No. 3 Supplement):357, 1964.
188. Gollnick, P. D., and King, D. W.: Effect of Exercise and Training on Mitochondria of Rat Skeletal Muscle. Amer. J. Physiol. *216*:1502, 1969.
189. Gollnick, P. D., and King, D. W.: Energy Release in the Muscle Cell. Medicine and Science in Sports *1*:23, 1969.
190. Gordon, B., Levine, S. A., and Wilmaers, A.: Observations on a Group of Marathon Runners. Arch. Intern. Med. *33*:425, 1924.
191. Gould, G. M., and Pyle, W. L.: Anomalies and Curiosities of Medicine. Philadelphia, W. B. Saunders Company, 1897.
192. Kraf, O.: Zur Frage der spezifischen Wirkung der Cola auf die körperliche Leistungsfähigkeit. Arbeitsphysiol. *2*:474, 1930.
193. Grande, F.: Nutrition and Energy Balance in Body Composition Studies. *In* Techniques for Measuring Body Composition. Edited by Brožek, J., and Henschel, A. Proceedings of a Conference. Washington, D.C., National Academy of sciences – National Research Council, 1961.
194. Granit, R., Holmgren, B., and Merton, P. A.: The Two Routes for Excitation of Muscle and Their Subservience to the Cerebellum. J. Physiol. *130*:213, 1955.
195. Gray, J. S.: The Multiple Factor Theory of the Control of Respiratory Ventilation. Science *103*:739, 1946.
196. Greene, M. M.: The Energy Cost of Track Running and Swimming. Unpublished Master's Thesis, Springfield College, Springfield, Massachusetts, 1930.
197. Grollman, A. N.: The Effect of Variation in Posture on the Output of the Human Heart. Amer. J. Physiol. *86*:285, 1928.
198. Grosse-Lordemann, H., and Müller, E. A.: Der Einfluss der Leistung und der Arbeitsgeschwindigkeit auf das Arbeitsmaximum und Wirkungsgrad beim Radfahren. Arbeitsphysiol. *9*:454, 1936.
199. Gullichsen, R., and Soisalon-Soininen, J. L.: Uber die Kohlenstoffabgabe des Menschen beim Fechten und Ringen. Skand. Arch. f. Physiol. *41*:188, 1921.
200. Guyton, A. C.: Textbook of Medical Physiology, 3rd ed. Philadelphia, W. B. Saunders Company, 1966.
201. Haggard, H. W., and Greenberg, L. A.: Diet and Physical Efficiency. New Haven, Yale University Press, 1935.
202. Haldi, J., and Wynn, W.: The Effect of Low and High Carbohydrate Meals on the Blood Sugar Level and on Work Performance in Strenuous Exercise of Short Duration. Amer. J. Physiol. *145*:402, 1946.
203. Haldi, J., and Wynn, W.: Industrial Efficiency as Affected by Food Intake during Mid-Morning and Mid-Afternoon Rest Periods. J. Appl. Physiol. *2*:268, 1949.
204. Hale, C. J.: The Effect of Preliminary Massage on the 440-Yard Run. Unpublished Master's Thesis, Springfield College, Springfield, Massachusetts, 1949.
205. Hall, D. M., Cain, R. L., and Tipton, C. M.: Keeping Fit: A 23-Year Study of Evaluation of Physical-fitness Tests. Urbana, University of Illinois Cooperative Extension Service, 1965.
206. Hambly, W. D., Pembrey, M. S., and Warner, E. O.: The Physical Fitness of Men Assessed by Various Methods. Guy's Hosp. Rep., London *75*:388, 1925.
207. Hammond, P. H., Merton, P. A., and Sutton, G. G.: Nervous Gradation of Muscular Contraction. Brit. Med. Bull. *12*:214, 1956.
208. Hannisdahl, B.: Der Einfluss von Muskelarbeit auf die Blutsenkung. Arbeitsphysiol. *11*:165, 1940.
209. Hansen, J. E., Stelter, G. P., and Vogel, J. A.: Arterial Pyruvate, Lactate, pH, and

P_{CO_2} during Work at Sea Level and High Altitude. J. Appl. Physiol. *23*:523, 1967.
210. Hansen, J. E., Vogel, J. A., Stelter, G. P., and Consolazio, C. F.: Oxygen Uptake in Man during Exhaustive Work at Sea Level and High Altitude. J. Appl. Physiol. *23*:511, 1967.
211. Harrison, W., Calhoun, J., and Harrison, T.: Afferent Impulses as a Cause of Increased Ventilation during Muscular Exercise. Amer. J. Physiol. *100*:68, 1932.
212. Hartman, F. A., Waite, R. H., and Powell, E. F.: The Relation of the Adrenals to Fatigue. Amer. J. Physiol. *60*:255, 1922.
213. Havard, R. E., and Reay, G. A.: The Influence of Exercise on the Inorganic Phosphates of the Blood and Urine. J. Physiol. *61*:35, 1926.
214. Hawkins, C.: The Effects of Conditioning and Training upon the Differential White-Cell Count. Dissertation, New York University, 1937.
215. Heath, B. H.: Need for Modifications of Somatotype Methodology. Amer. J. Phys. Anthrop. *21*:227, 1963.
215A. Heath, B. H., and Carter, J. E. L.: A Modified Somatotype Method. Amer. J. Phys. Anthrop. *27*:57, 1968.
216. Hellebrandt, F. A.: Cross Education. Ipsilateral and Contralateral Effects of Unimanual Training. J. Appl. Physiol. *4*:136, 1951.
217. Hellebrandt, F. A., Houtz, S. J., and Krikorian, A. M.: Influence of Bimanual Exercise on Unilateral Work Capacity. J. Appl. Physiol. *2*:446, 1950.
218. Hellebrandt, F. A., Houtz, S. J., Partridge, M. J., and Walters, C. E.: Tonic Neck Reflexes in Exercises of Stress in Man. Amer. J. Phys. Med. *35*:144, 1956.
219. Hellebrandt, F. A., Parrish, A. M., and Houtz, S. J.: Cross Education. The Influence of Unilateral Exercise on the Contralateral Limb. Arch. Phys. Med. *28*:76, 1947.
220. Hellebrandt, F. A., Rork, R., and Brogdon, E.: Effect of Gelatin on Power of Women to Perform Maximal Anaerobic Work. Proc. Soc. Exp. Biol. Med. *43*:629, 1940.
221. Hellebrandt, F. A., Schade, M., and Carns, M. L.: Methods of Evoking the Tonic Neck Reflexes in Normal Human Subjects. Amer. J. Phys. Med. *41*:89, 1962.
222. Hellebrandt, F. A., and Waterland, J. C.: Expansion of Motor Patterning under Exercise Stress. Amer. J. Phys. Med. *41*:56, 1962.
223. Henderson, Y., and Haggard, H. W.: The Maximum Power and its Fuel. Amer. J. Physiol. *72*:264, 1925.
223A. Henderson, Y., Haggard, H. W., and Dolley, F. S.: The Efficiency of the Heart and the Significance of Rapid and Slow Pulse Rates. Amer. J. Physiol. *82*:512, 1927.
224. Henry, F. M., and DeMoor, J.: Metabolic Efficiency of Exercise in Relation to Work Load at Constant Speed. J. Appl. Physiol. *2*:481, 1950.
225. Henry, F. M., and Fitzhenry, J. R.: Oxygen Metabolism of Moderate Exercise with Some Observations on the Effects of Tobacco Smoking. J. Appl. Physiol. *2*:464, 1950.
226. Henry, F. M., and Rogers, D. E.: Increased Response Latency for Complicated Movements and a "Memory Drum" Theory of Neuromotor Reaction. Res. Quart. Amer. Ass. Health Phys. Educ. *31*:448, 1960.
227. Henschel, A., Taylor, H. L., and Keys, A.: Performance Capacity in Acute Starvation with Hard Work. J. Appl. Physiol. *6*:624, 1954.
228. Herbst, R., and Schellenberg, P.: Cocain und Muskelarbeit; weitere Untersuchungen ueber die Beeinflussung des Gasstoffwechsels. Arbeitsphysiol. *4*:203, 1931.
229. Hermanson, L.: Anaerobic Energy Release. Medicine and Science in Sports *1*:32, 1969.
230. Hermanson, L., Hultman, E., and Saltin, B.: Muscle Glycogen during Prolonged Severe Exercise. Acta Physiol. Scandinav. *71*:129, 1967.
231. Hermanson, L., and Saltin, B.: Blood Lactate Concentration during Exercise at Acute Exposure to Altitude. *In* Exercise at Altitude. Edited by Margaria, R. Amsterdam, Excerpta Medica Foundation, 1967.
232. Hermanson, L., and Saltin, B.: Oxygen Uptake during Maximal Treadmill and Bicycle Exercise. J. Appl. Physiol. *26*:31, 1969.

233. Herxheimer, H.: Zur Wirkung des Kaffeins auf die Sportliche Leistung. Münch. Med. Wchnschr. *69*:1339, 1922.
234. Hettinger, T. W.: Physiology of Strength. Springfield, Illinois, Charles C Thomas, 1961.
235. Hettinger, T. W., and Müller, E. A.: Muskeltraining. Arbeitsphysiol. *15*:111, 1953.
236. Hewitt, J. E., and Callaway, E. C.: Alkali Reserve of Blood in Relation to Swimming Performance. Res. Quart. Amer. Ass. Health Phys. Educ. 7:83, 1936.
237. Hill, A. V.: The Physiological Basis of Athletic Records. Scient. Monthly *21*:409, 1925.
238. Hill, A. V.: Muscular Movement in Man. New York, McGraw-Hill Book Company, 1927.
239. Hill, A. V.: The Series Elastic Component of Muscle. Proc. Roy. Soc. [Biol.] *137*:273, 1950.
240. Hill, A. V., and Howarth, J. V.: The Reversal of Chemical Reactions in Contracting Muscle during an Applied Stretch. Proc. Roy. Soc. [Biol.] *151*:169, 1959.
241. Hill, A. V., and Lupton, H.: Muscular Exercise, Lactic Acid, and the Supply and Utilization of Oxygen. Quart. J. Med. *16*:135, 1923.
242. Hill, L., and Flack, M.: The Influence of Oxygen on Athletes. J. Physiol. *38*:28, 1909.
243. Hilsendager, D., and Karpovich, P. V.: Ergogenic Effect of Glycine and Niacin Separately and in Combination. Res. Quart. Amer. Ass. Health Phys. Educ. *35*:389, 1964.
244. Himwich, H. E., and Castle, W. B.: Studies in the Metabolism of Muscle. I. The Respiratory Quotient of Resting Muscle. Amer. J. Physiol. *83*:92, 1927.
245. Himwich, H. E., and Rose, M. I.: Studies in the Metabolism of Muscle. Amer. J. Physiol. *88*:663, 1929.
246. Hipple, J.: Warm-Up and Fatigue in Junior High School Sprints. Res. Quart. Amer. Ass. Health Phys. Educ. *26*:246, 1955.
246A. Hodgkins, J.: Influence of Unilateral Endurance Training on Contralateral Limb. J. Appl. Physiol. *16*:991, 1961.
247. Holloszy, J. D.: Biochemical Adaptations in Muscle. J. Biol. Chem. *242*:2278, 1967.
248. Holloszy, J. D., and Oscai, L. B.: Effect of Exercise on α-Glycerophosphate Dehydrogenase Activity in Skeletal Muscle. Arch. Biochem. *130*:653, 1969.
249. Holmgren, A.: Cardiorespiratory Determinants of Cardiovascular Fitness. Canad. Med. Ass. J. *96*:697, 1967.
250. Hoogerwerf, A., and Hoitink, A. W. J. H.: The Influence of Vitamin C Administration on the Mechanical Efficiency of the Human Organism. Int. Z. Angew. Physiol. *20*:164, 1963.
251. Hornbein, T., and Roos, A.: Effect of Polycythemia on Respiration. J. Appl. Physiol. *12*:86, 1958.
252. Hörnicke, E.: Breathing and Physical Efficiency. Münch. Med. Wchnschr. *71*:1569, 1924.
253. Hough, T.: Ergographic Studies in Muscular Soreness. Amer. J. Physiol. 7:76, 1902.
254. Houssay, B. A.: Human Physiology. New York, McGraw-Hill Book Company, 1951.
255. Howell, L., and Coupe, K.: Effect of Blood Loss upon Performance in Balke-Ware Treadmill Test. Res. Quart. Amer. Ass. Health Phys. Educ. *35*:156, 1964.
256. Hull, C. L.: Hypnosis and Suggestibility (An Experimental Approach). New York, Appleton-Century, 1933.
257. Huxley, H. E.: The Mechanism of Muscular Contraction. Sci. Amer. *21*:18, 1965.
258. Hyde, I. H., Root, C. B., and Curl, H.: A Comparison of the Effects of Breakfast, of No Breakfast and of Caffeine on Work in an Athlete and a Non-Athlete. Amer. J. Physiol. *43*:371, 1917.
259. Hyman, A. S.: Practical Cardiology. New York, Landsberger Medical Books, 1958.
260. Hyman, A. S.: The Cardiac Athlete. Medicina Sportiva *13*:313, 1959.
261. Iampietro, P. F., Vaughan, J. A., Goldman, R. F., Kreider, M. B., Masucci, F., and Bass, D. E.: Heat Production from Shivering. J. Appl. Physiol. *15*:632, 1960.
262. Ikai, M.: Étude du effect d'entraînement sur la force d'un muscle par unité de aire

en coupe transversale au moyen de la measure ultrasonore. Paper given before FIEP Congress, Strasbourg, France, July, 1969.
263. Ikai, M., and Fukunaga, T.: Calculation of Muscle Strength per Unit Cross-Sectional Area of Human Muscle by Means of Ultrasonic Measurement. Int. Z. Angew. Physiol. *26*:26, 1968.
264. Ikai, M., and Steinhaus, A.: Some Factors Modifying the Expression of Human Strength. J. Appl. Physiol. *16*:157, 1961.
265. Ikai, M., and Yabe, K.: Training Effect of Muscular Endurance by Means of Voluntary and Electrical Stimulation. Int. Z. Angew. Physiol. *28*:55, 1969.
266. Ikai, M., Yabe, K., and Ischii, K.: Muskelkraft und Ermudung bei willkurlicher Anspannung und Elektrischer Reizung des Muskels. Sportartz und Sportmedizin *5*:197, 1967.
267. Issekutz, B., Jr., Birkhead, N. C., and Rodahl, K.: Use of Respiratory Quotients in Assessment of Aerobic Work Capacity. J. Appl. Physiol. *17*:47, 1962.
268. Itallie, T. B. Van, Sinisterra, L., and Stare, F. J.: Nutrition and Athletic Performance. J.A.M.A. *162*:1120, 1957.
269. Iwata, M.: Athletics and Health of Girls. J.A.M.A. *101*:723, 1933.
270. Jacobson, E.: Progressive Relaxation. Chicago, University of Chicago Press, 1938.
271. Johnson, R. E., and Robinson, S.: Selection of Men for Physical Work in Hot Weather. Appendix I. CMR. OSRD., Dept. 16. Harvard Fatigue Laboratory, February 15, 1943.
272. Jokl, E.: A Medical Theory of Gymnastics. Clin. Proc., So. Africa *2*:1, 1943.
273. Jokl, E.: Syncope in Athletes. Manpower, So. Africa *1&2*:1, 1947.
274. Jokl, E., and Jokl, P. (editors): Exercise and Altitude. Basel, Switzerland and New York, S. Karger, 1968.
275. Jones, H. E.: The Relationship of Strength to Physique. Amer. J. Phys. Anthrop. *5*:29, 1947.
276. Jungmann, H.: Studies on the Course and Duration of Acclimatization to an Altitude of 2,000 m. (6562 ft.). *In* The Effects of Altitude on Physical Performance. Edited by Goddard, R. F. Chicago, Illinois, The Athletic Institute, 1967.
277. Kamon, E., and Gormley, J.: Muscular Activity Pattern for Skilled Performance and During Learning of a Horizontal Bar Exercise. Ergonomics *11*:345, 1968.
278. Kandel, G. E.: Evaluating Body Composition. Aerospace Med. *40*:486, 1969.
279. Kao, F., Schlig, B., and Brooks, C.: Regulation of Respiration during Induced Muscular Work in Decerebrate Dogs. J. Appl. Physiol. *7*:379, 1955.
280. Karpovich, P. V.: Water Resistance in Swimming. Res. Quart. Amer. Ass. Health Phys. Educ. *4*:21, 1933.
281. Karpovich, P. V.: Effect of Oxygen Inhalation on Swimming Performance. Res. Quart. Amer. Ass. Health Phys. Educ. *5*:24, 1934.
282. Karpovich, P. V.: Analysis of the Propelling Force in the Crawl Stroke. Res. Quart. Amer. Ass. Health Phys. Educ. *6*:49, 1935.
283. Karpovich, P. V.: Physiological and Psychological Dynamogenic Factors in Exercise. Arbeitsphysiol. *9*:626, 1937.
284. Karpovich, P. V.: Textbook Fallacies Regarding Child's Heart. Res. Quart. Amer. Ass. Health Phys. Educ. *8*:33, 1937.
285. Karpovich, P. V.: Respiration in Swimming and Diving. Res. Quart. Amer. Ass. Health Phys. Educ. *10*:3, 1939.
286. Karpovich, P. V.: Ergogenic Aids in Work and Sport. Res. Quart. Amer. Ass. Health Phys. Educ. *12*:432, 1941.
287. Karpovich, P. V.: Relation Between Bends and Physical Fitness. Air Surgeon's Bulletin *1*:5, 1944.
288. Karpovich, P. V.: Relation Between Breath Holding and Endurance in Running and the Harvard Step-Up Test Score. Fed. Proc. *5*:53, 1946.
289. Karpovich, P. V.: Breath Holding as a Test of Physical Endurance. Amer. J. Physiol. *149*:720, 1947.
290. Karpovich, P. V.: A Frictional Bicycle Ergometer. Res. Quart. Amer. Ass. Health Phys. Educ. *21*:210, 1950.
291. Karpovich, P. V.: Physical Fitness—Why, How Much, and How to Acquire. Ind. Med. Surg. *25*:372, 1956.
292. Karpovich, P. V.: Warming-Up and Physical Performance. J. Lancet *77*:87, 1957.

293. Karpovich, P. V.: The Mighty Muscle. Proceedings of the American Academy of Physical Education, 1958.
294. Karpovich, P. V.: Effect of Amphetamine Sulfate on Athletic Performance. J.A.M.A. *170*:558, 1959.
295. Karpovich, P. V., and Gollnick, P. D.: Electrogoniometric Study of Locomotion and of Some Athletic Movements. Fed. Proc. *21*:313, 1962.
296. Karpovich, P. V., and Hale, C.: Effect of Warming-Up Upon Physical Performance. J.A.M.A. *162*:1117, 1956.
297. Karpovich, P. V., and Hale, C. J.: Tobacco Smoking and Athletic Performance. J. Appl. Physiol. *3*:616, 1951.
298. Karpovich, P. V., and Ikai, M.: Relation Between Reflex and Reaction Time. Fed. Proc. *19*:300, 1960.
299. Karpovich, P. V., and Karpovich, G. P.: An Improved Lever Arm for an Electric Dynamometer. J. Appl. Physiol. *27*:906, 1969.
300. Karpovich, P. V., and Millman, N.: Athletes as Blood Donors. Res. Quart. Amer. Ass. Health Phys. Educ. *13*:166, 1942.
301. Karpovich, P. V., and Millman, N.: Vitamin B_1 and Endurance. New Eng. J. Med. *226*:881, 1942.
302. Karpovich, P. V., and Millman, N.: Energy Expenditure in Swimming. Amer. J. Physiol. *142*:140, 1944.
303. Karpovich, P. V., and Pestrecov, K: Effect of Gelatin upon Muscular Work in Man. Amer. J. Physiol. *134*:300, 1941.
304. Karpovich, P. V., and Pestrecov, K.: Mechanical Work Done and Efficiency in Swimming Crawl and Back Strokes. Arbeitsphysiol. *10*:504, 1939.
305. Karpovich, P. V., Starr, M. P., Kimbro, R. W., Stoll, C. G., and Weiss, R. A.: Physical Reconditioning after Rheumatic Fever. J.A.M.A. *130*:1198, 1946.
306. Kasch, F. W., Philips, W. H., Ross, W. D., Carter, J. E. L., and Boyer, J. L.: A Comparison of Maximal Oxygen Uptake by Treadmill and Step-test Procedures. J. Appl. Physiol. *21*:1387, 1966.
307. Katch, F. I., and Michael, E. D., Jr.: Prediction of Body Density from Skinfold and Girth Measurements of College Females. J. Appl. Physiol. *25*:92, 1968.
308. Kay, H., and Karpovich, P. V.: Effect of Smoking upon Recuperation from Local Muscular Fatigue. Res. Quart. Amer. Ass. Health Phys. Educ. *20*:251, 1949.
309. Keller, L. F.: The Relation of "Quickness of Bodily Movement" to Success in Athletics. Res. Quart. Amer. Ass. Health Phys. Educ. *13*:146, 1942.
310. Kelliher, M. S.: A Report on Kraus-Weber Test in East Pakistan. Res. Quart. Amer. Ass. Health Phys. Educ. *31*:34, 1960.
311. Kennedy, T. F.: Report on an Investigation of Energy Expended on Exercises of the Physical Training Tables for Recruits of All Arms. J. Royal Army Med. Corps *61*:108, 185, 257, 1933.
311A. Kennedy, T. F.: Some Figures on Effects of Smoking on Endurance. J. Royal Army Med. Corps 57:451, 1931.
312. Keys, A., and Brožek, J.: Body Fat in Adult Man. Physiol. Rev. *33*:245, 1953.
313. Keys, A., and Taylor, H.: The Behavior of the Plasma Colloids in Recovery from Brief Severe Work and the Question as to the Permeability of the Capillaries to Proteins. J. Biol. Chem. *109*:55, 1935.
314. King, D. W., and Gollnick, P. D.: Ultrastructure of Rat Heart and Liver after Exhaustive Exercise. Amer. J. Physiol. *218*:1150, 1970.
315. Klausen, K.: Cardiac Output in Man in Rest and Work during and after Acclimatization to 3,800 m. J. Appl. Physiol. *21*:609, 1966.
316. Klausen, K.: Exercise under Hypoxic Conditions. Medicine and Science in Sports *1*:43, 1969.
317. Klausen, K., Robinson, S., Michael, E. D., and Myhre, L. G.: Effect of High Altitude on Maximal Working Capacity. J. Appl. Physiol. *21*:1191, 1966.
318. Klissouras, V., and Karpovich, P. V.: Electrogoniometric Study of Jumping Events. Res. Quart. Amer. Ass. Health Phys. Educ. *38*:41, 1967.
319. Knehr, C. A., Dill, D. B., and Neufeld, W.: Training and Its Effects on Man at Rest and Work. Amer. J. Physiol. *136*:148, 1942.
320. Knuttgen, H. G.: Oxygen Debt, Lactate, Pyruvate, and Excess Lactate after Muscular Work. J. Appl. Physiol. *17*:639, 1962.

321. Knuttgen, H. G.: Oxygen Uptake and Pulse Rate While Running with Undetermined and Determined Stride Lengths at Different Speeds. Acta Physiol. Scandinav. *52*:366, 1961.
322. Knuttgen, H. G.: Comparison of Danish and American School Children. Res. Quart. Amer. Ass. Health Phys. Educ. *32*:190, 1961.
323. Koveshnikova, V. A., quoted from Yakevlev, N. N., Korobkov, A., and Yananis, S. V.: Physiological and Biochemical Foundations of Theory and Method of Sports Training. 2nd ed. Fizcultura i Sport, Moscow, 1960 (in Russian).
324. Kozar, A. J.: Telemetered Heart Rates Recorded during Gymnastic Routines. Res. Quart. Amer. Ass. Health Phys. Educ. *34*:102, 1963.
325. Kozlowski, S.: Physiologic Mechanism of Active Rest. Physical Education and Sport *5*:241, 1962 (in Polish).
326. Kral, J.: Les Competitions Sportives dans la Jeunesse. Atti dell' VIII Congresso Internazionale di Medicina Sportiva. Firenze, Montecatini 28, Maggio, 1950.
327. Kramer, P. J.: Developmental Trends of Selected Motor Performance Items in Mongoloid Males. Unpublished Master's Thesis, Springfield College, Springfield, Massachusetts, 1969.
328. Kraus, H., and Hirschland, R.: Minimum Muscular Fitness Tests in School Children. Res. Quart. Amer. Ass. Health Phys. Educ. *25*:178, 1954.
329. Kraut, H. A., and Müller, E. A.: Calorie Intake and Industrial Output. Science *104*:495, 1946.
330. Krestovnikoff, A., Danilov, A., and Kogan, G.: Die Wirkung von Monophosphaten auf das Blut und den Blutkreislauf bei körperlich Arbeit. Arbeitsphysiol. *8*:13, 1934.
331. Krestovnikoff, A.: Fiziologic Sporta. Fizkultura i Sport, Moscow, 1939.
332. Krogh, A., and Lindhard, J.: The Relative Value of Fats and Carbohydrates as Sources of Muscular Energy. Biochem. J. *14*:290, 1920.
333. Kroll, W.: Patellar Reflex Time and Reflex Latency under Jendrassik and Crossed Extensor Facilitation. Amer. J. Phys. Med. *47*:292, 1968.
334. Krustev, K., Iliev, I., and Purvanov, B.: Survey on an Elder Group of Tourists Taking Part in an Important Tourist Traverse. Bulgarian Olympic Committee, Bulletin d'Information *8*:24, 1963.
335. Lahiri, S., Milledge, J. S., Chattopadhyay, H. P., Bhattacharyya, A. K., and Sinha, A. K.: Respiration and Heart Rate of Sherpa Highlanders during Exercise. J. Appl. Physiol. *23*:545, 1967.
336. Lanoue, F.: Some Facts on Swimming Cramps. Res. Quart. Amer. Ass. Health Phys. Educ. *21*:153, 1950.
337. Laubach, L. L., and McConville, J. T.: Muscle Strength, Flexibility, and Body Size of Adult Males. Res. Quart. Amer. Ass. Health Phys. Educ. *37*:384, 1966.
338. Laubach, L. L., and McConville, J. T.: Relationship Between Flexibility, Anthropometry, and the Somatotype of College Men. Res. Quart. Amer. Ass. Health Phys. Educ. *37*:241, 1966.
339. Lautenbach, R., and Tuttle, W. W.: Relationship Between Reflex Time and Running Events in Track. Res. Quart. Amer. Ass. Health Phys. Educ. *3*:138, 1932.
340. Lefrancois, R., Gautier, H., Pasquis, P., and Vargas, E.: Factors Controlling Respiration During Muscular Exercise at Altitude. Fed. Proc. *28*:1296, 1969.
341. Lehman, H. C.: The Most Proficient Years at Sports and Games. Res. Quart. Amer. Ass. Health Phys. Educ. *9*:3, 1938.
342. Leithauser, D. J.: Early Ambulation and Related Procedures in Surgical Management. Springfield, Illinois, Charles C Thomas, 1946.
343. Lietzke, M.: Relation Between Weight Lifting Total and Body Weight. Science *124*:486, 1956.
344. Lind, A. R.: Cardiovascular Responses to Static Exercise. Circulation *51*:173, 1970.
345. Lind, A. R., and McNicol, G. W.: Muscular Factors Which Determine the Cardiovascular Responses to Sustained and Rhythmic Exercise. Canad. Med. Ass. J. *96*:706, 1967.
346. Lippold, O. C. J.: The Relation Between Integrated Action Potentials in a Human Muscle and Its Isometric Tension. J. Physiol. *117*:492, 1952.
347. Loewy, A., Eysern, A., and Oprisescu, S.: Untersuchungen bei körperlichen Hochstleistungen. Arbeitsphysiol. *4*:298, 1931.

348. Luft, U. C.: Aviation Physiology—The Effects of Altitude. *In* Handbook of Physiology, Section 3; Respiration, Volume II. Edited by Fenn, W. O., and Rahn, H. Washington, D.C., American Physiological Society, 1965.
349. Lukin, L., and Ralston, H. J.: Oxygen Deficit and Repayment in Exercise. Int. Z. Angew. Physiol. *19*:183, 1962.
350. Lundegren, H. M.: Changes in Skinfold and Girth Measures of Women Varsity Basketball and Field Hockey Players. Res. Quart. Amer. Ass. Health Phys. Educ. *39*:1020, 1968.
351. Lundsgaard, C., and Möller, E.: Immediate Effects of Heavy Exercise. J. Biol. Chem. *55*:477, 1923.
352. Lyon, R. S.: A Mathematical Analysis of World's Free Style Swimming Records. Unpublished Master's Thesis, Springfield College, Springfield, Massachusetts, 1952.
353. MacDonald, F. W., and Stearns, W. J.: A Mathematical Analysis of the Dolphin-Butterfly and Breast Strokes. Unpublished Master's Thesis, Springfield College, Springfield, Massachusetts, 1969.
354. Magel, J. R., and Faulkner, J. A.: Maximum Oxygen Uptakes of College Swimmers. J. Appl. Physiol. *22*:929, 1967.
355. Mahadeva, K., Passmore, R., and Woolf, B.: Individual Variations in Metabolic Cost of Standardized Exercises: Effects of Food, Age, Sex and Race. J. Physiol. *121*:225, 1953.
356. Maison, G. L., and Broeker, A. C.: Training in Human Muscles Working with and without Blood Supply. Am. J. Physiol. *132*:390, 1941.
357. Majdrakoff: Cited by Baur in Arch. f. Exper. Path. u. Pharmakol. *184*:51, 1936.
358. Malarecki, I.: Investigation on Physiological Justification of So-Called Warming Up. Acta Physiol. Pol. *5*:543, 1954.
359. Malhotra, M. S., Gupta, J., and Rai, R. M.: Pulse Count as a Measure of Energy Expenditure. J. Appl. Physiol. *18*:994, 1963.
360. Marbe, K.: Ueber der vermeintliche Leistungssteigerung durch Recresal und Natrium bicarbonicum. Arch. f. Exper. Path. u. Pharmakol. *167*:404, 1932.
361. Margaria, R. (editor): Exercise at Altitude. Amsterdam, Excerpta Medica Foundation, 1967.
362. Margaria, R., Cerretelli, P., Aghemo, P., and Sassi, G.: Energy Cost of Running. J. Appl. Physiol. *18*:367, 1963.
363. Margaria, R., Cerretelli, P., and Mangili, F.: Balance and Kinetics of Anaerobic Energy Release During Strenuous Exercise in Man. J. Appl. Physiol. *19*:623, 1964.
364. Margaria, R., Edwards, H. T., and Dill, D. B.: The Possible Mechanism of Contracting and Paying the Oxygen Debt and the Role of Lactic Acid in Muscular Contraction. Amer. J. Physiol. *106*:689, 1933.
365. Margaria, R., Milic, E. G., Petit, J. M., and Cavagana, G.: Mechanical Work of Breathing During Muscular Exercise. J. Appl. Physiol. *15*:354, 1960.
366. Margaria, R., Oliva, R. D., Di Prampero, D. E. A., and Cerretelli, P.: Energy Utilization in Intermittent Exercise of Supramaximal Intensity. J. Appl. Physiol. *26*:752, 1969.
367. Massey, B. H., Johnson, W. R., and Kramer, G. F.: Effect of Warm-up Exercise Upon Muscular Performance Using Hypnosis to Control the Psychological Variable. Res. Quart. Amer. Ass. Health Phys. Educ. *32*:63, 1961.
368. Mathews, P. B. C.: Muscle Spindles and Their Motor Control. Physiol. Rev. *44*:219, 1964.
369. Mazess, R. B.: Exercise Performance at High Altitude in Peru. Fed. Proc. *28*:1301, 1969.
370. McArdle, W. D., Foglia, G. F., and Patti, A. V.: Telemetered Cardiac Response to Selected Running Events. J. Appl. Physiol. *23*:566, 1967.
371. McCormick, W. J.: Vitamin B_1 and Physical Endurance. Medical Record *152*:439, 1940.
372. McCurdy, J. H., and Larson, L. A.: The Physiology of Exercise. Philadelphia, Lea and Febiger, 1939.
373. Meyer, F.: Energieumsatz und Wirkungsgrad des Alkoholgewohnten unter dem Einfluss von Alkohol. Arbeitsphysiol. *4*:433, 1931.

374. Miasnikov, A. L., quoted from Simonson, E.: Russian Physiology (Cardiovascular Aspects). Ann. Rev. Physiol. *20*:123, 1958.
375. Michael, E. D., Jr., and Katch, F. I.: Prediction of Body Density from Skinfold and Girth Measurements of 17-Year-Old Boys. J. Appl. Physiol. *25*:6, 1968.
376. Missiuro, W., Dill, D. B., and Edwards, H. T.: Effects of Adrenal Cortical Gland in Rest and Work. Amer. J. Physiol. *121*:549, 1938.
377. Missiuro, W., and Perlberg, A.: Effect of Gymnastics upon the Metabolism. Arbeitsphysiol. 7:62, 1934.
378. Mitchell, H., Sproule, B. J., and Chapman, C. B.: The Physiological Meaning of the Maximal Oxygen Intake Test. J. Clin. Invest. *34*:538, 1958.
379. Miyama, A.: The Effect of Oxygen upon Bodily Exercise. Acta Sch. Med. Univ. Kioto *14*:73, 1931.
380. Montoye, H. J., Spata, P. J., Pinckney, V., and Barron, L.: Effects of Vitamin B_{12} Supplementation on Physical Fitness and Growth of Young Boys. J. Appl. Physiol. 7:589, 1955.
381. Montoye, H. J., Van Huss, W. D., and Nevai, J. W.: Longevity and Morbidity of College Athletes: A Seven-Year Follow-Up Study. J. Sports Med. *2*:133, 1962.
382. Montoye, H. J., Van Huss, W. D., Olson, H. W., Hudec, A. G., and Mahoney, E.: Study of the Longevity and Morbidity of College Athletes. J.A.M.A. *162*:1132, 1956.
383. Morehouse, L.: Basal Metabolism of Athletes in Training. Unpublished Master's Thesis, Springfield College, Springfield, Massachusetts, 1937.
384. Morehouse, L. E., and Miller, A. T.: Physiology of Exercise, 5th ed. Saint Louis, The C. V. Mosby Company, 1967.
385. Morpurgo, B.: Ueber Activitats-Hypertrophie der willkürlichen Muskeln. Virchows Arch. Path. Anat. *150*:522, 1897.
386. Mostyn, E. M., Helle, S., Gee, J. B. L., Bentivoglio, L. G., and Bates, D. V.: Pulmonary Diffusing Capacity of Athletes. J. Appl. Physiol. *18*:687, 1963.
387. Moudgil, R., and Karpovich, P. V.: Duration of a Maximal Isometric Contraction. Res. Quart. Amer. Ass. Health Phys. Educ. *40*:3, 1969.
388. Müller, E. A.: The Regulation of Muscular Strength. J. Ass. Phys. Ment. Rehab. *11*:41, 1957.
389. Müller, E. A., and Hettinger, T.: Die Bedeutung des Trainingserlaufes für die Trainingsfestigkeit von Muskeln. Arbeitsphysiol. *15*:452, 1954.
390. Murray, J., and Karpovich, P. V.: Weight Training in Athletics. Englewood Cliffs, New Jersey, Prentice-Hall, Inc., 1956.
391. Newman, E. V., Dill, D. B., Edwards, H. T., and Webster, F. A.: The Rate of Lactic Acid Removal in Exercise. Amer. J. Physiol. *118*:457, 1937.
392. Newton, J. L.: The Assessment of Maximal Oxygen Uptake. J. Sports Med. *3*:164, 1963.
393. New York Herald Tribune, p. 14, July 1, 1952.
394. Nielsen, M., and Hanse, O.: Maximale körperliche Arbeit bei Atmung O_2-reich. Luft. Skand. Arch. f. Physiol. *76*:37, 1937.
395. O'Connell, A. L., and Gardner, E. B.: The Use of Electromyography in Kinesiological Research. Res. Quart. Amer. Ass. Health Phys. Educ. *34*:166, 1963.
396. Paillard, J.: The Patterning of Skilled Movements. *In* Handbook of Physiology, Section 1; Neurophysiology, Vol. 3. Edited by Field, J., Magoun, H. W., and Hall, V. E., Washington, D.C., American Physiological Society, 1960.
397. Palladin, A., and Ferdmann, D.: The Influence of Muscle Training on Creatine Content. Hoppe-Seyler's Ztschr. Physiol. Chem. *174*:284, 1928.
398. Pampe, W.: Hyperglykämie und körperliche Arbeit. Arbeitsphysiol. *5*:342, 1932.
399. Pařízková, J.: Longitudinal Study of the Development of Body Composition and Body Build in Boys of Various Physical Activity. Hum. Biol. *40*:212, 1968.
400. Pařízková, J., and Eiselt, E.: Body Composition and Anthropometric Indicators in Old Age and the Influence of Physical Exercise. Hum. Biol. *38*:351, 1966.
401. Pařízková, J., and Eiselt, E.: Longitudinal Study of Changes in Anthropometric Indicators and Body Composition of Old Men of Various Physical Activity. Hum. Biol. *40*:331, 1968.
402. Parnell, R. W.: Aetiology of Coronary Heart Disease. Brit. Med. J. *1*:232, 1959.

403. Parnell, R. W.: Behaviour and Physique. London, Edward Arnold (Publishers) Ltd., 1958.
404. Parsonnet, A. E., and Bernstein, A.: Heart Strain: A Critical Review. The Development of Physiologic Concept. Ann. Intern. Med. *16*:1123, 1942.
405. Passmore, R., and Durnin, J. V. G. A.: Human Energy Expenditure. Physiol. Rev. *35*:801, 1955.
406. Patty, W. W., and Van Horn, T. J.: Health of High School Athletes. J. Health & Phys. Educ. *6*:26, 1935.
407. Peachy, L. D.: Muscle. Ann. Rev. Physiol. *30*:401, 1968.
408. Perbix, J. A.: Relationship Between Somatotype and Motor Fitness in Women. Res. Quart. Amer. Ass. Health Phys. Educ. *25*:84, 1954.
409. Petren, T., Sjostrand, T., and Sylven, B.: Der Einfluss des Trainings auf die Haufigkeit der Capillaren in Herz- und Skeletmuskulatur. Arbeitsphysiol. *9*:376, 1936.
410. Pollock, M. L., Cureton, T. K., and Greninger, L.: Effects of Frequency of Training on Working Capacity, Cardiovascular Function and Body Composition of Adult Men. Medicine and Science in Sports *1*:70, 1969.
411. Press, H.: Uber die Wirkung des Sympatols auf Blutdruck und Puls an Normalen, Hypotonikern und Asthenikern bei körperlicher Arbeit. Ztschr. f. d. Ges. Exper. Med. *80*:66, 1931.
412. Pugh, L. G. C. E.: Athletes at Altitude. J. Physiol. *192*:619, 1967.
413. Pugh, L. G. C. E.: Cardiac Output in Muscular Exercise at 5,800 m. (19,000 ft.). J. Appl. Physiol. *19*:441, 1964.
414. Pugh, L. G. C. E.: Muscular Exercise on Mount Everest. J. Physiol. *141*:233, 1958.
415. Pugh, L. G. C. E., Gill, M. D., Lahiri, S., Milledge, J. S., Ward, M. P., and West, J. B.: Muscular Exercise at Great Altitudes, J. Appl. Physiol. *19*:431, 1964.
416. Pugh, L. G. C. E., and Owen, R.: Report of Medical Research Project into Effects of Altitude in Mexico City in 1965. British Olympic Association, 12 Buckingham Street, London, W.C. 2, 1966.
417. Pugh, L. G. C. E., Edholm, O. G., Fox, R. H., Wolff, H. S., Hervey, G. R., Hammond, W. H., Tanner, J. M., and Whitehouse, R. H.: A Physiological Study of Channel Swimming. *In* Medical Research on Swimming. Edited by Firsov, S., and Jokl, E. Medical Committee, Federation Internationale Natation Amateur, 1968.
418. Puni, A.: Der Einfluss von Monophosphaten auf einige psychische und psychomotorische Prozesse während der Erholungsperiode nach Muskelarbeit. Arbeitsphysiol. *8*:20, 1934.
419. Radloff, E. M.: The Oxygen Pulse in Athletic Girls During Rest and Exercise. Amer. J. Physiol. *96*:126, 1931.
420. Ralston, H. J.: Comparison of Energy Expenditure During Treadmill Walking and Floor Walking. J. Appl. Physiol. *15*:1156, 1960.
421. Rao, S.: Cardiovascular Responses to Head-Stand Posture. J. Appl. Physiol. *18*:987, 1963.
422. Rapaport, S. E., Fetcher, E. S., Shaub, H. G., and Hall, J. F.: Control of Blood Flow to the Extremities at Low Ambient Temperature. J. Appl. Physiol. *2*:61, 1949.
423. Rasch, P. J., Pierson, W. R., and Brubaker, M. L.: The Effect of Amphetamine Sulfate and Meprobamate on Reaction Time and Movement Time. Int. Z. Angew. Physiol. *18*:280, 1960.
424. Rathbone, J. L.: Teach Yourself to Relax. New York, Prentice-Hall, Inc., 1957.
425. Rathbun, E. M., and Pace, N.: The Determination of Total Body Fat by Means of the Body Specific Gravity. J. Biol. Chem. *158*:667, 1945.
426. Ray, G. B., Johnson, J. R., and Taylor, M. M.: Effect of Gelatine on Muscular Fatigue. Proc. Soc. Exp. Biol. Med. *40*:157, 1939.
427. Reid, J. G.: Static Strength Increase and its Effect upon Triceps Surae Reflex Time. Res. Quart. Amer. Ass. Health Phys. Educ. *38*:691, 1967.
428. Riabuschinsky, N. P.: The Effect of Phosphate on Work and Respiratory Exchange. Ztschr. f. d. Ges. Exper. Med. *72*:20, 1930.
429. Ringer, L. B., and Adrian, M. J.: An Electrogoniometric Study of the Wrist and

Elbow in the Crawl Arm Stroke. Res. Quart. Amer. Ass. Health Phys. Educ. *40*:353, 1969.

430. Robinson, S.: Experimental Studies in Physical Fitness in Relation to Age. Arbeitsphysiol. *10*:251, 1938.
431. Robinson, S., Edwards, H. T., and Dill, D. B.: New Records in Human Power. Science *85*:409, 1937.
432. Robinson, S., and Harmon, P. M.: The Lactic Acid Mechanism and Certain Properties of the Blood in Relation to Training. Amer. J. Physiol. *132*:757, 1941.
433. Robinson, S., Robinson, D. L., Mountjoy, R. J., and Bullard, R. W.: Influence of Fatigue on the Efficiency of Men During Exhausting Runs. J. Appl. Physiol. *12*:197, 1958.
434. Robinson, S., Turrell, E. S., Belding, H. S., and Horvath, S. M.: Rapid Acclimatization of Men to Work in Hot Climates. Amer. J. Physiol. *140*:168, 1943.
435. Rochelle, R. H.: Blood Plasma Cholesterol Changes During a Physical Training Program. Res. Quart. Amer. Ass. Health Phys. Educ. *32*:538, 1961.
436. Rochelle, R. H., Skubic, V., and Michael, E. D.: Performance as Affected by Incentive and Preliminary Warm-up. Res. Quart. Amer. Ass. Health Phys. Educ. *31*:499, 1960.
437. Ronkin, R. R.: Further Studies of the Harvard Step-Up Test. Report 2, Project 148. A.A.F. School of Aviation Medicine, August 17, 1944.
438. Rose, J. E., and Mountcastle, V. B.: Touch and Kinesthesis. *In* Handbook of Physiology, Section 1; Neurophysiology, Vol. 1. Edited by Field, J., Magoun, H. W., and Hall, V. E., Washington, D.C., American Physiological Society, 1959.
439. Rose, K. D., Schneider, P. J., and Sullivan, G. F.: A Liquid Pregame Meal for Athletes. J.A.M.A. *178*:1, 1961.
440. Roush, E. S.: Strength and Endurance in the Waking and Hypnotic State. J. Appl. Physiol. *3*:404, 1951.
441. Rowell, L. B.: Circulation. Medicine and Science in Sports *1*:15, 1969.
442. Rowell, L. B., Brengelmann, G. L., Blackmon, J. R., Twiss, R. D., and Kusumi, F.: Splanchnic Blood Flow and Metabolism in Heat-Stressed Man. J. Appl. Physiol. *24*:475, 1968.
443. Rowell, L. B., Blackmon, J. R., and Bruce, R. A.: Indocyanine Green Clearance and Estimated Hepatic Blood Flow During Mild to Maximal Exercise in Upright Man. J. Clin. Invest. *43*:1677, 1964.
444. Rowell, L. B., Taylor, H. L., and Wang, Y.: Limitations to Prediction of Maximal Oxygen Intake. J. Appl. Physiol. *19*:919, 1964.
445. Royce, J.: Isometric Fatigue Curves in Human Muscle with Normal and Occluded Circulation. Res. Quart. Amer. Ass. Health Phys. Educ. *29*:204, 1958.
446. Royce, J.: Re-evaluation of Isometric Training Methods and Results, a Must. Res. Quart. Amer. Ass. Health Phys. Educ. *35*:215, 1964.
447. Ruch, T. C., and Patton, H. D. (editors): Physiology and Biophysics, 19th ed. Philadelphia, W. B. Saunders Company, 1965.
448. Rusk, H. A.: Army Air Corps' New Convalescent Program. J. Indiana Med. Ass. *36*:127, 1943.
449. Sage, J. N.: Effects of Differing Breakfast Conditions and Habit Patterns on Performance in an Endurance Activity. Res. Quart. Amer. Ass. Health Phys. Educ. *40*:799, 1969.

449A. Saiki, H., Margaria, R., and Cuttica, F.: Lactic Acid Production in Submaximal Work. Int. Z. Angew. Physiol. *24*:57, 1967.

450. Saltin, B., and Åstrand, P.-O.: Maximal Oxygen Uptake in Athletes. J. Appl. Physiol. *23*:353, 1967.
451. Sargent, R. M.: Relation Between Oxygen Requirement and Speed in Running. Proc. Roy. Soc. [Biol.] *100*:10, 1926.
452. Schade, M., Hellebrandt, F. A., Waterland, J. C., and Carns, M. L.: Spot Reducing in Overweight College Women: Its Influence on Fat Distribution as Determined by Photography. Res. Quart. Amer. Ass. Health Phys. Educ. *33*:461, 1962.
453. Schaefer, K. E.: Oxygen Toxicity Studies in Underwater Swimming. J. Appl. Physiol. *8*:524, 1956.
454. Schellong, F.: Das Verhalten des Diastolischen Blutdrucks Nach Körperarbeit und seine klinische Bedeutung. Klin. Wchnschr. *9*:1340, 1930.

455. Scherrer, J., and Bourguignon, A.: Changes in the Electromyogram Produced by Fatigue in Man. Amer. J. Phys. Med. *38*:148, 1959.
456. Schirlitz, K.: Ueber Kaffein bei ermudender Muskelarbeit. Arbeitsphysiol. *2*:273, 1930.
457. Schmid, L.: Increasing Bodily Output by Warming-Up. Casop. Lek. Cesk. *86*:950, 1947.
457A. Schmidt, G.: Uber kolloidchemische Veränderungen bei der Ermüdung des Warmblutermuskels. Arbeitsphysiol. *1*:136, 1928.
458. Schneider, E. C.: Observations on Holding the Breath. Amer. J. Physiol. *94*:464, 1930.
459. Schneider, E. C., and Collins, R.: Venous Pressure Responses to Exercise. Amer. J. Physiol. *121*:574, 1938.
460. Schneider, E. C., and Foster, A. O.: The Influence of Physical Training on the Basal Metabolic Rate of Man. Amer. J. Physiol. *98*:595, 1931.
461. Schneider, E. C., and Havens, L. C.: Changes in the Blood after Muscular Activity and During Training. Amer. J. Physiol. *36*:239, 1915.
462. Schneider, E. C., and Ring, G. C.: The Influence of a Moderate Amount of Physical Training on the Respiratory Exchange and Breathing During Physical Exercise. Amer. J. Physiol. *91*:103, 1929.
463. Schneider, E. C., and Truesdell, D.: A Statistical Study of the Pulse Rate and the Arterial Blood Pressures in Recumbency, Standing and after a Standard Exercise. Amer. J. Physiol. *61*:429, 1922.
464. Schneider, E. C., and Truesdell, D.: Daily Variations in Cardiovascular Conditions and a Physical Efficiency Rating. Amer. J. Physiol. *67*:193, 1923.
465. Schorn, M.: Ueber die Wirkung des Recresals auf die körperliche und geistige Leistungsfähigkeit. Münch. Med. Wchnschr. *79*:371, 1932.
466. Schroll, W.: Ueber Veränderungen der Fahigkeit Askorbinsaure zu Oxygieren und Dehydroaskorbinsaure zu reduzieren im Training. Arch. f. d. Ges. Physiol. *240*:642, 1938.
467. Schwartz, L., Britten, R. H., and Thompson, L. R.: Studies in Physical Development and Posture. I. The Effect of Exercise on the Physical Condition and Development of Adolescent Boys. Pub. Health Bull. No. 179, 1928.
468. Scott, E. L., and Hastings, A. B.: Sugar and Oxygen Relationship in Blood of Dogs During Exercise. Proc. Soc. Exp. Biol. Med. *17*:120, 1920.
469. Sedgwick, A. W., and Whalen, H. R.: Effect of Passive Warm-up on Muscular Strength and Endurance. Res. Quart. Amer. Ass. Health Phys. Educ. *35*:45, 1963.
470. Sheldon, W. H., Dupertis, C. W., and McDermott, E.: Atlas of Men. New York, Gramercy Publishing Company, 1954.
471. Sheldon, W. H., Stevens, S. S., and Tucker, W. B.: The Varieties of Human Physique. New York, Harper and Brothers, 1940.
472. Sherrington, C. S.: Integrative Action of the Nervous System, 8th ed. New Haven, Yale University Press, 1926.
473. Shuck, G.: Effects of Athletic Competition on the Growth and Development of Junior High School Boys. Res. Quart. Amer. Ass. Health Phys. Educ. *33*:288, 1962.
474. Sills, F. D., and Mitchem, J.: Prediction of Performance on Physical Fitness Tests by Means of Somatotype Ratings. Res. Quart. Amer. Ass. Health Phys. Educ. *28*:64, 1957.
475. Sills, F. D., and Olson, A. L.: Action Potentials in Unexercised Arm When Opposite Arm is Exercised. Res. Quart. Amer. Ass. Health Phys. Educ. *29*:213, 1958.
476. Simonson, E.: Effect of Alcohol and Thyroidin on the Metabolism. Arch. Exp. Path. u. Pharmakol. *119*:259, 1927.
477. Simonson, E., and Enzer, N.: Physiology of Muscular Exercise and Fatigue in Disease. Medicine *21*:345, 1942.
478. Singh, M., and Karpovich, P. V.: Effect of Eccentric Training of Agonists on Antagonistic Muscles. J. Appl. Physiol. *23*:742, 1967.
479. Singh, M., and Karpovich, P. V.: Isotonic and Isometric Forces of Forearm Flexors and Extensors. J. Appl. Physiol. *21*:1435, 1966.

480. Singh, M., and Karpovich, P. V.: Strength of Forearm Flexors and Extensors in Men and Women. J. Appl. Physiol. *25*:177, 1968.
481. Sinning, W. E., and Forsyth, H. L.: Lower-Limb Actions While Running at Different Velocities. Medicine and Science in Sports *2*:28, 1970.
482. Siri, W. E.: Body Volume Measurement by Gas Dilution. *In* Techniques for Measuring Body Composition. Edited by Brožek, J., and Henschel, A. Proceedings of a Conference. Washington, D.C., National Academy of Sciences—National Research Council, 1961.
483. Skubic, V., and Hilgendorf, J.: Anticipatory Exercise and Recovery Heart Rates of Girls as Affected by Four Running Events. J. Appl. Physiol. *19*:833, 1964.
484. Slack, D.: Pythagorean Theorem Applied to Swimming the Crawl with Foot Flippers. Unpublished Master's Thesis, Springfield College, Springfield, Massachusetts, 1957.
485. Sloan, A. W.: Estimation of Body Fat in Young Men. J. Appl. Physiol. *23*:311, 1967.
486. Sloan, A. W., Burt, J. J., and Blyth, C. S.: Estimation of Body Fat in Young Women. J. Appl. Physiol. *17*:967, 1962.
487. Smith, G. M., and Beecher, H. K.: Amphetamine Sulfate and Athletic Performance. J.A.M.A. *170*:542, 1959.
488. Spaeth, R. A.: An Experimental Investigation of the Supposed Relation Between Good Physical Condition and Mutual Resistance to Infection. Amer. J. Hyg. *5*:839, 1925.
489. Spain, D. M., Nathan, D. J., and Gellis, M.: Weight, Body Type and the Prevalence of Coronary Atherosclerotic Heart Disease in Males. Amer. J. Med. Sci. *245*:63, 1963.
490. Specht, H., Goff, L. G., Brubach, H. F., and Bartlett, R. G.: Work Efficiency and Respiratory Response of Trained Underwater Swimmers Using a Modified Self-Contained Underwater Breathing Apparatus. J. Appl. Physiol. *10*:376, 1957.
491. Staton, W. M.: The Influence of Soya Lecithin on Muscular Strength. Res. Quart. Amer. Ass. Health Phys. Educ. *22*:201, 1951.
492. Steinhaus, A. H.: Chronic Effects of Exercise. Physiol. Rev. *13*:103, 1933.
492A. Steinhaus, A. H.: Exercise and Basal Metabolism in Dogs. Amer. J. Physiol. *83*:568, 1928.
493. Strydom, N. B., Morrison, J. F., Booyens, J., and Peter, J.: Comparison of Oral and Rectal Temperatures During Work in Heat. J. Appl. Physiol. *8*:406, 1956.
494. Stuart, D. G., and Collings, W. D.: Comparison of Vital Capacity and Maximum Breathing Capacity of Athletes and Nonathletes. J. Appl. Physiol. *14*:507, 1959.
495. Symposium: Abuse of Bed Rest. J.A.M.A. *125*:1075, 1944.
496. Tanner, J. M.: The Physique of the Olympic Athlete. London, George Allen and Unwin Ltd., 1964.
497. Taylor, C.: Some Properties of Maximal and Submaximal Exercise with Reference to Physiological Variation and the Measurement of Exercise Tolerance. Amer. J. Physiol. *142*:200, 1944.
498. Taylor, H. L.: Exercise and Metabolism. *In* Science and Medicine of Exercise and Sports. Edited by Johnson, W. R. New York, Harper and Brothers, 1960.
499. Taylor, H. L., Brožek, J., Henschel, A., Michelsen, O., and Keys, A.: The Effect of Successive Fasts on the Ability to Withstand Fasting During Hard Work. Amer. J. Physiol. *143*:148, 1945.
500. Taylor, H. L., Buskirk, E., and Henschel, A.: Maximal Oxygen Intake as an Objective Measure of Cardio-Respiratory Performance. J. Appl. Physiol. *8*:73, 1955.
501. Taylor, H. L., Henschel, A. F., and Keys, A.: Cardiovascular Adjustments of Man in Rest and Work during Exposure to Dry Heat. Amer. J. Physiol. *139*:583, 1943.
502. Taylor, H. L., Henschel, A., Michelsen, O., and Keys, A.: The Effect of the Sodium Chloride Intake in the Work Performance of Man during Exposure to Dry Heat and Experimental Heat Exhaustion. Amer. J. Physiol. *140*:439, 1943.
503. Taylor, H. L., Wang, Y., Rowell, L., and Blomqvist, G.: The Standardization and Interpretation of Submaximal and Maximal Test of Working Capacity. Pediatrics *32*:703, 1963.

504. Tcheng, T. K.: Predictability of Body Weight of Iowa High School Wrestlers Through Anthropometric Assessment. Unpublished Doctoral Thesis, University of Iowa, Iowa City, Iowa, 1969.
505. Teräslinna, P., Ismail, A. H., and MacLeod, D. F.: Nomogram by Åstrand and Ryhming as a Predictor of Maximum Oxygen Intake. J. Appl. Physiol. *21*:513, 1966.
506. Thiel, D., and Essig, B.: Cocain und Muskelarbeit; der Einfluss auf Liestung und Gasstoffwechsel. Arbeitsphysiol. *3*:287, 1930.
507. Thompson, C. W., Buskirk, E. R., and Goldman, R. F.: Changes in Body Fat Estimated from Skinfold Measurements of College Basketball and Hockey Players During a Season. Res. Quart. Amer. Ass. Health Phys. Educ. *27*:418, 1956.
508. Tillett, W. S.: The Needs for Physiological Knowledge: Civilian Medicine. Fed. Proc. *3*:190, 1944.
509. Time Magazine: Battle of the Skies. Reflex. *43*:22, 1944.
510. Time Magazine: Genetics. *90*:(3)70, 1967.
511. Tipton, C. M., and Francis, P. R.: Influence of a Weight Training Program on Quadriceps Reflex Time. Medicine and Science in Sports *1*:2, 1969.
512. Tipton, C. M., and Karpovich, P. V.: Exercise and the Patellar Reflex. J. Appl. Physiol. *21*:15, 1966.
513. Tipton, C. M., Tcheng, T. K., and Paul, W. D.: Evaluation of the Hall Method for Determining Minimum Wrestling Weights. J. Iowa Med. Soc. 571, July, 1969.
514. Turner, A. H.: The Adjustment of Heart Rate and Arterial Pressures in Healthy Young Women During Prolonged Standing. Amer. J. Physiol. *81*:197, 1927.
515. Turner, A. H.: Personal Character of the Prolonged Standing Circulatory Reaction and Factors Influencing It. Amer. J. Physiol. *87*:667, 1929.
516. Tuttle, W. W.: The Effect of Weight Loss by Dehydration and the Withholding of Food on the Physiologic Responses of Wrestlers. Res. Quart. Amer. Ass. Health Phys. Educ. *14*:159, 1943.
517. Tuttle, W. W., and Dickinson, R. E.: A Simplification of the Pulse-Ratio Technique for Rating Physical Efficiency and Present Condition. Res. Quart. Amer. Ass. Health Phys. Educ. *9*:73, 1938.
518. Tuttle, W. W., and Salit, E. P.: The Relation of Resting Heart Rate to the Increase in Rate Due to Exercise. Amer. Heart J. *29*:594, 1945.
519. Tuttle, W. W., Wilson, M., and Daum, K.: Effect of Altered Breakfast Habits on Physiologic Response. J. Appl. Physiol. *1*:545, 1949.
520. Vernon, H. M.: Industrial Fatigue in Relation to Atmospheric Conditions. Physiol. Rev. *8*:130, 1928.
521. Voegler, R. F., and Ferguson, V. W.: The Effect of Sugar upon Athletic Performance. Res. Quart. Amer. Ass. Health Phys. Educ. *3*:54, 1932.
522. Vogel, J. A., Hansen, J. E., and Harris, C. W.: Cardiovascular Responses in Man During Exhaustive Work at Sea Level and High Altitude. J. Appl. Physiol. *23*:531, 1967.
523. Wahlund, H.: Determination of Physical Working Capacity. Acta Med. Scandinav. (Supplement) *132*:1, 1948.
524. Wakabayashi, Y.: Liver Glycogen and Training. Z. Physiol. Chem. *179*:79, 1928.
525. Waterland, J. C., and Munson, N.: Reflex Association of Head and Shoulder Girdle in Nonstressful Movements of Man. Amer. J. Phys. Med. *43*:98, 1964.
526. Waterland, J. C., and Hellebrandt, E. A.: Involuntary Patterning Associated with Willed Movement against Progressively Increasing Resistance. Amer. J. Phys. Med. *43*:13, 1964.
527. Weiss, R. A., and Karpovich, P. V.: Energy Cost of Exercise for Convalescents. Arch. Phys. Med. *28*:447, 1947.
528. Welham, W. C., and Behnke, A. R., Jr.: The Specific Gravity of Healthy Men. J.A.M.A. *118*:498, 1942.
529. Wells, E. W.: The Microanatomy of Muscle. *In* Structure and Function of Muscle, Volume I. Edited by Bourne, G. H. New York and London, Academic Press, 1960.
530. Wells, J. G., Balke, B., and Van Fossan, D. D.: Lactic Acid Accumulation During Work. A Suggested Standardization of Work Classification. J. Appl. Physiol. *10*:51, 1957.

531. West, E. S., and Todd, W. R.: Textbook of Biochemistry, 3rd ed. New York, MacMillan Company, 1961.
532. West, H.: Clinical Studies on the Respiration. VI. A Comparison of Various Standards for the Normal Vital Capacity of the Lungs. Arch. Intern. Med. *25*:306, 1920.
533. Westerland, J. H., and Tuttle, W. W.: The Relationship Between Running Events and Reaction Time. Res. Quart. Amer. Ass. Health Phys. Educ. *2*:95, 1931.
534. Whip, B. J., Seard, C., and Wasserman, K.: Oxygen Deficit–Oxygen Debt Relationships and Efficiency of Anaerobic Work. J. Appl. Physiol. *28*:452, 1970.
535. Whipple, C. H.: The Hemoglobin of Striated Muscle. I. Variations due to Age and Exercise. Amer. J. Physiol. *76*:693, 1926.
536. White, H. L.: Circulatory Responses to Exercise in Man and Their Bearing on the Question of Diastolic Heart Tone. Amer. J. Physiol. *69*:410, 1924.
537. White, H. L., Barker, P. S., and Allen, D. S.: Venous Pressure Responses to Exercise in Patients with Heart Disease. Amer. Heart J. *1*:3, 1925.
538. White, H. L., and Moore, R. M.: Circulatory Responses to Static and Dynamic Exercise. Amer. J. Physiol. *73*:636, 1925.
539. White, J. R.: Effects of Eating a Liquid Meal at Specific Times Upon Subsequent Performances in the One-Mile Run. Res. Quart. Amer. Ass. Health Phys. Educ. *39*:206, 1968.
540. Widimsky, J., Berglund, E., and Malmber, R.: Effect of Repeated Exercise on the Lesser Circulation. J. Appl. Physiol. *18*:983, 1963.
541. Wilder, R. M.: Glycine in Myasthenia Gravis. Proc. Staff Meet., Mayo Clinic *9*:606, 1934.
542. Wilkie, D. R.: The Mechanical Properties of Muscle. Brit. Med. Bull. *12*:177, 1956.
543. Wilkie, D. R.: The Relation Between Force and Velocity in Human Muscle. J. Physiol. *110*:249, 1950.
544. Williams, C. G., Wyndham, C. H., Kok, R., and von Rahden, M. J. E.: Effect of Training on Maximum Oxygen Intake and on Anaerobic Metabolism in Man. Int. Z. Angew. Physiol. *24*:18, 1967.
545. Wilmore, J. H., and Behnke, A. R.: An Anthropometric Estimation of Body Density and Lean Body Weight in Young Men. J. Appl. Physiol. *27*:25, 1969.
546. Wilmore, J. H., and Behnke, A. R.: Predictability of Lean Body Weights through Anthropometric Assessment in College Men. J. Appl. Physiol. *25*:349, 1968.
547. Wolffe, J. B.: The Heart of the Athlete. J. Sports Med. *2*:20, 1962.
548. Woodbury, J. W.: The Cell Membrane: Ionic and Potential Gradients and Active Transport. *In* Physiology and Biophysics, 19th ed. Edited by Ruch, T. C., and Patton, H. D. Philadelphia, W. B. Saunders Company, 1965.
549. Wyndham, C. H., Strydom, N. B., Maritz, J. S., Morrison, J. F., Peter, J., and Potgieter, Z.: Maximum Oxygen Intake and Maximum Heart Rate During Strenuous Work. J. Appl. Physiol. *14*:927, 1959.
550. Yaglon, C. P.: Temperature, Humidity and Air Movement in Industries: The Effective Temperature Index. Amer. J. Physiol. *58*:439, 1927.
551. Young, E.: Hygiene of the School Age. *In* Pediatrics, Volume 1. Edited by Abt, I. A. Philadelphia, W. B. Saunders Company, 1923.
552. Zorbas, W. S., and Karpovich, P. V.: The Effect of Weight Lifting upon the Speed of Muscular Contractions. Res. Quart. Amer. Ass. Health Phys. Educ. *22*:145, 1951.
553. Zoth, O.: Ergographie und Ergometrie in Abderhalden's Handbuch der Biologischen Arbeitsmethoden. Abt. V. Teil 5A, p. 171, 1922-1931.
554. Zuntz and Schumburg: Studien zu einer Physiologie des Marches. Berlin, Hirschwald, 1901.

INDEX

Page numbers in *italics* indicate illustrations.
Page numbers in *italics* followed by *t* indicate tables.

A-band, 3
Acetylcholine, neuromuscular transmission and, 11
Acetyl coenzyme A, 72
Acid-base balance, 164–166
  second wind and, 160
Actin, 1, 6, 12
Action potential, 11, *11*
  muscle tone and, 29
Active rest, 67
Actomyosin, 6
Adenosine diphosphate, 12, 72–77
Adenosine triphosphate, 12
  synthesis of, 72–77
Aerobic capacity. See *Maximum oxygen intake.*
Aerobic reactions, 12
  during exercise, 76, 77
  efficiency of, 114
Age, athletic performance and, 278, *278t*
  blood pressure and, 217
  body composition and, 310
  growth and, 276
  pulse rate and, 199, *200*
  reaction time and, 64
  work capacity and, 278
Air, composition of, 105, *106*
Air conditioning, 251
Air embolism, 263
Air hunger, cause of, 117
Alactacid oxygen debt, 96
  intermittent work and, 100
Alcohol, as ergogenic aid, 323
Alkalies, as ergogenic aids, 324
Alkaline reserve, lactic acid and, 174
Allergic reaction to exercise, 273
Alpha-efferent fibers, 51
Alpha-ketoglutaric acid, 72, *73*
Altitude, 255–262
  acclimatization to, 260
  air changes and, 255, *256t*
  cardiovascular responses and, 257
Altitude, (*Continued*)
  oxygen consumption and, 255–256
  performance potential and, 259
  respiration and, 256
  sensory effects of, 259
Alveolar air, 153–154
  altitude and, 256
  composition of, 153, *153t*
    during exercise, 154
  hyperventilation and, 264
Alveoli, gas exchange in, 145, *146*
Amphetamine, as ergogenic aid, 325
Ampullae, 48, *49*
Anaerobic glycolysis, *74*, 75
Anaerobic reactions, 12
  altitude and, 255
  during exercise, 76, 77
  effect of training on, 27
  efficiency of, 76, 114
Anemia, protracted exercise and, 176
Anisotropic bands, 3
Annulospiral endings, 8, 52, *52*, *53*
Anthropometry, body composition and, 306–310
  somatotyping and, 298
Aortic bodies, respiratory regulation and, 156
Aortic valves, disease of, 192
Arterial blood pressure, 216–225
  age and, 217
  altitude and, 258
  anticipatory rise in, 218
  arm exercise and, 221
  during exercise, 218–221, *219*
  epinephrine and, 329
  factors determining, 216
  fatness and, 217
  following exercise, 222
  in athletes, 219
  isometric contractions and, 223
  leg exercise and, 221
  measurement of, 216

Arterial blood pressure, (*Continued*)
  menopause and, 217
  normal values for, 217
  physical fitness tests and, 283
  regulation during exercise, 221, 229
  stroke volume and, 187
  training and, 225, *226t*
  weight lifting and, 223
Arteries, function of, 215
  peripheral resistance and, 220
Arterio-venous oxygen difference, oxygen intake and, 187
  training and, 190, *190*
  work rate and, *188*
Åstrand-Ryhming test of maximum oxygen intake, 285–287, *286*
Ataxia, 42
Atherosclerosis, 194
Athlete(s), age and performance in, 278, *278t*
  blood pressure in, 219
  body composition of, 310–311
  children as, 276–278, *277t*
  diet of, 86
  heart of, 192
  hemoglobin in, 176
  longevity of, 275
  pulse rate of, 213, *213t*
  reaction time of, 64
  somatotypes of, 300–304, *302–303*
  tobacco and, 333
Athlete's heart, 192
Autonomic nervous system, function of, 39
  pulse rate and, 210
  vasomotor nerves of, 215

Basal ganglia, 39
Basal metabolic rate, 121
  training effects on, 141
Bed rest, exercise and, 315
Behnke's estimation of lean body weight, 306–307, *306t*
"Bends," 263
Benzedrine, 325
Beta-oxidation of fats, 76
Betz cells, 55
Bicarbonates, acid-base balance and, 165
Bicycle ergometer, 116
  Åstrand-Ryhming test and, 285
  maximum oxygen intake and, 111
Bicycling, air resistance and, 137
  energy cost of, 136
  mechanical efficiency of, 136
  pedaling rate of, 136
Blood, acid-base balance of, 164
  alkalinity of, fruit juices and, 327
  buffers in, 233
  carbon dioxide in, 170, 233
Blood, (*Continued*)
  circulation time of, 189, *189*
  composition of, 164–182. See also *Blood corpuscles.*
  flow of, local control of, 228
    nervous control of, 228
  lactic acid in, 172–174, *172*
  oxygen saturation of, 167
    altitude and, 257
    during exercise, 232
    tobacco and, 167
  pH of, 165
  platelets, 178
  pooling in muscles by, 222
  specific gravity of, 179
  temperature of, 170
    respiration and, 156
Blood circulation, active rest and, 67
  heart and, 183–197
  heat and, 247
  rate of, 189, *189*
  warm-up and, 30
Blood corpuscles, red, 175–177
    in urine, 234
    increase during exercise, 174–175
    sedimentation rate of, 176
  white, classes of, 177
    after muscular activity, 177–178
Blood donation, athletes and, 180–182
  work capacity and, 181, *181*
Blood pressure, 215–227. See also *Arterial blood pressure* and *Venous blood pressure.*
  bed rest and, 316
Blood sugar, 85
  effect of exercise on, 179–180
Blood transfusion, oxygen transport and, 176
Blood volume, heat and, 248
  of blood vessels, 216
  red corpuscle concentration and, 174
Body composition, 304–313
  aging and, 310
  anthropometric assessment of, 306–310
  exercise and, 310
  growth and, 310
  measurement of, 305–310
  weight changes and, 311
Body constitution, 295–304. See also *Somatotype* and *Somatotyping.*
  and body composition, 304, 314
Body densitometry, 305
Body fluids, bed rest and, 316
Body surface area, vital capacity and, 151
Body weight, blood pressure and, 217
  changes in body composition and, 311
  "ideal," in children, 307
  in wrestlers, 312
  strength and, 22
  weight charts and, 310
Bone marrow, training and, 176

Boredom, 242
Brain, nuclei of, 39, *40–41*
Brain stem, 41
Breath holding, as physical fitness test, 284
  dangers while swimming, 264
  on starting line, 148
  oxygen inhalation and, 330
  pulse rate and, 204
  sprinting and, 150
  swimming and, 150
  vital capacity and, 152
Breathing. See also *Pulmonary ventilation* and *Respiration.*
  depth of, 146–150, 153, 154
  efficiency of, 150
  nasal versus mouth, 154
  rate of, 146–150
    fatigue and, 150
  shallow, 150
  venous blood pressure and, 226
BTPD, 106
BTPS, 106
Buffer substances, 164

Caffeine, as ergogenic aid, 326
Calisthenics, energy cost of, 134
Calorimetry, methods of, 103–105, *104, 105, 106*
Capillarization, training and, 27
Carbohydrate(s), blood sugar and, 180
  computing use of, 81, 82
  energy expenditure and, 106
Carbon dioxide, diffusion of, 171
  dissociation curve of, 171, *171*
  respiratory quotient and, 80
  transport of, 170
Cardiac acceleratory center, 229
Cardiac inhibitory center, 229
Cardiac output, altitude and, 257
  during exercise, 186–189
    in heat, 251
  exercise rate and, 187
  measurement of, 183
  oxygen transport and, 232
  postural effects on, 184, *186*
  stroke volume and, 186
Cardio-body index, 193
Cardiovascular disease, exercise and, 195
  isometric exercise and, 225
  somatotype and, 301
Carotid bodies, respiration and, 156
Cerebellum, control of muscle contraction by, 55
  divisions of, 42, *43*
  peduncles of, 42
  stretch reflex and, 54
Cerebrum, cocaine and, 326
  hemispheres of, 39, *40*
Cheering, ergogenic effects of, 68
Chemoreceptors, respiration and, 156
Cheyne-Stokes breathing, 162
  altitude and, 259
Children, in athletics, 276, *277t*
  physical fitness of, 279
Chlorides, fatigue and, 238
Cholesterol, 195
Chromosomes, and sex determination, 35
Climate, work capacity and, 252
Cocaine, as ergogenic aid, 326
Cold, clothing and, 254, *254*
  work capacity and, 254
Concentric contraction, 13
  electromyogram during, 16, *15, 16*
  force during, 117–120, *119*
  positive work and, 103
Conditioned reflex, 62
Connective tissue, of muscle, 4, *5*
Contraction, of muscle. See *Muscle contraction.*
Convalescents, physical activity for, 315–320
  physical fitness and, 270
  pulse rates of, 214
  red corpuscles of, 177
  testing of, 319, *320*
Coordination of physiological functions, 228–235
Corpora quadrigemina, 41, *41, 42*
Corticospinal tracts, 55, *56*
Cramps, abdominal, 34
  during exercise, 152
  heat, 249
  in muscle, 33
Creatine, glycine and, 327
Crest load, 91
Crista ampullaris, 48, *49*
Cross training, 68
Cytochrome system, *74,* 75

Dead air space, 153
Deaf-mutes, 61
Deamination, 76
Decompression sickness, 263
Decussation of pyramids, 55, *56*
Dehydration, heat and, 248
Depolarization, of nerve and muscle, 10
Detraining, and weight training, 26
Diastolic pressure, 216, 217. See also *Arterial blood pressure.*
Diathermy, and warm-up, 30
Diffusion, of carbon dioxide and oxygen, 170

Digestion, during exercise, 233
pulse rate and, 201
Dissociation curve, of carbon dioxide, 171, *171*
of oxygen, 166, *167*
Diving, breath holding and, 150
hyperventilation before, 264
pulse rate during, 204
underwater swimming and, 262–265
Douglas bag, 105
Dreyer's formula, 151
Dynamic exercise, blood pressure during, 224, *224*
venous blood pressure during, 226
Dynamogenic effect, 67
Dynamometer, 117–120, *118*

Eccentric contraction, 13
electromyogram during, *15, 16*
force during, 117–120. *119*
negative work and, 103
Ectomorphy, 296, *296*
Efficiency, aerobic work and, 114
anaerobic work and, 114
gross, 113
learning effects on, 124
measurement of, 113
net, 113
of fats, 83
of running, 127
skill and, 130
training and, 141
Elasticity, of muscle, 17
Electric phenomena, in muscular contraction, 10, 14–18
Electrogoniometer, 63, *63*, 118
Electromyography, 14–17, 63
learning and, 59
motor unit summation and, 15
muscle contraction and, 15, *15*
muscle force and, 15, *15*
Electron transport, *74*, 76
Elgon, *63*, 118
Endolymph, 48
Endomorphy, 296, *296*
Endomysium, 4
End plate. See *Motor end plate.*
Endurance, 26, 28
alcohol and, 323
altitude effects on, 259
amphetamine and, 325
caffeine and, 326
cocaine and, 327
hypnosis and, 70
limitations to, 231
of respiratory muscles, 162
psychological limits and, 65, *66*
sugar and, 84
Endurance exercise, effect on resting pulse rate, 212, *213t*
Energy, basal requirements of, 121
carbohydrate as source of, 84, 85
fat as source of, 83
phosphocreatine and, 12, 77, 96
protein as source of, 80
release in the cell, 72–76, *73, 74*
role of ATP, 12, 72
Energy cost, 121–143
blood lactic acid and, 110
calculation of, 106
gross, 108
learning and, 124
measurement of, 107–110
methods of expressing, 110
net, 108
of bicycling, 136
of calisthenics, 134, *135*
of football, 134
of housework, 140, *140t*
of pack carrying, 125
of posture, 122, *122t*
of rowing, 134
of running, 126–128, *128*
of skiing, 129
of snowshoeing, 129
of snow shoveling, 138, *139, 139t*
of stair climbing, 126
of swimming, 130–134, *131*
of walking, 122–125
of weight lifting, 137
of wrestling, 138
Environmental effects. See *Altitude, Cold, Diving,* and *Heat.*
Enzymes, role in metabolic reactions, 72
training and, 78
Epimysium, 4
Epinephrine, blood sugar and, 85, 329
fatigue and, 239
glycogen and, 75
physiological regulation and, 229
pulse rate and, 211
Ergogenic aids, 321–338
alcohol, 323
alkalies, 324
amphetamine, 325
athletic performance and, 321
caffeine, 326
cocaine, 326
ethics of, 322
fruit juices, 327
gelatin, 327
glycine, 327
hormones, 329
lecithin, 329
oxygen, 330
phosphates, 331
purposes of, 323
sodium chloride, 332

Ergogenic aids, (*Continued*)
sugar, 332
tobacco, 333
ultraviolet rays, 335
vitamins, 335
Ergogenic effects, of cheering and music, 68, 69
Ergograph, leg, *158*
Ergometer(s), 114
selection of, 116
Erythrocytes. See *Blood corpuscles, red.*
Excitation-contraction coupling, 11
Excitement, ergogenic effects of, 69
pulse rate and, 202
Exercise effects. See listings for specific items. See also *Training.*
Exhaustion, collapse and, 274
heat and, 253
Extrafusal fibers, 8, *52*
Extrapyramidal system, 56, 58
tracts of, 55

Fainting, cardiac output and, 151
post-exercise, 222
Fasciculus cuneatus, 46
Fasciculus gracilis, 46
Fasting, work capacity and, 87
Fat(s), as energy source, 83
caloric equivalent of, 126
computation of, in body, 305
computing use of, 81, 82
energy expenditure and, 106
lecithin and, 329
of muscle, 7
per cent in men, 304
per cent in women, 304
skinfolds and, 307
Fatigue, 236–241
blood pressure and, 219
blood sugar and, 179
boredom and, 242
causes of, 237
cellular changes and, 80
electromyogram and, 16
ergogenic aids and, 323
glycine and, 328
industrial, 241
isometric contractions and, 225
nervous breakdown and, 243
of muscle, 16
reaction time and, 64
seats of, 239–241, *240*
symptoms of, 237
types of, 236
vitamin C and, 336
Fiber, muscle. See *Muscle cell.*
Fick method (cardiac output), 183
Fitness. See *Physical fitness.*

Flarimeter, 284
Flavoproteins, *74*, 75
Flexor reflex, 50
Flower spray endings, 8, 53, *55*
Football, energy cost of, 134
Forebrain, structure of, 39–40, *40*
Fruit juice, as ergogenic aid, 327

Gamma efferent fibers, 51, 52
Gamma motor system, function of, 54
Gas(es), partial pressure of, 166
transportation of, 166–171
Gelatin, as ergogenic aid, 327
Gelatinous layer, of macula, 48, *49*
Globus pallidus, 39, *40*
Glucagon, and glycogen, 75
Glycine, as ergogenic aid, 327
niacin and, 328
Glycogen, diet and, 85
in cell, *78*
of liver, 85
of muscle, 4, 7, 85
synthesis of, 73
training and, 27
vitamin B complex and, 336
Golgi tendon organ, 8, 46
stress reflex and, 53, *53*
Gray, and multiple factor theory of respiration, 159
Guiding role of the head, 60
Gymnastics, pulse rate during, 205
respiratory, 162

Hair cells, of membranous labyrinth, 48, *49*
Halle's classification of body types, 295
Hall's estimation of "ideal" body weight, 307-309, *309*
measurements for, *309*
Harvard step-up test, 289–292, *291t, 292t*
modifications of, 291
Head position, movement and, 61
Health, and physical fitness, 266–267, *267*
definition of, 266
Heart, diastolic filling of, 187
ergogenic aids and, 323
hypertrophy of, 191
in prepubescent child, 195, *196*
measurement of, size of, 192, *193*
strain of, 194
training and, 191
valvular diseases of, 192
volume of and growth, 196, *196*
Heart disease, and exercise, 194
Heart output. See *Cardiac output.*
Heart rate. See *Pulse rate.*
Heart-body ratio, training and, 191

Heat, acclimatization to, 253
blood circulation and, 247, 250
blood volume and, 248
body temperature and, 246, 248
cramps, 249
sodium chloride and, 332
dehydration and, 248
exercise in, 248–254
exhaustion and, 253
loss of, from body, 246, *246t*
metabolism and, 247
pulse rate and, 247, 250, *250*
humidity and, 252, *252*
respiration and, 247, 251
salt intake and, 249
working capacity in, 251
Hematocrit, altitude and, 260
Hemoglobin, altitude and, 260
as buffer, 165
dissociation of oxygen from 160, 167, *167*
during protracted exercise, 175
in urine, 175
oxygen debt and, 96
oxygen transport and, 166, *167*
Hindbrain, structure of, 41–44, *40, 42, 43, 44*
Hippocrates' classification of body types, 295
Hormones, as ergogenic aids, 328
Housework, energy cost of, 140, *140t*
Hyperventilation, before diving, 264
blood carbon dioxide and, 170
Hypnosis, 70
Hypothalamus, 39, *40*
body temperature and, 230

I-band, 3
Immunity, physical fitness and, 272
Inhibitory interneurons, 54
Injuries, warm-up and, 32
Innervation of muscle, 8
Innervation ratio, 9
Interfilamentous bridge, 3, 12
Intermittent exercise, 99, 114
Intestines, acid-base balance and, 166
Intrafusal muscle fibers, 8, 51, *52*, 54
Ischemia, and isometric and isotonic contractions, 27
Isometric contraction, 13
blood pressure and, 223–225, *224*
cross training and, 68
joint angle and, 118, *119*
strength gain and, 21, 22
venous blood pressure during, 226
Isotonic contraction, 13
concentric, 13, 118–119
cross training and, 68
Isotonic contraction, (*Continued*)
eccentric, 13, 118–119
joint angle and, 118, *119*
Isotropic bands, 3, *2*

Kidneys, acid-base balance and, 166
function during exercise, 233–234
lactic acid removal and, 174
Kinesthesia, 46. See also *Sensation.*
Kraus-Weber test, 293
Krebs cycle, 12, 73, *74*
Kretschmer's classification of body types, 295

Labyrinthine structures. See *Semicircular canals.*
Lactacid oxygen debt, 96
Lactic acid, acid-base balance and, 165
aerobic capacity and, 98
as criteria for maximum oxygen intake, 111
blood carbon dioxide and, 171
concentration during exercise, 77
during recovery, 94, *96*
exercise limitations and, 231
exercise rate and, 172
fatigue and, 237, 238
in blood, 172–174, *172*
intermittent work and, 100
muscle contraction and, 12
oxygen inhalation and, 330
physical fitness and, 173
ratio to oxygen debt, 95
respiration and, 156, 157, 159
respiratory stimulation and, 232
work classification and, 110
Latent period, 13, *14*
Lean body mass, 304. See also *Body composition.*
weight changes and, 312
Lean body weight, 305. See also *Body composition* and *Body weight.*
Learning, and training effects on respiration, 162
electromyogram and, 59
"memory drum" and, 58
motor skills and, 58
sensory "engrams" and, 58
Lecithin, as ergogenic aid, 329
Lemnisci, 46
Length-regulation, of muscle, 54
Liminal zone. See *Threshold zone.*
Limits to performance, physiological and psychological, 65, 66, *66*
Liquid food, as pre-game meal, 87

Liver, glycogen in, 85
  side pain and, 152
Lung(s), acid-base balance and, 166
  air space divisions of, 153
  blood oxygen saturation and, 232
  dead space of, 153
  of marathon runners, 151
  physical fitness testing and, 284, 285
  regulation of, 229
  reserve inspiratory air of, 153
  residual volume of, 153
  tidal air of, 153
  training and, 161
  ventilation of. See *Pulmonary ventilation.*
  vital capacity of, 151–152, 161
Lymph ducts, 8

Making weight, 312–313
Marathon runners, resting pulse rate in, 212, *213t*
  vital capacity of, 151, 161
Marathon running, blood sugar and, 179
  maximum oxygen intake and, 98
  oxygen debt during, 98
  tobacco and, 333
  white corpuscles and, 178
Massage, fatigue and, 67
  warm-up and, 30
Maximum oxygen intake, 90, 91
  altitude and, 255
    acclimatization and, 260
  cardiac output and, 187, *188*
  measurement of, 111
  performance capacity and, 97, 98
  prediction of, 285–287, *286*
  running and, 127
  stroke volume and, 187, *188*
  training and, *92*
McCurdy-Larson test, 292
Meat, in diet, 86
Medulla oblongata, 41, *40, 43, 44*
  physiological controls and, 229
Memory drum, theory of motor learning, 58
Menopause, blood pressure and, 217
Menstruation, muscle strength and, 35
Metabolism. See also *Basal Metabolic Rate.*
  fatigue and, 238
  heat and, 247
  measurement of, 103–107
  oxygen consumption and, 90
  rate of, 90–92
  blood oxygen and, 168
  thyroxin and, 329
  work classification and, 110, 237
Midbrain, structure of, 40, *40, 42*
"Milking action" of muscles, 8, 230
  venous blood pressure and, 226
Minute-volume, of the heart. See *Cardiac output.*
  of the lungs. See *Pulmonary ventilation.*
Mitochondria, in muscle cell, 4
  role in metabolic reactions, 75
  training and, 78
Mongoloid children, reflex and reaction of, 64
Motivation, warm-up and, 31
Motor cortex, 39, *40*
  Betz cells in, 55
  cerebellum and, 57, *57*
  functional areas of, *48*
  muscular contraction and, 55, *56*
  pyramidal system and, 55
  respiratory regulation and, 158
  stretch reflex and, 55
Motor end plate, acetycholine in, 11
  fatigue and, 239, *240*
  muscle cell and, 8, *38*
  neuromuscular transmission and, 11
  training and, 27
Motor learning. See *Learning.*
Motor nerves. See *Nerve fibers.*
Motor neuron pools, cerebellum and, 57, *57*
  description of, 8
  efferent fibers from, 8
  gamma, 51
    reticular tracts and, 54, *53*
  innervation of, 55–57, *56*
  spinal reflexes and, 50–55, *50, 53*
  subthreshold zone of, 44
  threshold zone of, 44
Motor unit, 9
  recruitment during contraction, 18
Mountain-sickness, 259
Multiple factor theory of respiratory control, 159
Muscle
  acetylcholine in, 11
  blood supply in, 7
  cell structure of, 1, *2*
    changes during exercise, 78, *78, 79*
    dimensions of components, *3*
    fatigue and, 239, *240*
  chemical composition of, 6
  connective tissue of, 4, *5*
  contraction of, 13–17
    aerobic, 10
    anaerobic, 12
    blood pooling and, 222
    chemistry of, 12
    concentric, 13, 103
    control of, 57–59, *57*
    dynamic, 13
    dynamogenic effects of, 67
    eccentric, 103
    electromyography and, 15
    in vivo, 14–17

Muscle, (*Continued*)
isometric, 13
load-velocity relationship of, 18–19, *18*
neural integration of, 49–59
of excised muscles, 13
of respiratory muscles, 149, 150
phasic, 13
process involved in, 9
pulse rate and, 211
sliding filament theory of, *4*, 12
strength of different kinds, 118, 119
supraspinal controls of, 55
types of, 13
cramps in, 33
elasticity of, 17
electrical characteristics of, 10–12
fiber. See *Muscle cell.*
arrangement of, 5, *6*
intrafusal, 51, 54
length of, 5
force and cross section, 5
hypertrophy and training, 20
innervation of, 8–9
milking action of, 8
neuromuscular transmission and, 11
physical properties of, 17
red fibers in, 9
soreness of, 32
spindles, 47, *45*
structure of, *2*, 4–6
strength of. See *Strength.*
tone of, 29
training of, 20
effects of, 20–28
capillarization and, 28
chemical composition and, 27
endurance and, 26
fiber use and, 28
strength and, 21
structure and, 20
weight training and, 25
white fibers in, 9
Music, ergogenic effects of, 68
Myofibril, muscle cell and, 1, *2*, *3*, *4*
Myofilament, 1
Myogen, 6
Myoglobin, 9
intermittent exercise and, 99
training and, 27
Myosin, 1, 6, 12
Myotatic unit, 52

Negative work, eccentric contraction and, 103
muscle soreness and, 33
Nerve cell. See *Neuron.*
Nerve endings, annulospiral, 52
flower spray, 52
Nerve endings, (*Continued*)
Golgi tendon, 46, 47, 53, *53*
of vestibular apparatus, 48, *49*
pacinian, 46
primary, 52
secondary, 52
terminal, 38
Nerve fibers. See also *Tracts of Nervous System.*
cross training and, 68
efferent, 8
intercostal, 155
phrenic, 155
sympathetic, 9
Nervous breakdown, prevention of, 243
Nervous system, blood pressure regulation and, 221–222
coordination of organ functions and, 228–229
ergogenic aids and, 323
fatigue and, 239–240
integration of muscle contraction and, 49–59
highly skilled movements and, 58
spinal reflexes and, 50–55, *50*, *52*, *53*
supraspinal controls and, 56–58, *56*, *57*
pulse rate regulation and, 210–212
respiratory regulation and, 155–159
sensation of movement and, 46–49, *47*, *48*, *49*
structure of, 37–46
forebrain, 38, *40*
hindbrain, 41, *40*, *43*
midbrain, 40, *40*, *42*
spinal cord, 44, *45*, *46t*
Neural integration, 49–58
extrapyramidal system and, 56
length of muscle and, 54
motor cortex and, 55
muscle spindle and, 51
of skilled movements, 58
pyramidal system and, 55, 56
proprioceptive stimuli and, 57
sensory cortex and, 51
spinal reflexes and, 50–55. See also *Reflexes.*
servocontrol in, 53
servomechanism in, 54
thalamus and, 51
Neuromuscular transmission, 11
training and, 27
Neurons, fatigue and, 239–241, *240*
sensory, order of, 46, *47*
structure of, 37–39, *38*
Neutrophilic phase of leukocytosis, 143
Nicotinamide adenine dinucleotide (NAD), 75
Nitrogen narcosis, 263
Nociceptive reflex, 50, *50*
Norepinephrine, and glycogen, 75

Normal load, 91
Nuclei of brain, 39, *40–41*
caudate, 39
corpora quadrigemina, 41, *41, 42*
cuneatus, 42
fastigial, 42, *43*
gracilis, 42
lentiform, 39
pontine, 41
red, 41, *41, 42*
substantia nigra, 41, *41, 42*
subthalamic, 40, *40*
vestibular, 42, *43*

Olympics, altitude effects at, 259
Orienteering, 91
Otoconia, 48, 49
Over load, 92
Oxaloacetic acid, 72
Oxidative enzymes, 75, 76
reactions, 72–77
Oxygen, arterio-venous difference of, 187, 190, *188, 190*
as ergogenic aid, 330
blood saturation with, 232
caloric expenditure and, 106
cell use of, 75, 76, 77, *72*
demand for, 90
dissociation curve of, 166, *167*
hemoglobin and, 166
in blood and lungs, 90
partial pressure of, 166
respiratory quotient and, 81
transport of, 166–169
exercise limitation and, 232–233
unloading of by muscles, 232
utilization of, during exercise, 232
venous saturation of, 168, *168*
Oxygen debt, 92–94, *94*
advantages of, 76
alactacid, 96
alcohol and, 324
alkalies and, 324
altitude and, 255
anaerobic reactions of, 73, 76, 77, *74*
determination of, 93
efficiency and, 96
exercise limitation and, 97, 231
intermittent work and, 100
lactacid, 94–97
phosphocreatine and, 96
pulse rate recovery and, 208
running and, 127
running time and, 97–99
skiing and, 143
training effects on, 143, *143*
values for, 93, *94*
Oxygen deficit, 93, *95*
Oxygen intake, altitude and, 255
as limitation to exercise, 97
cardiac output and, 186, *186*
factors determining rate of, 91, 100
intermittent work and, 100
maximum. See *Maximum oxygen intake.*
measurement of, 104
pulmonary ventilation and, 147, *147*
altitude and, 256, *258*
pulse rate and, 109
resting rate of, 90
running time and, 97, 98
steady state of, 90, 91
tobacco and, 334
training and, 143, *143*
Oxygen poisoning, diving and, 264, *263*
Oxygen pulse, 169
following exercise, 208, *209*

Pacemaker of heart, body temperature and, 211
Pacinian receptors, in joints, 46
Pack carrying, energy cost of, 125
Pain, in side, 152–153
Palleocerebellum, 42, *43*
Partial pressure, of carbon dioxide, 170
of oxygen, 166
Peduncles, 42
Perimysium, 4
Peripheral vascular resistance, altitude and, 258
blood pressure and, 216, 220, *221*
Phosphates, as ergogenic aids, 331
in exercise, 180
Phosphocreatine, in resynthesis of ATP, 12, 77
oxygen debt and, 96
training and, 180
Phosphoric acid, 165
Physical conditioning. See *Training.*
Physical fitness, 268
aging and, 275
circulatory adjustments and, 185
degree needed, 268
health and, 267–273, *267*
hormones and, 329
immunity to disease and, 272
intelligence and, 271
lactic acid and, 173
learning and, 271
of American children, 279
pulse rate and, 201
somatotype and, 301
sources of, 279
tests of, 268, 281–294
blood pressure and, 283
breath holding, 284
classification of, 281

Physical fitness, (*Continued*)
for heart, 282
Harvard step-up test, 289
McCurdy-Larson test, 292
of oxygen use, 285
of physical work capacity, 287
pulse rate and, 205, 283
respiratory, 284
selection of, 293–294
Tuttle pulse-ratio test, 288
vitamins and, 337
Ponderal index, 297
Pons, 41, *40, 43, 44*
Positive work, concentric contraction and, 103
muscle soreness and, 33
Posture, blood pressure and, 217–218
circulatory adjustments to, 185
energy cost of, 122
Pre-game meal, 87
Pre-motor area, 39, *40*
Proprioceptive stimuli, 57
Protein(s), as buffers, 165
as source of energy, 80
energy expenditure and, 106
Proteinuria, 234
Psychological factors, in exercise, bed rest and, 316
blood donation and, 181
excitement, 69
gelatin and, 328
in strength and endurance, 65, *66*
physical rehabilitation and, 317
pulse rate recovery and, 207
Pulmonary ventilation, altitude and, 256, 260, *258*
blood pH and, 159, 165
exercise and, 148–150
heat and, 247, 251
in sedentary people, 146
maximum rate of, 147
minute-volume of, 146–150
oxygen consumption and, 147, *147, 148t*
regulation of. See *Respiration, regulation of.*
Pulse pressure, 216
during exercise, 220
Pulse rate, 198–214
age and, 199, *200*
digestion and, 201
diving and, 204
during exercise, 202–204, *203*
in heat, 252, *252*
emotions and, 201
energy expenditure and, 109, *109*
exercise acceleration of, 210
gymnastics and, 205
heart catheterization and, 203
heat and, 247, 250, *250,* 253
Pulse rate, (*Continued*)
kind of activity and, 204
maximum oxygen intake and, 109, *110*
altitude and, 258, 260
oxygen debt and, 208
oxygen intake and, 109
oxygen pulse and, 169
physical fitness tests and, 283
postural changes and, 199
recovery of, following exercise, 207
regulation of, 210
resting, before exercise, 202
measurement of, 210
normal values for, 198, *198*
resting-recovery relationship, 209
step-up tests and, 205
training effects on, 212–214
weight lifting and, 204
Putamen, 39
Pyramidal system, 55, *56*
tracts, 55
Pyramids, 42

Quadriceps reflex, 53
Quotient, respiratory. See *Respiratory quotient.*

"Rapture of the deep," 263
Reaction time, 61–65, *65t*
amphetamine and, 326
testing of, 62
Reciprocal inhibition, 51
Reconditioning, kinesiological versus physiological, 318
Recovery, following exercise, 94–97
blood oxygen and, 169
blood pressure during, 222
lactic acid and, 172, 173, *172*
oxygen debt and, 93
platelet count during, 178
pulse rate during, 207–208, *208*
white corpuscles and, 177
Red blood corpuscles. See *Blood corpuscles, red.*
Reflex(es), 50
crossed extensor, 51
labyrinthine, 60
movement patterns and, 61
muscle tone and, 29
patellar tendon, 53
pulse rate response and, 211
quadriceps, 53
respiratory control and, 157, 158
spinal cord centers, 44
spinal level, 50–55, *50*
time, 61–65
testing of, 62, *62*

Reflex(es), (*Continued*)
tonic neck, 60
versus chemical effects, 231
visual, 58
Relaxation period, of muscle, 13, *14*
Renshaw loop, 54
Reserve inspiratory air, 153
Residual volume, 153
Respiration, 145–163. See also *Pulmonary ventilation.*
alveolar air in. See *Alveolar air.*
aortic bodies and, 156
caffeine and, 326
carotid bodies and, 156
chemoreceptors and, 156
Cheyne-Stokes, 162, 259
depth of, 146
ergogenic aids, and, 323
exercise limitations and, 232
heat and, 247, 251
intolerable, 232
minute volume of, 146
nasal versus mouth, 154
physical fitness tests and, 284
second wind and, 160
training and, 161
venous pressure and, 226
Respiratory center, 155
Respiratory gymnastics, 162
Respiratory muscles, fatigue and, 150
Respiratory quotient, 80
during exercise, 82, 83
foodstuffs used and, 81–83, *82*
in calorimetry, 106
maximum oxygen intake and, 111
spurious, 83
Resting metabolism, following exercise, 96, 108
Resting potential, of nerve and muscle, 10
Reticular activating system, 43
Reticular formation, 43, *44*
Reverberating circuit, 50, 51
Rheumatic fever, physical activity and, 317
Rhythmical movements, and muscle elasticity, 17
Rowing, energy cost of, 134, *135*
Running. See also *Marathon running* and *Sprinting.*
alcohol and, 323
energy cost of, 126–128, *128*
measurement of energy expenditure, 105
prediction of time, 98, *98*
pulse rate during, 204
step length of, 128
sugar and, 332
vitamin B_{12} and, 336
Runners, somatotypes of, 300, *302*

Sarcolemma, 1
Sarcomere, 3
Sarcoplasm, 1
Sarcoplasmic reticulum, 1
Scuba diving, 262
"Second wind," 159
Sedimentation rate, of red blood corpuscles, 176
Semicircular canals, 43, 48, *49*
Sensation, of movement, 46–49
of touch and pressure, 47, *47*
Sensorimotor cortex, in forebrain, 39, *40*
kinesthesia and, 46
pain and, 51
reflexes and, 51
sensation and, 46
sensory areas of, 46, *48*
skilled movement and, 58
tracts to, 47
Sensory cortex. See *Sensorimotor cortex.*
Sensory "engrams," 58
Septa, of muscle, 4
Servocontrol, 53
Servomechanism, 54
Sensory end organs. See *Nerve endings.*
Sex, and energy cost of walking, 123
determination of, 35
Shivering, 254
Shortening period, of muscle, 13, *14*
Sinoauricular node, 211
Skating, maximum oxygen intake during, 112
Skiing, energy cost of, 129, *129t, 128, 130*
Skilled movements, alcohol and, 324
efficiency and, 124, 130
learning of, 58
neural control of, 58, 59
Skinfolds, body composition and, 307
in somatotyping, 298
measurement of, *308*
training effects on, 311
Sliding filament theory, 11
Smoking, 333–335
Snorkels, 262
Snowshoeing, energy cost of, 129
Snow shoveling, energy cost of, 138, *139, 139t*
Sodium chloride, as ergogenic aid, 332
fatigue and, 238
in sweat, 249
Somatotype, cholesterol and, 304
classifications of, 296, *296*
coronary heart disease and, 301
exercise physiology and, 300–304
motor performance and, 301
of football players, 300
of sprinters, 300
of swimmers, 300

Somatotype, (*Continued*)
physical fitness and, 301
serum lipids and, 304
tendency to exercise and, 304
Somatotyping, Heath-Carter method, 298–300, *299t*
Kretschmer's method, 295
Parnell's method, 297
Sheldon's method, 295–297, *296*
Somesthetic area. See *Sensorimotor cortex.*
Specific gravity, of blood, 179
Speed, amphetamine and, 326
training and, 28
Spinal cord, as reflex center, 44
motor pools of, 44
neurons in, 44
reflexes of, 50–55, *50, 53*
structure of, 44–46, *45*
tracts of, 44, 46, 56, *45, 46t, 56*
Spinothalamic tracts, 51
Spleen, red corpuscles in, 174
Sportherz, 192
Spot reduction, 126
Sprinting, altitude and, 259
blood donation and, 181
body type and, 300
caffeine and, 326
effect on resting pulse rate, 212, *213t*
energy for, 84
lactic acid and, 172
limitations to, 231
white corpuscles and, 178
Stair climbing, energy cost of, 126
Staleness, 242–243
glycine and, 328
Starling's Law of the Heart, and venous return, 230
Static exercise. See *Isometric contraction.*
"Steady state," 90
breathing and, 148
oxygen consumption during, 108
Step-up tests, pulse rate and, 205, *206*, 210
to measure maximum oxygen intake, 112
"Stitch," 152–153
STPD, 106
Strength, body weight and, 22
cross training and, 68
dynamogenic effects and, 67
glycine and, 328
head position and, 61
hypnosis and, 70
measurement of during isotonic contraction, 117–119, *118*
physiological limit of, 65
psychological limit of, 65
tobacco and, 334
training of, athletics and, 25
capillarization and, 28
chemical changes and, 27
connective tissue and, 20
endurance gains and, 26
Strength, (*Continued*)
fiber use and, 28
isometric exercise and, 21
isotonic exercise and, 22
muscle size and, 20
nerve transmission and, 27
speed and, 28
stretch reflex and, 64
weight exercises and, 25
Stretch receptor, 8, 51, 52, *52*
Stretch-stress reflex, 51–55, *53*
Striate body, 39, *40*
Stroke volume, blood pressure and, 187
cardiac output and, 186–189
during exercise, 186–189
heart rate and, 187
in athletes, 190, 233, *190*
oxygen pulse and, 169
training and, 190, *190*
Subliminal zone, 44. See also *Motor neuron pool(s).*
Substantitia nigra, 41, *41, 42*
Subthreshold zone, 44. See also *Motor neuron pool(s).*
Sugar. See also *Blood sugar* and *Glycogen.*
as ergogenic aid, 332
endurance and, 84
Summation, muscle force and, 14, 15, *14*
Supraspinal controls of muscle contraction, 55–58. See also *Neural integration.*
Sweating, heat acclimatization and, 254
heat cramps and, 249
heat elimination and, 230
kidney function and, 234
Swimmers, alveolar gas transfer in, 145
pulmonary diffusion capacity of, 162
somatotype of, 300
Swimming, alcohol and, 323
breathing and, 148
efficiency of, 130
energy cost of, 130–134
maximum oxygen intake during, 112
measurement of energy expenditure of, 105
oxygen inhalation and, 330
Pythagorean theorem and, 133, *133*
sugar and, 333
underwater, 262–265
use of legs, 133
vitamin B and, 336
water resistance in, *132t*
work during, 130
Synapse, 38, *38*
fatigue and, 240
Systolic pressure, 216–217

Tapering off, 26
Temperature, body, atmospheric conditions and, 246, *246t*

Temperature(s) (*Continued*)
environmental temperature and, 246
heat and exercise effects on, 248
pulse rate and, 211
regulation of, 230
environmental, effects of, 245–255
of blood, 156, 170
Thalamus, 39, 51, *40*
Threshold stimulus for muscle contraction, 13, *14*
Threshold zone, 44
Thrombocytes, 178
Thyroxin, metabolism and, 329
Tidal air, 153
during exercise, 154
Tobacco, as ergogenic aid, 333
sensitivity to, 335
Tract(s), association, 39
commissural, 39
extrapyramidal, 55
of nervous system, 55, *56*
projection, 39
pyramidal, 55
reticulospinal, 56
spinothalamic, 46, *45, 47*
Training, alkaline reserve and, 165
altitude and, 260
arterial blood pressure and, 225, *226t*
arterio-venous oxygen difference and, 190, *190*
at high altitude, 261
basal metabolism and, 141
body composition and, 310
bone marrow and, 176
cardiac output and, 191
cell function and, 77
chemical changes and, 27
effect on blood oxygen of, 168, *168*
efficiency and, 141
endurance and, 26, 28
heart and, 191
lactic acid and, 98, 173, *173*
maximum oxygen intake and, 98, *92*
muscle structure and, 20
oxygen consumption and, 143, *143*
oxygen debt and, 143, *143*
oxygen pulse and, 169
phosphocreatine and, 180
pulse rate and, 212–214, *213t*
red corpuscles and, 175
respiration and, 161
skinfolds and, 311
speed and, 28
stroke volume and, 190, *190*
strength and, 21
vital capacity and, 161
white corpuscles and, 178
work capacity and, 141, *142*
Transmembrane potential, 10
Transmitter substance, 11
Treadmill, energy cost of walking on, 124, *125*
per cent grade of, 111
T-system, 1
Tuttle pulse-ratio test, 288

Ultraviolet rays, as ergogenic aids, 335
Underwater swimming, 262–265
Underwater weighing, 305
Urine, exercise and, 234
erythrocytes in, 234
hemoglobin in, 175
lactic acid in, 174
pH of, 166
protein in, 234
Urticaria, after exercise, 226
Utricle, 43, 48, *49*

Valsalva phenomenon, 222–223
Vasoconstriction, 215
blood pressure and, 221
caffeine and, 326
Vasodilatation, 215
Vasomotor center, 229
Vasomotor regulation, peripheral resistance and, 216
posture and, 185
Vegetarian diet, 86
Venous blood pressure, 225–227
regulation during exercise, 229
Venous return, during exercise, 230
Ventilation, pulmonary. See *Pulmonary ventilation.*
Vestibular apparatus, 48, 58
Vital capacity, 151–152
training and, 161
Vitamins, as ergogenic aids, 335
deficiency of, 336

Walking, energy cost of, 122–125, *123*
Warm-up, 29–32, *32t*
alcohol and, 324
types of, 30
Water resistance, *132, 132t*
Weight. See *Body Weight.*
Weight lifting, 21, 22
blood pressure during, 223–225
energy cost of, 137
motor performance and, 138
predicting records in, 23, *24, 24t*
pulse rate during, 204
training athletes and, 25

Weight training, 26. See also *Weight lifting.*
White blood corpuscles. See *Blood corpuscles, white.*
Withdrawal reflex, 50
Work, 102, 103
  classification of, 110
  during swimming, 130
  measurement of, 114–117
  negative, 103
  positive, 103
  pulse rate and, 203
Work capacity, air conditioning and, 251, *251*
  climate and, 252
  cold and, 254, *254*
  effect of fasting on, 88
Work capacity, (*Continued*)
  glycine and, 328
  glycogen and, 85
  heat and, 251–253
  phosphates and, 331
  training effects on, 141, *142*
Work classification, white corpuscles and, 177
  fatigue and, 237
Wrestler's weights, 313
Wrestling, energy cost of, 138
  Valsalva effect and, 223

Z-lines, 3